Dietary Fiber in Health and Disease

GWUMC Department of Biochemistry
Annual Spring Symposia

Series Editors:
Allan L. Goldstein, Ajit Kumar, and George V. Vahouny
The George Washington University Medical Center

DIETARY FIBER IN HEALTH AND DISEASE
Edited by George V. Vahouny and David Kritchevsky

Dietary Fiber in Health and Disease

Edited by

George V. Vahouny

The George Washington University Medical Center
Washington, D.C.

and

David Kritchevsky

The Wistar Institute
Philadelphia, Pennsylvania

Plenum Press • New York and London

Library of Congress Cataloging in Publication Data

Main entry under title:

Dietary fiber in health and disease.

(GWUMC department of biochemistry annual spring symposia)
Includes bibliographical references and index.
1. Fiber deficiency diseases. 2. High-fiber diet. I. Vahouny, George V., 1932-
II. Kritchevsky, David, 1920- . III. Series. [DNLM: 1. Dietary fiber—Congresses. WB 427 D565]
RC627.F5D54 616.3′96 82-7549
ISBN 978-1-4615-6852-0 ISBN 978-1-4615-6850-6 (eBook) AACR2
DOI 10.1007/978-1-4615-6850-6

Softcover reprint of the hardcover 1st edition 1982

A Division of Plenum Publishing Corporation
233 Spring Street, New York, N.Y. 10013

Contributors

MARGARET J. ALBRINK
Department of Medicine
West Virginia University Medical Center
Morgantown, West Virginia 26505

R. ALI
Nutritional Research and Development
Bristol-Myers International Division
New York, New York 10154

JAMES W. ANDERSON
Endocrine–Metabolic Section
Veterans Administration Medical Center
and Departments of Medicine and Clinical Nutrition
University of Kentucky College of Medicine
Lexington, Kentucky 40511

LUIS ARIAS-AMADO
Centro de Investigación en Ciencas de la Alimentación
Facultad de Medicina Veterinaria y Zootecnia
Universidad de Guadalajara
Guadalajara, Mexico

J. D. BAIRD
Wolfson Gastrointestinal Laboratory
Metabolic Unit, University Department of Medicine
Department of Clinical Chemistry
Western General Hospital
Edinburgh EH4 2XU, Scotland

P. C. BOYLE
Department of Nutrition and Health Sciences
General Foods Technical Center
Tarrytown, New York 10591

W. J. BRANCH
Dunn Clinical Nutrition Centre
Old Addenbrookes Hospital
Cambridge CB2 1QE, England

W. G. BRYDON
Wolfson Gastrointestinal Laboratory
Metabolic Unit, University Department of Medicine
Department of Clinical Chemistry
Western General Hospital
Edinburgh EH4 2XU, Scotland

MARIE M. CASSIDY
Department of Physiology
The George Washington University
School of Medicine and Health Sciences
Washington, D.C. 20037

JOHN H. CUMMINGS
Dunn Clinical Nutrition Centre
Old Addenbrookes Hospital
Cambridge CB2 1QE, England

N. T. DAVIES
Rowett Research Institute
Bucksburn, Aberdeen AB2 9SB, Scotland

M. A. EASTWOOD
Wolfson Gastrointestinal Laboratory
Metabolic Unit, University Department of Medicine
Department of Clinical Chemistry
Western General Hospital
Edinburgh EH4 2XU, Scotland

HUGH JAMES FREEMAN
University of British Columbia Faculty of Medicine (Gastroenterology) and Cancer Research Center
Vancouver, B.C., Canada V5Z 1L3

PEDRO M. GARCÍA L.
Centro de Investigación en Ciencias de la Alimentación
Facultad de Medicina Veterinaria y Zootecnia
Universidad de Guadalajara
Guadalajara, Mexico

PEDRO GARZÓN
Laboratorio del Bioquímica
División de Biología del Desarrollo
Unidad de Investigaciones Biomédicas de Occidente
Instituto Mexicano del Seguro Social
Guadalajara, Mexico

M. HARTOG
University Department of Medicine
Bristol Royal Infirmary
Bristol BS2 8HW, England

K. W. HEATON
University Department of Medicine
Bristol Royal Infirmary
Bristol BS2 8HW, England

S. HELLIWELL
Wolfson Gastrointestinal Laboratory
Metabolic Unit, University Department of Medicine
Department of Clinical Chemistry
Western General Hospital
Edinburgh EH4 2XU, Scotland

C. L. HENRY
University Department of Medicine
Bristol Royal Infirmary
Bristol BS2 8HW, England

M. J. HILL
Bacterial Metabolism Research Laboratory
PHLS Centre for Applied Microbiology and Research
Salisbury SP4 0JG, England

MICHAEL J. KELLEY
Department of Foods and Nutrition
Purdue University
West Lafayette, Indiana 47907

JUNE L. KELSAY
Carbohydrate Nutrition Laboratory
Beltsville Human Nutrition Research Center
Human Nutrition, Science and Education Administration
U.S. Department of Agriculture
Beltsville, Maryland 20705

DAVID KRITCHEVSKY
The Wistar Institute of Anatomy and Biology
Philadelphia, Pennsylvania 19104

ANTHONY RICHARD LEEDS
Department of Nutrition
Queen Elizabeth College
University of London
London W8 7AH, England
and Department of General Medicine and Endocrinology
Central Middlesex Hospital
London NW10 7NS, England

G. A. LEVEILLE
Department of Nutrition and Health Sciences
General Foods Technical Center
Tarrytown, New York 10591

FRED G. LIGHTFOOT
Department of Anatomy
The George Washington University
School of Medicine and Health Sciences
Washington, D.C. 20037

A. MANHIRE
University Department of Medicine
Bristol Royal Infirmary
Bristol BS2 8HW, England

JUAN M. MUNOZ
Fargo Clinic
Fargo, North Dakota 58123

J. L. PRITCHARD
Wolfson Gastrointestinal Laboratory
Metabolic Unit, University Department of Medicine
Department of Clinical Chemistry
Western General Hospital
Edinburgh EH4 2XU, Scotland

BANDARU S. REDDY
Naylor Dana Institute for Disease Prevention
American Health Foundation
Valhalla, New York 10595

JOHN G. REINHOLD
Centro de Investigación en Ciencas de la Alimentación
Facultad de Medicina Veterinaria y Zootecnia
Universidad de Guadalajara
Guadalajara, Mexico

BARBARA OLDS SCHNEEMAN
Departments of Nutrition and Food Science and Technology
University of California
Davis, California 95616

J. H. SMITH
West Granton Medical Group
Edinburgh EH4 4PL, Scotland

D. A. T. SOUTHGATE
Nutrition and Food Quality Division
Agricultural Research Council
Food Research Institute
Norwich NR4 7UA, England

GENE A. SPILLER
Department of Biology
Mills College
Oakland, California 94613

H. STAUB
Department of Nutrition and Health Sciences
General Foods Technical Center
Tarrytown, New York 10591

JON A. STORY
Department of Foods and Nutrition
Purdue University
West Lafayette, Indiana 47907

JAMES N. THOMAS
Department of Foods and Nutrition
Purdue University
West Lafayette, Indiana 47907

IRMA H. ULLRICH
Department of Medicine
West Virginia University Medical Center
Morgantown, West Virginia 26505

GEORGE V. VAHOUNY
Department of Biochemistry
The George Washington University
School of Medicine and Health Sciences
Washington, D.C. 20037

Preface

Dietary fiber is a topic that has burgeoned from an esoteric interest of a few research laboratories to a subject of international interest. This growth has been helped by the intense public interest in the potential benefits of adding fiber to the diet. The general popularity of fiber may have been helped by the perception that, for once, medicine was saying "do" instead of "don't." There has been a proliferation of excellent scientific books on dietary fiber. Why another?

The Spring Symposium on Dietary Fiber in Health and Disease was an outgrowth of our belief that informal discussion among peers—a discussion in which fact is freely interlaced with speculation—was the most effective way to organize our knowledge and direct our thinking. The normal growth progression of a discipline includes its branching into many areas. Soon the expertise, which was once general, is broken into many specialties. Intercommunication becomes increasingly difficult. It was our intent to provide a forum that would expose its participants to developments in areas related to their research interest. Free exchange under these conditions could not help but broaden everyone's knowledge and expand his horizons.

We feel that this symposium was singularly successful in achieving its goals. It resulted in a free and friendly exchange of knowledge and ideas. It helped to establish seeds for future collaborations based on mutual interest and friendship. The proceedings of this conference will serve as yet another basic resource in the fiber field.

The volume opens with a discussion of definitions and terminology—ours is a field in which we all recognize the subject of our research but cannot name it adequately.

Chapters on the effect of fiber on colonic function are followed by discussions of fiber and its effects on nutrients, with special reference to trace minerals. Metabolic aspects of fiber ingestion are covered in chapters on obesity and diabetes. These are followed by discussions of lipids and lipid metabolism. The closing chapters are devoted to dietary fiber and cancer. In all, this volume represents a state-of-the-art report as of the date of the conference. State-of-the-art quickly becomes obsolete in rapidly developing areas of research, but this meeting included enough basic background and future direction to provide a valuable, lasting reference.

We also feel that homage should be paid to the scientists who have done so much to popularize the metabolic aspects of dietary fiber and to establish its place in the field of nutrition. Surgeon Captain T. L. Cleave propounded much of the basic theory in his paper entitled "The Neglect of Natural Principles in Current Medical Practice," which appeared in the *Journal of the Royal Naval Medical Service* in 1956 (Vol. 42, p. 55). Cleave's hypothesis was refined, confirmed, and expanded, principally by Denis Burkitt, but also by Hugh Trowell and A. R. P. Walker. Clearly, dietary fiber is not a panacea, but its place in nutrition has been established through the efforts of many, including the scientists cited.

We are particularly grateful for the important suggestions and valuable contributions of several colleagues during the organization of the conference. These included Drs. M. M. Cassidy, M. J. Hill, R. Kay, D. Southgate and G. Spiller. The organizers of the conference are indebted to the George Washington University and to I. T. T. Continental Baking Company for their continued interest and support not only of this conference, but of other research and communication activities in the area of human nutrition and pathophysiology. We are also indebted to Mrs. Fran Nigro for assistance in the preparation of the index.

George V. Vahouny
David Kritchevsky

Contents

1

Definitions and Terminology of Dietary Fiber

D. A. T. SOUTHGATE

1. INTRODUCTION

In this chapter I outline a basis for the usage of the term *dietary fiber* and discuss the implications of this basis for the analysis of dietary fiber and for studies of its mechanism of action.

If we look back, say, ten years, I think that we would have not predicted the current scientific (and popular) interest in dietary fiber. For this we must thank the two "evangelists" of dietary fiber, Burkitt and Trowell, and in order to develop our thoughts on definitions and terminology, we must return to their original hypothesis.

2. THE DIETARY FIBER HYPOTHESIS

Despite the many thousands of words that have been spoken and written about dietary fiber, I do not recall seeing any formal statement of the hypothesis, and I would like to suggest the following, which is based on Burkitt and Trowell (1975) and Trowell (1976).

The hypothesis has two primary statements:

1. A diet that is rich in foods which contain plant cell walls (for example, high-extraction cereals, fruits, and vegetables) is protective against a

D. A. T. SOUTHGATE • Nutrition and Food Quality Division, Agricultural Research Council, Food Research Institute, Norwich NR4 7UA, England.

range of diseases, in particular those prevalent in affluent Western communities (for example, constipation, diverticular disease, large bowel cancer, coronary heart disease, diabetes, obesity, and gallstones).

2. In some instances a diet providing a low intake of plant cell walls is a causative factor in the etiology of the disease and in others it provides the conditions under which other etiological factors are more active.

It is quite clear that the original hypothesis relates to *types of diet* and argues that the essential difference between protective diets and nonprotective diets is the amounts of plant cell wall material they provide, and that the protective agent is, or is derived from, the plant cell walls in the diet.

Both Trowell (1976) and Heaton (1973) have advanced the view that the protective effect is due, in part, to the consumption of the plant foods in their natural state, that is, with the cellular structures largely intact and enclosing the other nutrients.

However, this view is essentially related to the mechanism of action of dietary fiber and is not essential to the hypothesis as such.

The diets rich in plant cell walls that are consumed by communities showing a low incidence of the disease against which dietary fiber is thought to be protective have a number of other characteristics which probably contribute to the postulated protective properties. They are, for instance low in fat, refined sucrose, and salt and high in starch (Perissé *et al.*, 1969)

3. DEFINITION OF DIETARY FIBER

Our usage of the term *dietary fiber* and its definition should be related to the original hypothesis.

Thus it is possible to argue that Hipsley's (1953) original use of the term and Trowell's (1972) later definition are the most acceptable. I think that they would have been, had not the word *fiber* been part of the term. This led to considerable confusion and to a series of naive studies in which cellulose and/or bran was regarded as equivalent to the plant cell wall. Furthermore, in an effort to study the mechanism of action of dietary fiber, studies with isolated polysaccharides were made and a debate ensued as to whether or not these were dietary fibers.

Spiller *et al.* (1976) suggested that a new term was needed to eliminate this confusion and they coined the word *plantix*. Trowell *et al.* (1976) proposed a definition that refined his earlier proposition in restricting dietary fiber to the sum of lignin and the plant polysaccharides that are undigested by the endogenous secretions of the human digestive tract. This has been criticized in that, first, it focuses attention on the indigestibility of dietary fiber and assumes knowledge of the course of digestion of polysaccharides, knowledge which is, in fact, surprisingly limited; and second, it is a definition that is difficult to translate into an analytical procedure. It has also been suggested that in omitting the other components of the plant cell wall (e.g., proteins, lipids, cutins), one is not referring to dietary fiber in a strict sense.

4. THE NATURE OF CELL WALL MATERIAL

The immature cell wall is predominantly a mixture of polysaccharides organized into a complex supramolecular arrangement (Albersheim, 1977) with minor amounts of protein and other constituents. As the plant matures, lignin is deposited in specific cell types and eventually makes a substantial contribution to the wall. The polysaccharides and lignin account for between 80 and 90% of the plant cell wall (Table I). Therefore measuring these will provide a good estimate of total cell wall material.

The polysaccharides in the cell wall include cellulose and a range of noncellulosic polysaccharides, ranging from those rich in uronic acids (for example, pectic substances) to those with only a few uronic acid residues (for example, xylans, arabinogalactans, galactomannans, and β-1-3-glucans). These polysaccharides have one common feature: they are not α-linked-glucan-polysaccharides and since the mammalian digestive tract is only known to secrete α-glucosidases, it is reasonable to argue that these polysaccharides will not provide the body with carbohydrates. This is the logic behind McCance and Lawrence's (1929) separation of food polysaccharides into available (starch and dextrins) and unavailable (the nonstarch polysaccharides) and the basis for the Trowell *et al.* (1976) definition and the derivative Southgate *et al.* (1978) version.

When the diet includes processed foods a range of non-α-glucan-polysac-

TABLE I. Composition of Plant Cell Walls[a]

	Percent of dry cell wall	
	Immature primary	Mature
Cellulose	24.8	38.3
Noncellulosic polysaccharides	60.0	43.4
Lignin	Tr	16.7
Waxes, etc.	4.2	—
Protein	9.5	3.2
Ash	—	1.4

[a] From Siegel (1968).

charides may be present, including pectin, guar and locust bean gum, the algal polysaccharides (e.g., alginates, carrageenins), and derivatives of cellulose. These contribute to the total noncellulosic polysaccharide fraction of the diet. In many cases they are structurally similar to components of the plant cell wall and are virtually impossible to distinguish analytically; therefore it is reasonable to regard them as part of the total dietary fiber. Individually, however, these polysaccharides are not *dietary fibers* and in this I support one basic thesis of Trowell *et al*. (1978).

5. TERMINOLOGY

I think that we should develop our terminology as summarized in Table II, distinguishing between dietary fiber and isolated polysaccharides as follows.

5.1. Dietary Fiber

The term *dietary fiber* should be used for the sum of lignin and the non-α-glucan-polysaccharides in food or diet. This will in most cases be derived mainly from cell wall material and therefore this usage is compatible with the original hypothesis.

TABLE II. Terminology in Dietary Fiber Studies

Hypothesis	Mixture in diet	Terminology when used as supplements or in metabolic studies		Analytical definition
Dietary fiber = plant cell wall	Protein Lipids Inorganic constituents Lignin Cellulose Noncellulosic polysaccharides	Cellulose Pectic substances	Systematic structural name where known	Dietary fiber
	Polysaccharide food additives	Pectin Gums Mucilages Algal polysaccharides Modified cellulose		

5.2. Isolated Polysaccharides

When used as a supplement or in a metabolic study isolated polysaccharides do not need to be called fiber at all—they should be referred to by their trivial or systematic (polysaccharide) names. Since, however, these supplements are often made to advance the understanding of the mechanism of the action of dietary fiber, the inclusion of studies using these isolated polysaccharides in dietary fiber bibliographies is essential (Trowell *el al.*, 1978).

6. ANALYTICAL IMPLICATIONS

The proposed definition can be used to form the basis of an analytical scheme. For practical purposes dietary fiber is the sum of lignin and the nonstarch polysaccharides.

This simplicity is, however, deceptive, since it has not resolved the analytical difficulty of defining the boundary between starch and the nonstarch polysaccharides. This can be done by selective extraction (Selvendran *et al.*, 1979) or by enzymatic means (Southgate, 1976). Specific enzymes provide the simpler approach, but the enzyme used and the conditions used for hydrolysis will always define an empirical boundary, since absolute boundaries rarely, if ever, exist in nature.

The important point of principle is that *dietary fiber* includes both water-soluble and water-insoluble substances and that many of these are nonfibrous in the accepted sense of the word.

7. IMPLICATIONS FOR THE STUDY OF THE MECHANISM OF ACTION OF DIETARY FIBER

The dietary fiber hypothesis relates to types of diet and has been extended to implicate the plant cell walls (dietary fiber) as the protective agent. Studies to test this hypothesis fall into two categories: first, epidemiological studies of the relation between intake and disease experience in populations, and second, studies of mechanism of action. The complex nature of dietary fiber has led many workers to study the physiological properties of isolated polysaccharides; these may not be relevant to the hypothesis as such, despite their intrinsic interest and the important physiological insights that stem from these studies.

The amounts of pectin or guar gum, for instance, that have been fed in these studies are more relevant to what has been described as the "pharmacology of polysaccharides."

Studies that have used cell wall preparations to increase dietary fiber intakes appear more relevant, but even this may be superficial, because the preparation of the cell wall material may involve physical damage to the cellular structures and the coprecipitation of protein *inter alia*. Furthermore, the addition of dietary fiber to a diet does not in itself change the nature of the other components of the diet.

Evaluating the dietary fiber hypothesis involves both evidence for epidemiological association and evidence for mechanisms of action under conditions that parallel the dietary compositional features seen in the epidemiological studies.

8. CONCLUSIONS

The dietary fiber hypothesis relates to *types of diets:* one of the most noticeable features of protective diets is a high intake of plant cell wall material. Dietary fiber is thus equivalent to the plant cell wall material in a diet. Because the cell wall is principally lignin and polysaccharide, measurement of these gives a good estimate of cell wall material. The polysaccharides in the cell wall are non-α-glucan-polysaccharides and the measurement of nonstarch polysaccharides plus lignin is therefore a convenient basis for analytical measurement.

Diets that contain processed foods may contribute noncellulosic polysaccharides of this type to the total dietary fiber. These polysaccharides should not be individually regarded as dietary fiber, but referred to as the specific polysaccharide, either by their trivial name or according to systematic polysaccharide nonmenclature where this is known. These materials can be described collectively as gums, mucilages, or algal polysaccharides, respectively, and it does not seem desirable to label these substances as "fibers," which they unquestionably are not.

Finally I come to the conclusion that the *dietary fiber* hypothesis would not be regarded as acceptable according to some of Karl Popper's (1959) criteria, because of the eminent difficulty of falsifying it. To end on a positive note, I think at this time there are a large number of exciting and potentially important physiological effects of isolated polysaccharides and plant cell wall material in the gastrointestinal tract of man to be studied and understood. Whether at the end of the work a clear link between dietary fiber intake and health and disease will emerge is unclear at present. I think that our guide should be the words of Modan in Bazil: "the prospects are too good to pass up!"

REFERENCES

Albersheim, P., 1977, The primary cell wall, in: *Plant Biochemistry* (J. Bonner and J. E. Varner, eds.), Academic Press, New York, pp. 266–277.

Burkitt, D. P., and Trowell, H. C., 1975, *Refined Carbohydrate Foods and Disease. Some Implications of Dietary Fiber,* Academic Press, London.

Heaton, K. W., 1973, Food fiber as an obstacle to energy intake, *Lancet* **ii:**1418–1421.

Hipsley, E. H., 1953, Dietary 'fiber' and pregnancy toxaemia, *Br. Med. J.* **2:**420.

McCance, R. A., and Lawrence, R. D., 1929, The Carbohydrate Content of Foods, Special Report Series, Medical Research Council, London, No. 135, HMSO, London.

Perissé, J., Sizaret, F., and Francois, P., 1969, The effect of income on the structure of the diet, *Nutrition Newsletter (FAO)* **7:**1–9.

Popper, K. R., 1959, *The Logic of Scientific Discovery,* Hutchinson, London.

Selvendran, R. R., Ring, S. G., and Du Pont, M. S., 1979, Assessment of procedures used for analysing dietary fiber and some recent developments, *Chem. Ind.* **1970**(7):225–230.

Siegel, S. M., 1968, The biochemistry of the plant cell wall, in: *Comprehensive Biochemistry,* Vol. 12A (M. Florkin and E. H. Stotz, eds.), Elsevier, Amsterdam, pp. 1–51.

Southgate, D. A. T., 1976, *Determination of Food Carbohydrates,* Applied Science, London.

Southgate, D. A. T., Hudson, G. J., and Englyst, H., 1978, The analysis of dietary fiber. The choices for the analyst, *J. Sci. Food Agric.* **29:**978–988.

Spiller, G. A., Fassett-Cornelius, G., and Briggs, G., 1976, A new term for plant fibers in nutrition, *Am. J. Clin. Nutr.* **29:**934–935.

Trowell, H. C., 1972, Ischemic heart disease and dietary fiber, *Am. J. Clin. Nutr.* **25:**926–932.

Trowell, H., 1976, Definition of dietary fiber and hypotheses that it is a protective factor in certain diseases, *Am. J. Clin. Nutr.* **29:**417–427.

Trowell, H., Southgate, D. A. T., Wolerer, T. M. S., Leeds, A. R., Gassull, M. A., and Jenkins, D. A., 1976, Dietary fibre redefined, *Lancet* **1:**967.

Trowell, H., Godding, E., Spiller, G., and Briggs, G., 1978, Fiber bibliographies and terminology, *Am. J. Clin. Nutr.* **31:**1489–1490.

2

Consequences of the Metabolism of Fiber in the Human Large Intestine

JOHN H. CUMMINGS

1. INTRODUCTION

The effect of fiber on bowel habit is widely known (Royal College of Physicians, 1980; Eastwood *et al.*, 1980), and there is little need to restate the evidence for this. Studies have shown, however, that the increase in fecal output, shortened transit time, and altered colonic metabolism due to fiber are subject to considerable variation, particularly in relation to the source of fiber and among the individuals under observation. Analysis of the reasons for this variation has led to new understanding of the role of fiber in the colon and of colonic physiology.

2. IMPORTANCE OF THE SOURCE OF FIBER

When equivalent amounts of fiber from different plant sources are fed to healthy subjects, not all have similar effects on bowel habit (Williams and Olmsted, 1936; Eastwood *et al.*, 1973; Baird *et al.*, 1977; Heller, 1977; Cummings *et al.*, 1978; Spiller *et al.*, 1980). While the methods used in these studies are not strictly comparable, in general the results concur, with fiber from wheat usually producing the greatest changes in stool weight. Colonic function is less influenced by pectin or guar gum, while intermediate effects are seen with

JOHN H. CUMMINGS • Dunn Clinical Nutrition Centre, Old Addenbrookes Hospital, Cambridge CB2 1QE, England.

fiber from carrot and cabbage. Every sort of fiber will produce some change in colonic function, but the extent of these changes varies with the source of fiber. In order to explain these contrasting effects of different fibers, it has been necessary to examine the way in which fiber acts in the colon.

2.1. Physical Properties

2.1.1. Water-Holding Capacity

The ability of fiber to hold water in varying amounts has been used as a ready explanation for the differing effects of fiber on fecal output. As early as 1941, Gray and Tainter suggested that the *in vitro* water-holding ability of bulk laxatives was an important determinant of their capacity to alter stool output. Their human feeding studies, however, failed to demonstrate the expected relationship between water-holding and fecal-bulking ability.

More recently, MacConnell *et al.* (1974) have applied tests of water-holding to a much wider range of fiber-containing materials and shown a greater than tenfold variation in the water-holding capacity of plant materials found in the normal diet. Cell wall preparations from potato, banana, and bran hold only 2 or 3 g of water per gram of material, while fiber from cucumber, carrot, and lettuce hold 20–24 g/g. They have gone on to show that within a given species, such as the carrot, water-holding can vary, dependeng on maturity (Robertson *et al.*, 1980), and that the treatment a food source receives can materially affect its capacity to hold water (Robertson and Eastwood, 1981).

Fewer studies comparing *in vitro* and *in vivo* effects of these materials have been reported. When Stephen and Cummings (1979) measured water-holding capacity for an assortment of plant materials and then fed these same materials to groups of healthy subjects, the opposite of what might have been predicted was found. Those preparations that held the least water *in vitro,* such as bran and bagasse, produced the largest changes in the stool weight, while gums such as pectin, which are very good at holding water *in vitro,* produced virtually no change in stool output.

If anything, these data suggest that materials that hold more water are likely to be least effective at increasing fecal bulk. This, of itself, is an important observation and does not exclude water-holding as one mechanism whereby fiber increases stool output. However, in order to explain the effects of fiber one must look to alternative, perhaps complementary, mechanisms.

2.1.2. Particle Size

A number of studies have been done that have shown that when fiber from a single source is fed at two different particle sizes then contrasting effects on bowel habits can be seen (Kirwan *et al.*, 1974; Brodribb and Groves, 1978; Heller *et al.*, 1980). The effects of changing the particle size of a fiber source on bowel habits are not large, but it is clear from these studies that the coarser the particle size of the fiber, the more effective it is at increasing stool output. Although the water-holding capacity of coarse bran is greater than that of fine bran (Brodribb and Groves, 1978; Heller *et al.*, 1980), it is by no means clear that this is the full explanation for their differing effects on the large bowel, since the digestibilities of these two materials also differ significantly (Heller *et al.*, 1980).

2.2. Chemical Composition

An alternative way of explaining the varied effects of different fiber sources on bowel function has been put forward based on the contrasting sugar composition of cell wall polysaccharides (Cummings *et al.*, 1978). When fiber from five different sources (carrot, cabbage, apple, bran, and guar gum) was fed to groups of volunteers in equal amounts and fecal output measured, one particular chemical component of the fiber was most closely related to the changes in stool weight. This component was the pentose fraction of the noncellulosic polysaccharides. The greater the amount of pentose in the fiber, the greater the increase in fecal weight observed. Data from other published studies where detailed information about the chemical composition of fiber has been reported fit well with this concept (Cummings, 1978). However, insufficient work has been done on the chemistry of dietary fiber to enable further generalizations about this to be made. Clearly, if a fiber is fed at two particle sizes and has different effects on colonic function, then the pentose hypothesis cannot be used as the explanation for the differences, since identical intakes of pentose would have been present in the fiber source. Nevertheless, it is important to bear in mind that the chemical in addition to the physical composition of fiber is important in determining changes in bowel function.

3. THE ROLE OF TRANSIT TIME

The effect of fiber on the gut is also modified by the transit time of the individual. Transit has been a topic of interest for many years, although its physiological significance has never been clearly demonstrated. It is believed that transit is the result rather than the cause of colonic events. However, evidence is now emerging that transit itself influences events in the human large intestine and that it may not be the result of a particular diet or environmental factor, but more importantly modify independently the effect of these substances on the gut. Some evidence to support the independent role of transit in determining

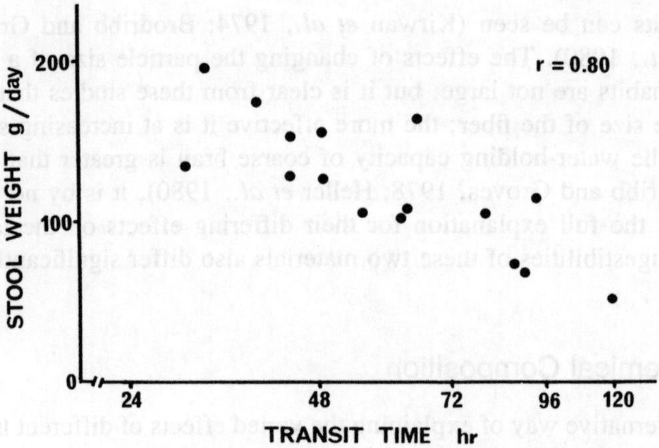

FIGURE 1. The relationship between mean daily stool weight and transit time in 16 healthy young men eating identical controlled diets. (From Cummings *et al.*, 1978.)

colonic function is shown in Fig. 1. Here transit and stool output have been measured in 16 normal subjects who were all eating identical diets, containing about 20 g of dietary fiber. Stool weight varied from 65 to 194 g/day and transit time from 31 to 117 hr. The threefold range in stool output cannot be explained on a dietary basis, since all were taking the same diet. Much of the variation can be accounted for by equally large differences in transit among the group, those subjects with the slowest transit having the smallest stool weight and vice versa.

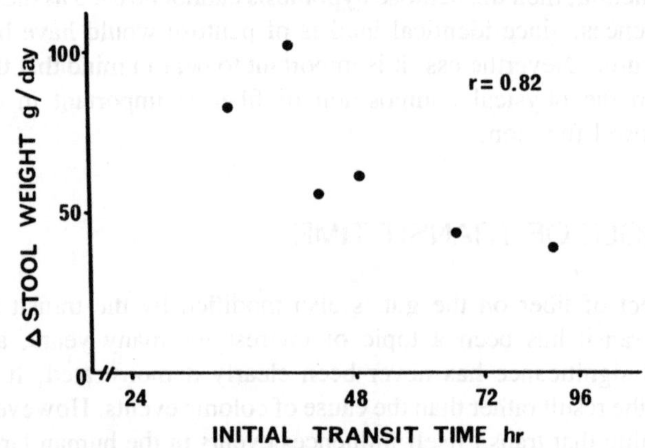

FIGURE 2. Increase in stool weight in grams per day in six healthy subjects when 18 g dietary fiber from apple was added to their controlled diets, in relation to their transit time while on the basal diet without apple fiber. (From Cummings *et al.*, 1978.)

If transit is an independent factor modifying colonic function, then it should be possible to demonstrate this by changing transit and showing that stool output and colonic physiology are altered. Stephen (1980) has done this by giving a constant diet to healthy subjects and then adding either senna or codeine phosphate to the diet and observing changes in colonic function. These studies showed quite clearly that by slowing down transit pharmacologically, stool weight was decreased, and conversely, by speeding it up, stool weight was increased. This supports the concept that transit itself modified the colonic response to diet.

Further evidence can be adduced from Fig. 2. This shows the change in fecal weight in six subjects when 18 g of dietary fiber from apple was added to their diet in relation to the transit time of the subject when on the control diet. Despite all subjects' taking an identical increment in dietary fiber, it is clear that the increase in stool weight is much greater in those subjects who had the fastest transit time initially. Thus transit determines how an individual will respond to a given amount of dietary fiber.

Transit is also important in controlling other aspects of fiber metabolism in the colon. When a group of five healthy subjects took a controlled diet it was clear from fecal analysis that each was excreting different amounts of cellulose, despite all having identical cellulose intakes. Furthermore, cellulose excretion in the stool varied from week to week for each subject. Figure 3 shows that cellulose was extensively degraded in the gut of these subjects most of the time. In those subjects whose transit time fell below 50 hr, however, cellulose diges-

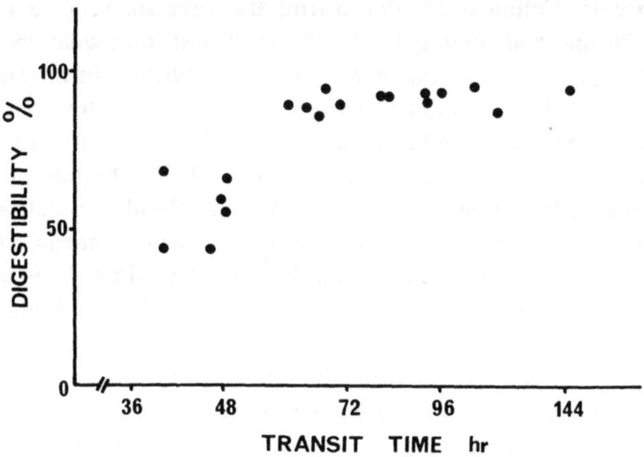

FIGURE 3. Relation of cellulose breakdown to transit time in five healthy subjects while consuming a standard Western-type diet with and without 36 g of added pectin. Each point represents a 4- to 7-day fecal collection and simultaneous transit measurement in a subject. Each subject was studied for 9 weeks and 3–4 fecal collections were analyzed from each person. (From Cummings *et al.*, 1979a.)

tibility declined. Similarly, Stephen (1980) has shown that the breakdown of dietary fiber in the large intestine can be related to transit time. Faster transit times are associated with incomplete metabolism of dietary fiber.

There is, therefore, much evidence that the effect of fiber on colonic function can be modified by a number of factors, including the source of the fiber, its preparation, and its physical and chemical form, and by factors in the individual, such as transit time. All of these factors must in some way influence the fate of fiber in the large intestine.

4. BREAKDOWN OF FIBER IN THE COLON

Many of the effects of fiber in the colon relate to the fact that it is extensively broken down in this part of the gut. In fact, fiber breakdown, which is an anaerobic process and should therefore be called *fermentation*, is probably the principal metabolic event occurring in the large bowel of man.

The main constituent of dietary fiber is polysaccharide, as recognized by Trowell (1972) in his original definition. Plant cell wall polysaccharides are a diverse group of substances chemically, but can be divided into two broad classes: (1) cellulose and (2) noncellulosic polysaccharides. Table I shows the results of a number of studies in which the breakdown of cellulose in man has been reported. It is clear from this evidence that cellulose is degraded extensively in man, on average to about 50%. Moreover, different sources of cellulose are fermented to varying extents. Cellulose present in fruit and vegetables, for example, as in the study of Prynne and Southgate (1979), is almost three-quarters fermented, while cellulose present in wheat bran or fed in a purified form (Heller, 1977; Van Soest *et al.*, 1978) is broken down to a much less extent. Cellulose is a relatively insoluble material and its breakdown by the colonic microflora requires that they gain access to its surface (Berg *et al.*, 1972). The rate of breakdown is related to the surface area of cellulose available. Highly crystaline celluloses isolated from wood are much less susceptible to enzymic breakdown than the cellulose present in fruit and vegetable cell walls. Cellulose digestion is also a relatively slow process. Rumen studies (Van Soest, 1973, 1975; Van Soest and Robertson, 1976) indicate that cellulose digestion may proceed for up to 48 hr, supporting a similar concept for man illustrated in Fig. 3.

The majority of the polysaccharides in the plant cell wall are in fact noncellulosic in type. These are chemically very different from cellulose in that they have a much more open molecular structure, are water-soluble, and are quite diverse in their properties. Table II shows evidence for their breakdown in the human. It will be noted that all studies show extensive breakdown and that in general these substances are metabolized to a greater extent than cellulose. Some water-soluble polysaccharides, such as pectin, are completely fermented in the

TABLE I. Cellulose Digestion in Man[a]

Subjects (*N*)	Sources	Intake (g/day)	Percent digested (range)	Reference
Young men (3)	Wheat bran	6.9	29 (24–33)	Williams and Olmsted (1936)
	Carrot	8.7	67 (62–72)	
	Peas	9.6	45 (40–49)	
	Cabbage	9.1	55 (43–61)	
	Celluflour	10.1	7 (0–11)	
Young men (16)	All bran	10–19	(0–6)	Hoppert and Clark (1945)
	Lettuce	5.6	29	
	Cabbage	4.1	42	
	Celery	8.0	29	
	Oranges	7.5	24	
	Apples	10.4	57	
Young men (12)	Mixed diet	8.0	15 (−7 to 29)	Southgate and Durnin (1970)
Young women (14)	Mixed diet	5.2	26 (6–40)	
Elderly men (11)	Mixed diet	7.9	44 (21–59)	
Elderly women (12)	Mixed diet	5.9	26 (−9 to 51)	
Adults (16)	Mixed diet	8.5	43 (15–87)	Milton-Thompson and Lewis (1971)
Adults (5)	Mixed diet	4.9	82 (69–91)	Southgate *et al*. (1976)
	Mixed diet + bran biscuit	7.7	64 (42–73)	
Young men (8)	Wheat bran			
	Coarse	2.7	6	Heller (1977)
	Fine	2.7	23	
Young men	Solka floc	—	20	Van Soest *et al*. (1978)
	Cabbage	—	81	
Adults (14)	Mixed diet	10.5	74	Farrel *et al*. (1978)
	Mixed diet + bran	18.7	63	
Adults (4)	Mixed diet	6.1	82	Cummings *et al*. (1979a)
Adults (4)	Mixed diet	4.9	73 (62–81)	Prynne and Southgate (1979)

[a] These studies were all performed as “balances” in which cellulose intake and fecal excretion were measured in subjects on controlled or known diets.

TABLE II. Noncellulosic Polysaccharide Digestion in Man[a]

Subjects (N)	Sources	Intake	Percent digested (range)	Reference
Young men (3)	Wheat bran	14.3	35 (32–39)	Williams and Olmsted (1936)
	Carrot	7.8	84 (80–89)	
	Peas	2.4	84 (80–87)	
	Cabbage	6.9	80 (79–80)	
	Agar	13.4	60 (55–65)	
Young men (12)	Mixed diet	13.5	72 (69–78)	Southgate and Durnin (1970)
Young women (14)	Mixed diet	11.0	77 (73–84)	
Elderly men (11)	Mixed diet	20.4	84 (80–87)	
Elderly women (12)	Mixed diet	15.0	85 (82–89)	
Adults (5)	Mixed diet	9.2	76 (60–90)	Southgate *et al.* (1976)
	Mixed diet + bran biscuit	19.5	72 (66–78)	
Young men (8)	Wheat bran			
	Coarse	9.0	50	Heller (1977)
	Fine	9.0	54	
Young men	Cabbage	—	53	Van Soest *et al.* (1978)
Adults (14)	Mixed diet	18.5	99	Farrell *et al.* (1978)
	Mixed diet + bran	21.4	42	
Adults (4)	Mixed diet	8.8	81	Cummings *et al.* (1979a)
	Mixed diet + pectin	39.6	93	
Adults (4)	Mixed diet	14.7	73	
	Mixed diet + ispaghula	37.4	87[b]	

[a] See footnote to Table I.
[b] Apparent digestibility of ispaghula only.

human gut by all subjects (Cummings *et al.*, 1979a), but this does not apply to all. For example, in the study of Prynne and Southgate (1979) one subject at least digested ispaghula, which is a water-soluble pentosan, very poorly. The findings fit well with animal studies and show that the breakdown of dietary fiber in man has many parallels with similar processes in other species.

Lignin is an important plant cell wall component and is present in the human diet in small but measurable quantities. Its breakdown is an aerobic process and therefore one would not expect much metabolism to occur in the anaerobic regions of the large intestine. Current evidence suggests that it is not broken down to any great extent in the gut, although loss of some side chains may occur (Gordon, 1978). Lignin, however, does influence the fermentation of the carbohydrate components of plant cell walls. In general, the more lignified a cell wall or plant structure, the less liable it is to complete fermentation in the gut. This can be illustrated by looking at the data for the breakdown of wheat bran in human studies. In Tables I and II it is clear that the nonstarch polysaccharides in wheat bran are broken down much less than those present in less lignified cell walls, such as cabbage or apples. How this inhibition of fermentation occurs is not clear, but it is a further example of how the detailed structure and composition of plant cell wall material alters its role in the human gut.

The most likely site for fermentation is undoubtedly the large intestine. The extent to which cell wall polysaccharides are metabolized in other parts of the gut remains to be established, but current evidence suggests that the colon is the primary location. The unique feature of polysaccharide metabolism in the colon is that it is an anaerobic process performed by the microorganisms of the large bowel. Its end products are different from those of aerobic metabolism in the rest of the body and so make intraluminal events in the colon especially important. The three main end products of fermentation are short-chain fatty acids, various gases, and energy.

5. SHORT-CHAIN FATTY ACID METABOLISM

Acetic, propionic, and butyric acids are the main anions in the large intestine of all mammalian species studied and similarly are the major anions in human feces. No values for other parts of the colon have as yet been obtained, due principally to the difficulties in obtaining material from this organ in *vivo*. Short-chain fatty acid (SCFA) concentrations in human feces average around 75 mmol/kg, with values over 100 mmol being reported (Rubinstein *et al.*, 1969; Bjork *et al.*, 1976; Zijlstra *et al.*, 1977; Cummings *et al.*, 1979b). The principal ion is acetate, which represents about 60% of total SCFA concentration. These levels of SCFA in the human are very similar to those found in animals, suggesting that a fermentation pattern in the human colon exists which is similar to that in the rumen and the colon of herbivorous species.

The main source of SCFA is carbohydrate and in man it is presumed that this is entirely plant cell wall carbohydrates. However, any carbohydrate would serve as a substrate for anaerobic metabolism, including starch and glycoprotein. Total production of SCFA in the human colon each day is at the present time unknown. However, an equation for fermentation in the human colon has been derived by Miller and Wolin (1979) based on the molar ratios of SCFA in feces and known production of carbon dioxide and methane:

$$34.4C_6H_{12}O_6 \rightarrow 64SCFA + 23.75CH_4 + 34.23CO_2 + 10.5H_2O$$

On the basis of this equation it can be calculated that if around 20 g of cell wall polysaccharides and other carbohydrates are broken down in the human colon each day, approximately 200 mmol of short chain fatty acids will be produced, of which 62 will be acetate, 25 propionate, and 16 butyrate. Clearly in those individuals who consume greater dietary fiber intakes, SCFA production will be correspondingly increased.

The fate of SCFA is to be absorbed from the human colon (McNeil *et al.*, 1978; McNeil, 1980; Ruppin *et al.*, 1980), as they are from the colon of all other species studied. The mechanism of transport remains to be worked out, but the effect of SCFA absorption is to stimulate sodium and water absorption from the colon (Crump *et al.*, 1980; Roediger and Moore, 1981) and the secretion of bicarbonate. Following absorption into the mucosal cell, SCFA provide an important source of energy for colonic mucosa, particularly butyrate (Roediger, 1980). Acetate and propionate are transported directly to the liver, where they are available for energy metabolism in the usual way. It can be seen from this, therefore, that the fermentation products of dietary fiber are absorbed and are available for energy metabolism in man and that dietary fiber, when digested, can be considered legitimately to have a calorie value. The role of SCFA in the human colon has probably been considerably underestimated and few studies of colonic metabolism have taken it into account. It is likely that SCFA are involved in every aspect of colonic function and so therefore is dietary fiber.

Other implications of SCFA production in the human colon relate to their effects on nitrogen metabolism, pH, and mucus production and their antibacterial effect (Cummings, 1981).

6. GAS PRODUCTION

A second inevitable consequence of fermentation in the colon is the production of gas. The gases derived from the breakdown of carbohydrate anaerobically are indicated in the equation. However, only 30–50% of humans, at least in Western countries, produce methane from fermentation (Bond *et al.*, 1971; Pitt *et al.*, 1980; Haines *et al.*, 1977), which is strikingly different from rumen fermentation. The main gases in man are hydrogen and carbon dioxide (Levitt and Bond, 1970). The importance of gas production lies mainly in the symptoms it causes in the gut. These include abdominal colic, feelings of distension, malodorous breath, and bowel disturbance, all of which people attribute to the presence of gas in the colon. The role of gas production in the etiology of colonic disease in man has never been considered in any great detail, although in individual cases it clearly leads to problems (Levitt *et al.*, 1976).

7. BACTERIAL METABOLISM

The third main by-product of the fermentation of cell wall polysaccharides in the human colon is the production of energy that is used by the microflora for cell growth and division. The presence of digestible fiber in the diet therefore stimulates microbial growth in the colon and fecal microbial cell mass. In people living on Western-type diets with relatively low dietary fiber intakes, the fecal microflora are the main component of stool. They represent approximately 55% of the dry weight of feces and up to 80% of the wet weight. When such subjects are given dietary fiber from cabbage for example, an increase in microbial cell excretion is seen and this contributes significantly to the 60% increase in stool output seen in these subjects with this change in diet (Stephen and Cummings, 1980). Not all types of dietary fiber stimulate microbial growth to this extent. The much less digestible wheat fiber produces only a small change in fecal colonic microbial excretion, presumably due to the fact that is is poorly digested compared with cabbage. Nor do all digestible types of fiber lead to big changes in stool output. Pectin is notable in its lack of effect on fecal bulk (Durrington *et al.*, 1976; Cummings *et al.*, 1979a). The reasons for this are unknown at present.

Other consequences of the stimulation of microbial growth by dietary fiber breakdown include alterations in nitrogen metabolism in the colon (Cummings *et al.*, 1981) and in the conversion of primary to secondary bile acids (Wicks *et al.*, 1978).

8. CONCLUSION

Fiber has important effects on colonic microbial metabolism. The microflora ferment fiber, leading to the production of SCFA, gas, and energy that they use for growth. The stimulation of microbial cell growth reroutes much of colonic nitrogen into bacterial protein, while the production of SCFA stimulates salt and water absorption from the colon and provides an energy substrate for epithelial growth. Other factors, however, such as transit time, contribute to the control of these aspects of colonic function, since marked individual variation is seen despite similar dietary intakes.

REFERENCES

Baird, I. M., Walters, R. L., Davies, P. S., Hill, M. J., Drasar, B. S., and Southgate, D. A. T., 1977, The effects of two dietary fiber supplements on gastrointestinal transit, stool weight and frequency, and bacterial flora, and fecal bile acids in normal subjects, *Metabolism* **26:**117–128.

Berg, B. B., Hofsten, B. Van, and Pettersson, G., 1972, Electron-microscopic observations on the degradation of cellulose fibres by Cellvibrio fulvus and sporocytophaga myxococcoides, *J. Appl. Bacteriol.* **35:**215–219.

Bjork, J. T., Soergel, K. H., and Wood, C. M., 1976, The composition of "free" stool water, *Gastroenterology* **70:**A6/864.

Bond, J. H., Engel., R. R., and Levitt, M. D., 1971, Factors influencing pulmonary methane excretion in man, *J. Exp. Med.* **133:**572–588.

Brodribb, A. J. M., and Groves, C., 1978, Effect of bran particle size on stool weight, *Gut* **19:**60–63.

Crump, M. H., Argenzio, R. A., and Whipp, S. C., 1980, Effects of acetate on absorption of solute and water from the pig colon, *Am J. Vet. Res.* **41:**1565–1568.

Cummings, J. H., 1978, Physiological effects of dietary fibre in man, in: *Topics in Gastroenterology 6* (S. C. Truelove and M. F. Hayworth, eds.), Blackwell, Oxford, pp. 49–62.

Cummings, J. H., 1981, Short chain fatty acids in the human colon, *Gut* **22:**763–779.

Cummings, J. H., Southgate, D. A. T., Branch, W. J., Houston, H., Jenkins, D. J. A., and James, W. P. T., 1978, Colonic response to dietary fibre from carrot, cabbage, apple, bran and guar gum, *Lancet* **1:**5–8.

Cummings, J. H., Southgate, D. A. T., Branch, W. J., Wiggins, H. S., Houston, H., Jenkins, D. J. A., Jivraj, T., and Hill, M. J., 1979a, The digestion of pectin in the human gut and its effect on calcium absorption and large bowel function, *Br. J. Nutr.* **41**(3):477–485.

Cummings, J. H., Hill, M. J., Bone, E. S., Branch, W. J., and Jenkins, D. J. A., 1979b, The effect of meat protein and dietary fiber on colonic function and metabolism. Part II. Bacterial metabolites in feces and urine, *Am. J. Clin. Nutr.* **32:**2094–2101.

Cummings, J. H., Stephen, A. M., and Branch, W. J., 1981, Implications of dietary fiber breakdown in the human colon, in: *Banbury Report No. 7, Gastrointestinal Cancer: Endogenous Factors* (R. Bruce, S. Tannenbaum, and P. Correa, eds.), Cold Spring Harbor Laboratory, Cold Spring Harbor, New York, pp. 71–81.

Durrington, P. N., Manning, A. P., Bolton, C. H., and Hartog, M., 1976, Effect of pectin on serum lipids and lipoprotein, whole-gut transit-time, and stool weight, *Lancet* **2:**394–395.

Eastwood, M. A., Kirkpatrick, J. R., Mitchell, W. D., Bone, A., and Hamilton, T., 1973, Effects of dietary supplements of wheat bran and cellulose on faeces and bowel function, *Br. Med. J.* **4:**392–394.

Eastwood, M. A., Brydon, W. G., and Tadess, K., 1980, Effect of fiber on colonic function, in: *Medical Aspects of Dietary Fiber* (G. A. Spiller and R. M. Kay, eds.), Plenum Press, New York, pp. 1–26.

Farrell, D. J., Girle, L., and Arthur, J., 1978, Effects of dietary fibre on the apparent digestibility of major food components and on blood lipids in men, *Austr. J. Exp. Biol. Med. Sci.* **56:**469–479.

Gordon, A. J., 1978, The chemical structure of lignin and quantitation and qualitation methods of analysis in foodstuffs, in: *Topics in Dietary Fiber Research* (G. A. Spiller, ed.), Plenum Press, New York, pp. 59–103.

Gray, H., and Tainter, M. L., 1941, Coloid laxatives available for clinical use, *Am. J. Dig. Dis.* **8:**130–139.

Haines, A., Metz, G., Dilawari, J., Blendis, L., and Wiggins, H. S., 1977, Breath-methane in patients with cancer of the large bowel, *Lancet* **2:**481–483.

Heller, S. N., 1977, The Effect of Particle Size of Dietary Wheat Bran on Colonic Function in Young Adult Men, Ph.D. Thesis, Cornell University, Ithaca, New York.

Heller, S. N., Hackler, L. R., Rivers, J. M., Van Soest, P. J., Roe, D. A., Lewis, B. A., and Robertson, J., 1980, Dietary fiber: The effect of particle size of wheat bran on colonic function in young adult men, *Am. J. Clin. Nutr.* **33:**1734–1744.

Hoppert, C. A., and Clark, A. J., 1945, Digestibility and effect on laxation of crude fibre and cellulose in certain common foods, *J. Am. Dietet. Assoc.* **21:**157–160.

Kirwan, W. O., Smith, A. N., McConnell, A. A., Mitchell, W. D., and Eastwood, M. A., 1974, Action of different bran preparations on colonic function, *Br. Med. J.* **4:**187–189.

Levitt, M. D., and Bond, J. H., 1970, Volume, composition and source of intestinal gas, *Gastroenterology* **59:**921–929.

Levitt, M. D., Lasser, R. B., Schwartz, J. S., and Bond, J. H., 1976, Studies of a flatulent patient, *N. Engl. J. Med.* **295:**260–262.

MacConnell, A. A., Eastwood, M. A., and Mitchell, W. D., 1974, Physical characteristics of vegetable foodstuffs that could influence bowel function, *J. Sci. Food Agric.* **25:**1457–1464.

McNeil, N. I., 1980, Short Chain Fatty Acid Absorption from the Human Colon, M.D. Thesis, University of Cambridge, Cambridge, England.

McNeil, N. I., Cummings, J. H., and James, W. P. T., 1978, Short chain fatty acid absorption by the human large intestine, *Gut* **19:**819–822.

Miller, T. L., and Wolin, M. J., 1979, Fermentations by saccharolytic intestinal bacteria, *Am. J. Clin. Nutr.* **32:**164–172.

Milton-Thompson, G. J., and Lewis, B., 1971, The breakdown of dietary cellulose in man, *Gut* **12:**853.

Pitt, P., De Bruijn, K. M., Beeching, M. F., Goldberg, E., and Blendis, L. M., 1980, Studies on breath methane: The effect of ethnic origins and lactulose, *Gut* **21:**951–959.

Prynne, C. J., and Southgate, D. A. T., 1979, The effects of a supplement of dietary fibre on faecal excretion by human subjects, *Br. J. Nutr.* **41:**495–503.

Robertson, J. A., and Eastwood, M. A., 1981, An examination of factors which may affect the water holding capacity of dietary fibre, *Br. J. Nutr.* **45:**83.

Robertson, J. A., Eastwood, M. A., and Yeoman, M. M., 1980, An investigation into the physical properties of fibre prepared from several carrot varieties at different stages of development, *J. Sci. Food Agric.* **31:**633–638.

Roediger, W. E. W., 1980, The colonic epithelium in ulcerative colitis: An energy-deficiency disease? *Lancet* **2:**712–715.

Roediger, W. E. W., and Moore, A., 1981, Effect of short chain fatty acid on sodium absorption in isolated human colon perfused through the vascular bed, *Dig. Dis. Sci.* **26:**100–106.

Royal College of Physicians, 1980, *Medical Aspects of Dietary Fibre,* Pitman Medical, London.

Rubinstein, R., Howard, A. V., and Wrong, O. M., 1969, *In vivo* dialysis of faeces as a method of stool analysis—The organic anion component, *Clin. Sci.* **37:**549–564.

Ruppin, H., Bar-Meir, S., Soergel, K. H., Wood, C. M., and Schmitt, M. G., 1980, Absorption of short chain fatty acids by the colon, *Gastroenterology* **78:**1500–1507.

Southgate, D. A. T., and Dernin, J. V. G. A., 1970, Calorie conversion factors. An experimental reassessment of the factors used in the calculation of the energy value of human diet, *Br. J. Nutr.* **24:**517–535.

Southgate, D. A. T, Branch, W. J., Hill, M. J., Drasar, B. S., Walters, R. L., Davies, P. S., and Baird, I. M., 1976, Metabolic responses to dietary supplements of bran, *Metabolism* **25:**1129–1135.

Spiller, G. A., Chernoff, M. C., Hill, R. A., Gates, J. E., Nassar, J. J., and Shipley, E. A., 1980, *Am. J. Clin. Nutr.* **33:**754–759.

Stasse-Wolthuis, M., Albers, H. F. F., van Jeveren, J. G. C., Wil de Jong, J., Hautvast, J. G. A. J., Hermus, R. J. J., Katan, M. B., Brydon, W. G., and Eastwood, M. A., 1980, Influence of dietary fiber from vegetables and fruits, bran or citrus pectin on serum lipids, fecal lipids, and colonic function, *Am. J. Clin. Nutr.* **33:**1745–1756.

Stephen, A. M., 1980, Dietary Fibre and Human Colonic Function, Ph.D. Thesis, University of Cambridge, Cambridge, England.

Stephen, A. M., and Cummings, J. H., 1979, Water holding by dietary fibre *in vitro* and its relationship to faecal output in man, *Gut* **20:**722–729.

Stephen, A. M., and Cummings, J. H., 1980, Mechanism of action of dietary fibre in the human colon, *Nature* **284:**283–284.

Thornton, J. R., and Heaton, K. W., 1981, Do colonic bacteria contribute to cholesterol gall-stone formation. Effects of lactulose on bile, *Br. Med. J.* **282:**1018.

Trowell, H., 1972, Ischaemic heart disease and dietary fibre, *Am. J. Clin. Nutr.* **25:**926–932.

Van Soest, P. J., 1973, The uniformity and nutritive availability of cellulose, *Fed. Proc.* **32:**1804–1808.

Van Soest, P. J., 1975, Physico-chemical aspects of fibre digestion, in: *Digestion and Metabolism in the Ruminant* (I. W. McDonald and A. C. I. Warner, eds.), University of New England Publishing Unit, Armidale, Australia, pp. 351–365.

Van Soest, P. J., and Robertson, J. B., 1976, Chemical and physical properties of dietary fibre, in: *Proceedings of the Miles Symposium, Nova Scotia, Canada,* pp. 13–25.

Van Soest, P. J., Robertson, J. D., Roe, D. A., Rivers, J., Lewis, B. A., and Hackler, I. R., 1978, The role of dietary fibre in human nutrition, in: *Proceedings, 1978 Cornell Nutrition Conference for Feed Manufacturers,* pp. 5–12.

Wicks, A. C. B., Yeates, J., and Heaton, K. W., 1978, Bran and bile: Time course of changes in normal young men given a standard dose, *Scand. J. Gastroenterol.* **13:**289–292.

Williams, R. D., and Olmsted, W. H., 1936, The effect of cellulose, hemicellulose and lignin on the weight of stool. A contribution to the study of laxation in man, *J. Nutr.* **11:**433–449.

Zijlstra, J. B., Beukema, J., Wolthers, B. G., Byrne, B. M., Groen, A., and Donkert, L., 1977, Pretreatment methods prior to gas chromatographic analysis of volatile fatty acids from faecal samples, *Clin. Chem. Acta.* **78:**243–250.

3

Dietary Fiber and Colon Function in a Population Aged 18–80 Years

M. A. EASTWOOD, J. D. BAIRD, W. G. BRYDON, J. H. SMITH, S. HELLIWELL, and J. L. PRITCHARD

1. INTRODUCTION

It has been suggested that a range of diseases may be linked to the intake of dietary fiber (Burkitt and Trowell, 1975). This hypothesis has been examined in a number of ways, but rarely in a normal population. In this study fecal constituents have been measured in a defined population aged 18–80 and the association of fecal constituents with age, diet, and serum lipids investigated.

2. SUBJECTS AND METHODS

The study took place over a 6-month period between December and June. An attempt was made to recruit volunteers from as small a geographical area as possible, living within a one-mile radius. These were selected by a general practitioner (JHS) from among individuals in his practice who were living at home and known to be free from gastrointestinal disease, as adjudged by questioning and their case records. None of the subjects took laxatives. Married couples were recruited to facilitate collections. It proved difficult to recruit younger people within this age group, since they were unwilling to collect feces.

M. A. EASTWOOD, J. D. BAIRD, W. G. BRYDON, S. HELLIWELL, and J. L. PRITCHARD • Wolfson Gastrointestinal Laboratory, Metabolic Unit, University Department of Medicine, Department of Clinical Chemistry, Western General Hospital, Edinburgh EH4 2XU, Scotland. J. H. SMITH • West Granton Medical Group, Edinburgh EH4 4PL, Scotland.

This younger group was augmented by others from the same area: five nurses from the hospital nursing school, 11 university students, two members of the clinical staff, and two research students. All were active and free from gastrointestinal disease.

A total of 83 volunteers with an age range between 18 and 80 years participated. Of these, 21 provided apparently incomplete stool collections as judged from the recovery of radioopaque markers. The final number considered to have participated satisfactorily in the study was 62, of whom 33 were males and 29 were females, and this group included 14 married couples.

All subjects were visited at home by MAE, who described the study, and subsequently by a dietician (SH), who recorded their height and weight, took a detailed diet history, and gave instructions for writing a diet diary for 7 days of the study. Also at this visit, the subjects swallowed 40 markers impregnated with barium in the presence of a dietician so that their mouth-to-anus transit time could be measured (Hinton *et al.*, 1969). During the 7 days of the study all stools passed were collected in plastic slings. These were then placed in sealed plastic boxes, which were collected daily. In the laboratory, the stools were weighed and the number of barium-impregnated markers counted under a fluoroscope. Transit time, where it could be measured, was regarded as the time taken to pass 80% of the pellets (Hinton *et al.*, 1969).

The stools were then mixed, and an aliquot freeze-dried (giving dry weight) and subsequently analyzed for bile acids (Evrard and Janssen, 1968), fat (Kamer *et al.*, 1949), and neutral sterols (Miettinen *et al.*, 1965, modified). The fecal contents of sodium, potassium, calcium, magnesium, and zinc were also estimated using flame photometry and atomic absorption spectrophotometry, after the stools had been charred with nitric acid, or ashed and dissolved in concentrated HCl (Aristar grade).

On the third day of the study, the subjects fasted from 6 p.m., and on the next morning a blood sample was taken (between 6 a.m. and 9 a.m.). The serum was analyzed for cholesterol (Roschlau *et al.*, 1974), triglycerides (Egstein, 1966), and phospholipids (Zilversmit and Davis, 1950).

The daily intake of calories, protein, fat (saturated and unsaturated), carbohydrate, alcohol, vitamin D, cholesterol, total dietary fiber, cellulose, lignin, and noncellulose polysaccharide was calculated from diet data collected using food tables (Southgate *et al.*, 1976; Paul and Southgate, 1978; D. A. T. Southgate, personal communication, 1978).

3. RESULTS

In 51 subjects recovery of pellets was 80% or greater. In 11 subjects with less than 80% recovery, pellets began to appear on or after day 4 and were still present in the final stool collected, indicating a prolonged transit time. Another 16 subjects had less than 80% recovery of pellets, with a peak recovery occurring in the first 3 days. Five subjects passed no pellets. Only the data from subjects who passed more than 80% of the markers or from subjects who were considered to have a delayed transit time were included in this paper.

The mean and standard deviation heights of male and female subjects were 1.75 ± 0.07 and 1.60 ± 0.07 m, respectively. The mean and standard deviation weights were 71 ± 10 and 63 ± 11 kg, respectively.

The mean and standard deviation percentage standard weights of males and females were 98 ± 12 and 101 ± 16, respectively.

Figure 1 shows the great variation in stool weight passed. The results range from the smallest mean fecal output, 19 g/24 hr, to the largest, 278 g/24 hr. The variation in stool weight for each subject is considerable.

In Table I information on fecal characteristics is presented. The concentration of neutral sterols was higher in females than in males ($p < 0.05$). In Table II the association between stool weight and fecal characteristics for males and females is presented. There was a significant inverse correlation between stool weight and transit time ($p < 0.05$). All fecal constituents correlated strongly with stool weight. Stool weight also increased with body weight ($p < 0.05$).

In Table III the mean and standard deviation food intakes of male and female subjects are presented. In Table IV the mean and standard deviation dietary intakes of male and female subjects are presented. Energy intake of males

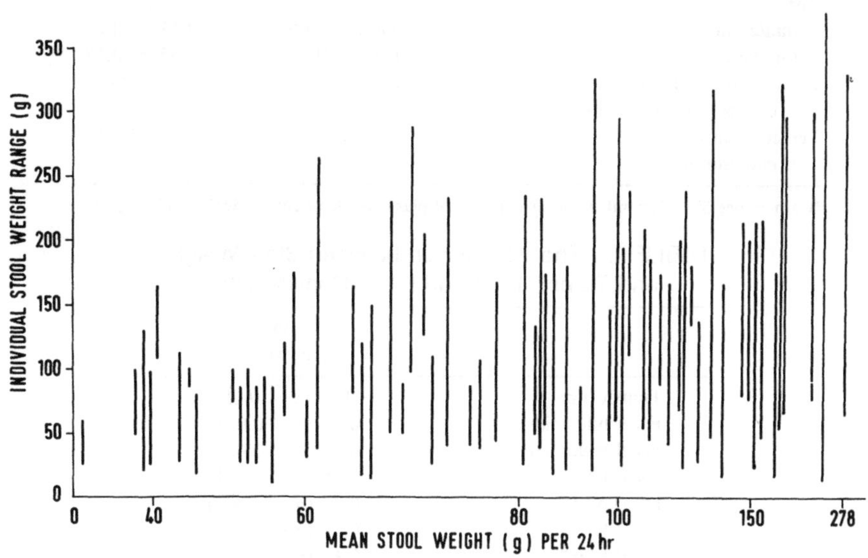

FIGURE 1. Variation in stool weight.

TABLE I. Fecal Characteristics of Subjects[a]

	Males (33)	Females (29)
Weight (wet), g/24 hr	98 ± 49	74 ± 34
Percent H_2O	73 ± 4.8	72 ± 6.3
Transit time	3.0 ± 1.5	3.6 ± 1.8
Bile acids		
mm/24 hr	0.71 ± 0.31	0.56 ± 0.35
mm/kg feces	8.0 ± 4.2	7.8 ± 3.3
Neutral sterols[b]		
mm/24 hr	1.52 ± 0.63	1.46 ± 0.82
mm/kg feces	16.9 ± 6.2	21.7 ± 10.0
Fat		
mm/24 hr	9.5 ± 5.3	8.2 ± 4.6
mm/kg feces	100 ± 42	114 ± 48
Na^+		
mm/24 hr	1.1 ± 0.9	1.0 ± 0.9
mm/kg feces	12.4 ± 10.8	13.4 ± 10.0
K^+		
mm/24 hr	9.6 ± 5.3	7.2 ± 2.7
mm/kg feces	99 ± 36	101 ± 32
Ca^{2+}		
mm/24 hr	17.1 ± 8.5	16.9 ± 9.2
mm/kg feces	183 ± 97	218 ± 96
Mg^{2+}		
mm/24 hr	6.4 ± 2.7	5.9 ± 3.9
mm/kg feces	70 ± 26	75 ± 31
Zn^{2+}		
mm/24 hr	0.13 ± 0.05	0.11 ± 0.04
mm/kg feces	1.44 ± 0.43	1.55 ± 0.58
Percent bile acids as chenodeoxy plus cholic[c]	8.1 (0–20.7)	8.2 (0–38.0)
Percent neutral sterol as cholesterol[c]	10.0 (2.2–97)	10.5 (0.5–75.4)

[a]Mean ± one SD. [b]Neutral sterols as coprostanol plus cholesterol only. [c]Median and range given.

TABLE II. The Association between Stool Weight and Transit Time and Fecal Constituents

	Correlation (r) with stool weight
Transit time	−0.25 ($p < 0.05$)
Fecal bile acids	0.40 ($p < 0.01$)
Fecal neutral sterols	0.37 ($p < 0.01$)
Fecal fat	0.50 ($p < 0.001$)
Fecal Na^+	0.43 ($p < 0.001$)
Fecal K^+	0.76 ($p < 0.001$)
Fecal Mg^{2+}	0.60 ($p < 0.001$)
Fecal Ca^{2+}	0.47 ($p < 0.001$)
Fecal Zn^{2+}	0.63 ($p < 0.001$)

TABLE III. Food Intake

	Intake,[a] g/24 hr	
	Males (33)	Females (29)
Beef	54 ± 36	45 ± 32
Pork	36 ± 35	29 ± 23
Lamb/poultry	22 ± 28	10 ± 10
Offal/processed food	31 ± 33	14 ± 16
Fish	17 ± 12	13 ± 8
Milk	367 ± 273	309 ± 133
Milk products	104 ± 55	74 ± 34
Egg/cheese dishes	7 ± 11	4 ± 9
Fruit	67 ± 84	91 ± 74
Potatoes	129 ± 99	90 ± 44
Other vegetables	85 ± 44	85 ± 54
Bread	103 ± 88	91 ± 64
Cake/puddings	49 ± 46	48 ± 40
Sugar/preserves	76 ± 54	44 ± 36
Other cereal foods	103 ± 90	72 ± 57

[a] Mean ± one SD.

TABLE IV. Daily Dietary Intake[a]

	Males (33)	Females (29)
Protein, g	81 ± 23	62 ± 12
Fat, g	109 ± 35	89 ± 21
Carbohydrate, g	275 ± 87	185 ± 56
Energy, MJ	10.5 ± 3.0	7.5 ± 1.6
Unsaturated fat, g	7 ± 4.7	5.4 ± 2.9
Cholesterol, g	0.46 ± 0.20	0.37 ± 0.13
Vitamin D, i.u.	105 ± 69	94 ± 56
Alcohol,[b] percent total calories	2.9 (0–36)	0 (0–8.3)
Calcium, g	1.10 ± 0.48	0.86 ± 0.25
Magnesium, g	0.31 ± 0.10	0.22 ± 0.08
Zinc, mg	10.6 ± 3.5	8.0 ± 1.8
Total fiber, g	15.9 ± 6.1	12.7 ± 4.6
Cellulose, g	4.0 ± 1.6	3.0 ± 1.0
Lignin, g	0.7 ± 0.6	0.7 ± 0.6
Noncellulose polysaccharide, g	11.3 ± 4.3	9.0 ± 3.3
Hexose residues, g	6.6 ± 2.6	4.7 ± 1.4
Pentose residues, g	2.3 ± 1.4	2.4 ± 1.6
Uronic acid residues, g	2.3 ± 1.1	2.0 ± 0.7

[a] Mean ± one SD.
[b] Median and range given.

TABLE V. Serum Lipids[a]

	Males (31)	Females (28)
Serum cholesterol, mmol/l	6.03 ± 1.43	6.60 ± 1.61
Serum triglyceride, mmol/l	1.21 ± 0.56	1.24 ± 0.48
Serum phospholipid, mmol/l	2.99 ± 0.52	3.29 ± 0.48

[a] Mean ± one SD. Two male and one female subjects did not provide a blood sample for serum lipids.

was greater than that of females. Females did, however, consume more fruit. In Table V the mean and standard deviation serum lipids of male and female subjects are shown.

In Table VI details of the regression coefficients of the association between age and fecal characteristics, diet, and serum lipids are presented. None of the fecal constituents correlated significantly with age, nor was there any significant change in the percentage of primary bile acids and cholesterol with age. There was a marked association between energy consumed by men and age. This was not observed in women. Serum lipid concentration increased up to age 65 in men but decreased beyond that age. This pattern was not seen in females, where all lipids showed a significant increase with age.

In Table VII the association between dietary constituents and fecal weight is presented. Dietary fiber, protein, carbohydrate, fat, and energy intake are significantly associated with stool weight. Of the dietary fiber fractions, both cellulose and noncellulose components correlated with stool weight, as did all individual residues of the noncellulose polysaccharide fraction.

TABLE VI. The Association between Age and Diet, Colon Function, and Serum Lipids

	Correlation (r) with age[a]	
	Males	Females
Energy	−0.52 ($p < 0.01$)	−0.13 (ns)
Protein	−0.58 ($p < 0.001$)	−0.06 (ns)
Carbohydrates	−0.50 ($p < 0.01$)	−0.06 (ns)
Fat	−0.36 ($p < 0.05$)	−0.06 (ns)
Alcohol	−0.35 ($p < 0.05$)	−0.06 (ns)
Total fiber	−0.49 ($p < 0.01$)	−0.37 ($p < 0.05$)
Stool weight	−0.31 (ns)	+0.12 (ns)
Transit time	−0.34 (ns)	+0.06 (ns)
Serum cholesterol	+0.10 (ns)	+0.41 ($p < 0.05$)
Serum triglyceride	+0.06 (ns)	+0.61 ($p < 0.001$)
Serum phospholipid	+0.15 (ns)	+0.58 ($p < 0.001$)

[a] ns, Not significant.

TABLE VII. The Association between Dietary Constituents and Stool Weight

	Correlation (r) with stool weight
Total fiber	0.48 ($p < 0.001$)
Cellulose	0.50 ($p < 0.001$)
Lignin	0.31 ($p < 0.05$)
Hemicellulose	0.42 ($p < 0.001$)
Hexose residues	0.44 ($p < 0.001$)
Pentose residues	0.43 ($p < 0.001$)
Uronic acid residues	0.35 ($p < 0.01$)
Protein	0.43 ($p < 0.001$)
Carbohydrate	0.37 ($p < 0.01$)
Fat	0.27 ($p < 0.05$)
Energy	0.36 ($p < 0.01$)

Dietary fiber intake correlated significantly with protein, fat, carbohydrate, and total energy intake ($p < 0.001$).

In Table VIII the effect of dietary fiber on colon function, serum lipids, and fecal constituents is presented. Transit time did not correlate significantly with dietary fiber intake. All serum lipids measured correlated significantly and inversely with total dietary fiber. Serum cholesterol correlated significantly and inversely with the intake of the hexose fraction of noncellulose polysaccharide

TABLE VIII. The Association between Dietary Fiber, Serum Lipids, and Fecal Constituents

	Correlation (r) with total dietary fiber[a]
Stool weight	0.48 ($p < 0.001$)
Transit time	(ns)
Serum cholesterol	−0.30 ($p < 0.05$)
Serum triglyceride	−0.29 ($p < 0.05$)
Serum phospholipid	−0.26 ($p < 0.05$)
Fecal bile acids	(ns)
Fecal neutral sterols	(ns)
Fecal fat	0.31 ($p < 0.05$)
Fecal Na^+	(ns)
Fecal K^+	0.44 ($p < 0.001$)
Fecal Mg^{2+}	0.44 ($p < 0.001$)
Fecal Ca^{2+}	(ns)
Fecal Zn^{2+}	0.43 ($p < 0.001$)

[a] ns, Not significant.

TABLE IX. The Association between Serum and Fecal Lipids

	Serum cholesterol	Serum triglycerides[a]	Serum phospholipids[a]
Fecal bile acids	r −0.29 ($p < 0.05$)	(ns)	(ns)
Fecal neutral			
Sterols	r −0.32 ($p < 0.05$)	(ns)	(ns)
Fecal fat (males)	r −0.58 ($p < 0.001$)	r −0.41 ($p < 0.05$)	r −0.51 ($p < 0.01$)

[a] ns, Not significant.

($p < 0.05$), and serum triglyceride correlated significantly and inversely with cellulose ($p < 0.05$) and the pentose fraction of noncellulose polysaccharide ($p < 0.05$). Fecal fat, potassium, and magnesium excretion were significantly increased by dietary fiber. Dietary fiber had no effect on bile acid or neutral sterol excretion expressed either as mmol/day or mmol/kg dry or wet feces. There was no significant correlation between relative weight and dietary fiber.

In Table IX the association between serum and fecal lipids is presented. There were weak but significant inverse correlations between serum cholesterol and both fecal neutral sterols and bile acids. There were significant inverse correlations between fecal fat and all serum lipids measured in males but not females.

Serum cholesterol, triglyceride, and phospholipids were not significantly related to dietary protein, fat, carbohydrate, or cholesterol.

4. DISCUSSION

The purpose of this study was to obtain information about fecal characteristics in a group of people free from intestinal disease and investigate the relationship of these characteristics with diet and serum lipids.

There is a high incidence of both colonic and ischemic heart disease in Scotland and similar studies in other areas might yield different results and clues to etiology.

The dietary intake in this group was similar to that observed in a Cambridgeshire village (Bingham *et al.*, 1981), with men consuming more total energy than females, but the dietary fiber intakes were higher for both males and females in the English group. Part of this difference may be ascribed to seasonal variation, since the English study was undertaken between May and August, and this study took place between December and June. The contribution of fruit to the dietary fiber intake in the Scottish group was very low (1–2 g/24 hr as fiber).

There is little information on fecal composition in a nonhospital population.

Studies on bowel function have, in general, relied on questionnaires or diaries (Hardy, 1945; Connell *et al.*, 1965; Milne and Williamson, 1972). Such reports have been of value in confirming the wide range of frequency of defecation both in working populations and the elderly. In general both old and young people have the same frequency. A study where stool weight was measured took place in a Federal Correction Institute in Texas, where 8267 stools were collected over 3 years from 115 males. Mean stool weight was 124 ± 40 g on 3500 cal/24 hr (Rendtorff and Kashgarian, 1967). The details of the diet are not available.

In this study there was a wide range of individual and mean stool weight. The range is in accord with the findings of Burkitt and Trowell (1975) and Wyman *et al.* (1978). The average fecal weight is about 90 g/24 hr. The percentage of water in the stool is constant over a small range (Table I). There was no effect of age on transit time, stool weight, and fecal constituents. However, fecal output is associated with an increase in fecal constituents. This suggests that merely increasing stool weight could result in an enhanced steroid and cation excretion.

However, in short-term experiments, the effect of increasing stool weight has a variable effect on fecal constituents (Cummings *et al.*, 1978).

Stool weight is strongly associated with diet, in particular dietary fiber, expecially the cellulose and noncellulose polysaccharide components. The association of stool weight with fiber is well documented (Eastwood *et al.*, 1973; Cummings *et al.*, 1976). Stool weight also correlates strongly with protein. It has recently been suggested that undigested protein should be considered part of the dietary fiber composite (Saunders and Betschart, 1980). The association between stool weight and other dietary components, i.e., protein, carbohydrate, and fat, is not surprising in a study of this kind. People eating their customary diet take main courses, which usually contain fiber in association with carbohydrate, protein, and fat, e.g., steak, vegetables, and potatoes will contain all four components, as will cereal, milk, and sugar. Further, in this population the major proportion of the fiber originated from cereal and bread sources, which are also rich in protein and carbohydrate.

Dietary fiber correlated weakly but significantly and inversely with serum lipid levels.

In experiments where fiber is added to the diet, pectin has been shown to be the most hypocholesterolemic of the fiber fractions (Judd *et al.*, 1976), whereas water-insoluble fiber from cereal sources has been shown overall to have little effect on serum cholesterol (Truswell and Kay, 1976). Stasse-Wolthuis *et al.* (1979) have suggested that when the feeding of a high-fiber diet is followed by a lowering of serum cholesterol, part of the fall could be attributed to high intakes of dietary fiber and the remainder to differences in fat intake. In this study there was no correlation between dietary fat and fiber intake.

Fecal fat excretion increased significantly with dietary fiber intake. Fiber may interfere with absorption from the small bowel (without causing steatorrhea) or may lead to an increased bacterial mass in the colon (Stephens and Cummings, 1980), which would lead to an increase in the measured fat in feces.

There is a significant correlation between dietary fiber intake and potassium, magnesium, and zinc excretion in the feces, but not with calcium and sodium. From *in vitro* studies fiber has been shown to have cation exchange properties (McConnell *et al.*, 1974). The selective effect of fiber on fecal cation excretion suggests a complex balance between intestinal absorption and fiber adsorption. Dietary fiber has been suggested as a possible causative factor in secondary zinc and magnesium deficiency situations (Reinhold *et al.*, 1976). Drews *et al.* (1979) have shown significant changes in fecal mineral loss (Mg, Zn, Cu) when augmenting the diet with fiber, especially hemicellulose.

There is no apparent influence of fiber on relative body weight. However, a study such as this cannot assess whether fiber is displacing dietary calories.

The effect of dietary fiber on serum lipids and colon function has usually been examined in short-term experiments where the fiber usually a single type, e.g., bran, has been added to the individual diets. In this study colon function and serum lipids have been examined without dietary intervention. There is a strong relationship between dietary fiber intake and stool weight, and a weak but significant inverse relationship between dietary fiber and serum lipids. It remains to be seen if such relationships are relevant to disease etiology.

ACKNOWLEDGMENTS. We are grateful to our subjects for their cheerful and good cooperation, to the Incorporated National Association of British and Irish Millers for generous financial support for this project and to the staff of the Metabolic Unit for serum lipid analysis. Dr. D. A. T. Southgate kindly gave us details of fiber analysis for fruit and vegetables not mentioned in Southgate *et al.* (1976). Mr. Alex Smith and Mrs. Susan Christie gave valuable technical assistance.

REFERENCES

Bingham, S., McNeil, N. I., and Cummings, J. H., 1981, A study of a randomly chosen cross section of British adults in a Cambridgeshire village, *Br. J. Nutr.* **45**:23–35.

Burkitt, D. P., and Trowell, H. L. (eds.), 1975, *Refined Carbohydrate Foods and Diseases*, Academic Press, London.

Connell, A. M., Hilton, C., Irvine, G., Lennard-Jones, J. E., and Misiewica, J. J., 1965, Variation of bowel habit in two population samples, *Br. Med. J.* **ii**:1095–1099.

Cummings, J. H., Hill, M. J., Jenkins, D. J. A., Pearson, J. R., and Wiggins, H. S., 1976, Changes in faecal composition and colonic function due to cereal fibre, *Am. J. Clin. Nutr.* **29**:1468–1473.

Cummings, J. H., Branch, W., Jenkins, D. J. A., Southgate, D. A. T., Houston, H., and James, W. P. T., 1978, Colonic response to dietary fibre from carrot, cabbage, apple, bran and guar gum, *Lancet* **1**:5–9.

Drews, L. M., Kies, C., and Fox, H. M., 1979, Effect of dietary fibre on copper, zinc and magnesium utilization by adolescent boys, *Am. J. Clin. Nutr.* **32**:1893–1897.

Eastwood, M. A., Kirkpatrick, J. R., Mitchell, W. D., Bone, A., and Hamilton, T., 1973, Effects of dietary supplements of wheat bran and cellulose on faeces and bowel function, *Br. Med. J.* **4**:392–394.

Egstein, M., 1966, Eine neue Bestimmung der Neutralfette im Blutserum und Gewebe. 2. Zuverlässigkeit der Methode, andere Neutralfette, Bestimmungen, Normalwert für Triglyceride und Glycerine im menschlichen Blut, *Klin. Wochenschr.* **44**:267–273.

Evrard, E., and Janssen, G., 1968, Gas liquid chromatographic determination of human fecal bile acids, *J. Lipid Res.* **9**:226–236.

Hardy, T. L., 1945, Order and disorder in the large intestine, *Lancet* **i**:519–524.

Hinton, J. M., Lennard-Jones, J. E., and Young, A. C., 1969, A new method for measuring gut transit using radio opaque markers, *Gut* **10**:842–847.

Judd, P. A., Kay, R. M., and Truswell, A. S., 1976, The cholesterol-lowering effect of pectin, *Nutr. Metab.* **21**:84–85.

Kamer, J. H. van de, Huinink, H. ten B., and Weyers, H. A., 1949, Rapid method for determination of fat in faeces, *J. Biol. Chem* **177**:347–355.

McConnell, A. A., Eastwood, M. A., and Mitchell, W. D., 1974, Physical characteristics of vegetable foodstuffs that could influence bowel function, *J. Sci. Food Agric.* **25**:1457–1464.

Miettinen, T. A., Ahrens, E. H. Jr., and Grundy, S. M., 1965, Quantitative isolation and gas liquid chromatographic analysis of total dietary and fecal neutral sterols, *J. Lipid. Res.* **6**:411–424.

Milne, J. S., and Williamson, J., 1972, Bowel habit in older people, *Gerentol. Clin.* **14**:56–60.

Paul, A. A., and Southgate, D. A. T., 1978, in: *The Composition of Foods* (R. A. McCance and E. M. Widdowson, eds.), HMSO, London.

Reinhold, J. G., Faradji, B., Abadi, P., and Ismail-Beigi, F., 1976, Decreased absorption of calcium, magnesium, zinc, and phosphorus by humans due to increased fibre and phosphorus consumption as wheat bread, *J. Nutr.* **106**:493–503.

Rendtorff, R. C., and Kashgarian, M., 1967, Stool patterns of healthy adult males, *Dis. Colon Rectum* **10**:222–228.

Roschlau, P., Bernt, E., and Gruber, W., 1974, Enzymatic determination of total cholesterol in serum, *Z. Klin. Chem. Biochem.* **12**:403–407.

Saunders, R. M., and Betschart, A. A., 1980, The significance of protein as a component of dietary fibre, *Am. J. Clin. Nutr.* **33**:960–961.

Stasse-Wolthuis, M., Hautvast, J. G. A. J., Hermus, R. J. J., Katan, M. B., Bausch, J. E., Rietberg-Brussard, J. H., Velema, J. P., Zondervan, J. H., Eastwood, M. A., and Brydon, W. G., 1979, The effect of a natural high-fibre diet on serum lipids, fecal lipids and colon function, *Am. J. Clin. Nutr.* **32**:1881–1888.

Stephen, A. M., and Cummings, J. H., 1980, Mechanisms of action of dietary fibre in the human colon, *Nature* **284**:283–284.

Southgate, D. A. T., Bailey, B., Collinson, E., and Walker, A. F., 1976, A guide to calculating intakes of dietary fibre, *J. Human Nutr.* **30**:303–313.

Truswell, A. S., and Kay, R. M., 1976, Bran and blood lipids, *Lancet* **1**:367.

Wyman, J. B., Heaton, K. W., Manning, A. P., and Wicks, A. C. B., 1978, Variability of colonic function in healthy adults, *Gut* **19**:146–150.

Zilversmit, D. B., and Davis, A. K., 1950, Microdetermination of plasma phospholipids by trichloracetic acid precipitation, *J. Lab. Clin. Med.* **35**:155–160.

4

Colonic Bacterial Activity

Effect of Fiber on Substrate Concentration and on Enzyme Action

M. J. HILL

1. INTRODUCTION

As currently used by most physicians, the term *dietary fiber* includes all of the nonstarch polysaccharides, lignins, sugar alcohols, and low-molecular-weight carbohydrates not metabolized by the enzymes of the human small intestinal mucosa. If this definition of dietary fiber as the indigestible residue of the carbohydrate fraction of the diet is accepted, then it should clearly also include a proportion of the dietary starch, the magnitude of which varies with the extent and type of cooking.

"Fiber" conjures up a very clear picture to the layman, and most members of the general public are very clear about what they think they have chosen when they elect to eat a high-fiber diet. Unfortunately, when they read (almost daily) in their newspapers of the many magical properties of dietary fiber, the physician who has painted the glowing picture has clearly included the effects of many carbohydrates that the layman would never consider to be fibrous.

In this chapter I will discuss the effect of dietary fiber on bacterial metabolism under three headings. First, I will consider the effect of dietary fiber on substrate concentration. Second, I will describe the effect of fiber on the intestinal physiology and the effect of these changes on bacterial enzyme activity. Third, I will discuss the effect of the products of fiber metabolism on bacterial enzyme action.

M. J. HILL • Bacterial Metabolism Research Laboratory, PHLS Centre for Applied Microbiology and Research, Salisbury SP4 0JG, England.

2. EFFECT OF FIBER ON SUBSTRATE CONCENTRATION

There is a growing body of data on the effect of various classes of dietary fiber on the substrates of bacterial enzymes; to describe the effects it is necessary to divide dietary fiber into (1) sequestering agents (such as lignin), (2) gel-forming fibers (such as pectin and guar), (3) readily fermentable fibers (such as lactulose), and (4) cereal fiber (such as wheat bran).

Some components of dietary fiber, such as lignin, are known to have anion binding properties *in vitro* (Eastwood and Hamilton, 1968; Story *et al.*, 1979) and it is assumed that these properties are also manifest *in vivo*. Dietary fibers rich in such components would be expected to have a cholestyraminelike action and increase the fecal loss of acid steroids and fatty acids while having no effect on the loss of neutral substances such as the neutral steroids. Such an effect was observed when healthy volunteers were given 10.5 g bagasse (sugar cane fiber) in their diet (Table I); the daily loss of bile acid was increased by 65% while the daily loss of neutral steroid was unchanged (Baird *et al.*, 1977). Since this fiber also caused a 67% increase in fecal bulk, the fecal concentration of bile acid was similar to the control value, while neutral steroid was diluted.

Pectic substances such as pectin and guar are gel-forming agents; if such gels were formed in the small intestine, it is reasonable to expect that they would inhibit the absorption of compounds trapped in the gel. The strength of the gel,

TABLE I. The Effect in Humans of Dietary Fiber on Fecal Loss per Day and on Fecal Concentration of Steroids[a]

	Acid steroids		Neutral steroids		
Dietary fiber, g/day	Daily loss	Concentration	Daily loss	Concentration	Reference
Bagasse, 10	165	99	100	60	Baird *et al*. (1977)
Lignin	—	153	—	—	Kay (1981)
Wheat bran					
16	100	60	100	60	Eastwood *et al*. (1973)
39	104	63	108	64	Walters *et al*. (1975)
54	91	55	97	58	Kay and Truswell (1977a)
100	171	57	168	56	Cummings *et al*. (1976)
Oat bran, 100	—	125	—	—	Kay (1981)
Pectin					
15	140	108	117	89	Kay and Truswell (1977b)
36	146	117	141	113	Thompson and Hill (1980)
Guar, 36	177	109	168	104	Thompson and Hill (1980)
Cellulose, 16	100	60	100	60	Eastwood *et al*. (1973)

[a] Both expressed as a percentage of the control value.

and therefore its ability to "trap" substrates, is of course related to the concentration of the pectin substance. This may be the mechanism by which pectin and guar increase the fecal loss of acid steroids, neutral steroids, fats, and amino acids such as tryptophan. The effect is dose-related and is nonspecific because the trapping action is nonspecific; consequently the magnitude of the increase in fecal loss of bile acids and of neutral steroids is similar (Table I). Dietary pectin and guar cause an increase in fecal bulk, but this is smaller than the increase in fecal loss of steroids and so there is an increase in fecal concentration of both acid and neutral steroids.

Totally fermentable "fiber" which is non-gel-forming does not "trap" or sequester molecules and so does not prevent their absorption from the small intestine. Such dietary fibers (e.g., lactulose, mannitol) do not increase daily fecal loss of steroids or fat; however, they are potent fecal bulking agents and so cause a nonspecific reduction of the fecal concentration of acid steroids, fats, and neutral steroids. The mechanism of the increased fecal bulking has been discussed elsewhere (Cummings, Chapter 2, this volume; Stephen and Cummings, 1980); via this mechanism we would expect that those substrates that can be used as nutrients for bacterial synthesis would be removed from the colon, and such substances include fatty acids and ammonia. This has been demonstrated; Agostini *et al.* (1972) showed that lactulose therapy led to a decreased fecal loss of ammonia, while Bown *et al.* (1974) showed that in persons on lactulose treatment the cecal pH was only 4.8 while the fecal pH was 6.6. This loss of short-chain fatty acids from the colon is probably due both to absorption from the colon and to synthesis of bacterial cell mass.

Cereal fiber appears to have a special place in its effect on substrate concentration. It is poorly digested during its transit through the gut (Stephen and Cummings, 1980), but appears to be a poor sequestering agent and does not have gel-forming properties, although it is a good stool-bulking agent. When given in low doses its only effect appears to be stool bulking; there is no increased daily loss of fecal steroids (Eastwood *et al.*, 1973) and so their concentration is reduced in proportion to the increase in stool bulk. At high doses, in contrast, there is a sequestering action manifest by the increased daily loss of bile acid (Cummings *et al.*, 1976); although the bulking action is greater, this is offset by the increased daily loss of steroid and so their fecal concentration is decreased no more than by small doses of bran. The above applies to wheat bran, but oatmeal behaves differently in that it has a greater effect on fecal bile acid loss (Kay, 1981); this has been thought to be due to the unsaturated fat content of oatmeal. It is of interest that the effect of wheat bran on the fecal loss of steroids was greatly enhanced by a high-protein diet (Cummings *et al.*, 1979).

All of the studies of the effects of fiber supplements have been short-term and there is an urgent need for longer-term studies.

3. PHYSIOLOGICAL EFFECTS OF FIBER ON ENZYME ACTION

Fiber affects intestinal physiology directly and via its metabolites; the effect of the metabolites on intestinal pH is discussed in Section 4. The major direct physiological effects are on intestinal transit time and on the multiphasic composition of the intestinal contents.

The intestinal transit time (ITT) represents the time available to the bacteria to carry out their metabolic reactions; this has been assumed to be the major factor in determining the extent to which a substrate is metabolized by bacterial enzymes. However, the situation is not simple. When the transit time is shorter than T_1 (Fig. 1) the colonic contents pass through so rapidly that little metabolism can take place. This is only partly because the time available is too short; other factors involved include the relatively low degree of anoxia that a very rapid transit might induce. When the transit time is between T_1 and T_2 the extent of metabolism of a substrate is proportional to the transit time, but when more time is available than T_2, then the extent of metabolism is curtailed by other factors, such as competing reactions and dehydration of the colonic contents. The times

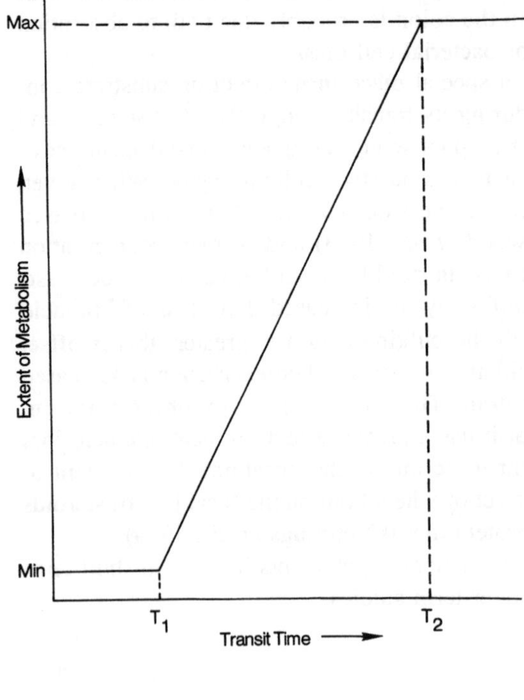

FIGURE 1. The relation between intestinal transit time (ITT) and the extent of metabolism of a substrate by the intestinal bacteria. T_1 and T_2 will vary with the nature of the substrate.

TABLE II. Relation between the Extent of Metabolism of Cholesterol and the Intestinal Transit Time in Persons on a Normal Diet and on One Supplemented with Bagasse

	Normal diet		Bagasse diet	
Subject	ITT, hr	Percent degradation	ITT, hr	Percent degradation
A	31	38	46	71
B	28	92	26	81
C	43	65	37	79
D	61	72	45	71
E	32	73	31	78
F	81	83	38	80
G	38	90	24	87
H	28	65	34	73
I	57	70	61	88
J	62	85	35	86

T_1 and T_2 will vary with the substrate, but we would only expect an effect of fiber on bacterial enzymic activity if the transit times during the control period and during the high-fiber period were between T_1 and T_2. This is clearly not so for cholesterol metabolism (Table II) since the extent of reduction to coprostanol was unchanged by dietary supplements of bagasse, bran, fibrogel pectin, and guar (Table III). Similarly there was little effect on the overall extent of bile acid dehydroxylation by bran supplements, even though these had a marked effect on transit time (Jivraj and Hill, unpublished results). With both of these substrates the ITT achieved by bran supplements (and supplements of the other fibers) was still higher than T_2. In contrast, there was a correlation between transit time and the amount of urinary volatile phenol produced by bacterial metabolism of tyrosine (Cummings *et al.*, 1979).

Undigested dietary components, including dietary fiber, are present in the

TABLE III. The Effect of Various Dietary Supplements on the Percentage Conversion of Cholesterol to Coprostanol and Coprostanone

		Percentage metabolism of cholesterol	
Dietary supplement	Number of persons	Test diet	Control
Bagasse	10	76	77
Bran	37	74	72
Fibrogel	37	73	72
Guar	5	79	82
Pectin	5	79	81

colon as solid particles and may form a matrix on which microbial reactions might take place. This would happen if the matrix was one to which the substrate was adsorbed. It is known that bacteria grow preferentially on solid surfaces rather than freely suspended in fluid and so at the matrix surface there would be high concentrations of both substrate and bacterial enzyme—a situation similar to that in classical catalysis. The effect of this increase in surface area in the colonic contents on bacterial enzyme activity has not been studied but would undoubtedly yield interesting results.

There have been suggestions that dietary fiber might modify bacterial enzyme activity by modifying the composition of the flora itself. This modification was monitored by measuring the activity of such "sentinel enzymes" as β-glucuronidase and azoreductase (Goldin and Gorbach, 1976; Mastromarino *et al.*, 1976). Positive reports of changes in sentinel enzyme activity have come mainly from animal studies; our own (unpublished) studies of glycosidases in human feces revealed a large day-to-day variation in enzyme activity but little evidence of any effect of diet.

4. METABOLISM OF DIETARY FIBER

The polysaccharide component of dietary fiber is metabolized by the colonic bacteria to mono- and disaccharides and then to CO_2, H_2, and short-chain fatty acids; the nature of the fatty acid products depends on the organism, but the principal ones are acetic, lactic, propionic, and butyric acids (in that order). From the work of Stephen and Cummings (1980) it is clear that a very high proportion of dietary fiber is degraded in this way, the fatty acid either being used in the synthesis of bacterial mass or being absorbed and used as nutrient by the host animal. When the dietary fiber is rapidly metabolized, as with lactulose, the local concentration of short-chain fatty acid is sufficient to cause a fall in the pH. A dose of 30 g lactulose in five volunteers gave a mean pH of 4.8 in the cecum and ascending colon; this is important because the right colon is the main site of bacterial metabolism of a range of substrates. Evidence for this comes from studies of patients who have had a right hemicolectomy; their metabolism of bile acids, cholesterol, and tyrosine is unimpaired (Table IV).

The effect of lactulose on ammonia metabolism has been investigated extensively because of the successful use of this disaccharide in the treatment of hepatic coma. Originally the rationale for lactulose therapy was that the pH achieved would provide a gradient against which ammonia (in the form of its cation NH_4^+) could not be absorbed (Elkington, 1970; Bircher *et al.*, 1971). The success of the treatment in preventing ammonia intoxication and the demonstrable fall in intestinal pH was thought to confirm this mechanism. However, if ammonia were trapped in the colon in this way, then the fecal concentration of ammonia

TABLE IV. Metabolism of Bile Acids, Cholesterol, and Tyrosine by the Intestinal Bacteria in Persons with Hemicolectomy, Complete Colon, and Total Colectomy

	Total colectomy	Hemicolectomy		Normal colon
		Right	Left	
Number of patients	7	2	12	12
Percent cholesterol remaining undegraded	>85%	58%	29%	23%
Extent of dehydroxylation of fecal bile acids	−	+	+ + +	+ + +
Urinary volatile phenols produced from tyrosine, mg/day	<10	—	50–70	50–100

should be increased. Agostini *et al.* (1972) demonstrated that this is not so, and in later publications Vince and Burridge (1980) showed that lactulose inhibits ammonia production from nonurea sources. Stephen and Cummings (1980) have suggested, alternatively, that the products of lactulose metabolism are utilized in the synthesis of a new bacterial mass and that this results in the utilization of large amounts of colonic ammonia. It is possible that all three mechanisms are, in fact utilized; the synthesis of nonurea ammonia may be impaired, and the absorption of the ammonia generated from such sources as urea may be inhibited by the pH gradient, with this inhibition allowing it to be utilized in the synthesis of bacterial mass, and the overall effect being a very low concentration of ammonia in feces.

It is known that the bacterial enzymes metabolizing steroids tend to have pH optima close to 7; in particular, the 7α-dehydroxylase that produces lithocholic acid from chenodeoxycholic acid and deoxycholic acid from cholic acid has greatly reduced activity at acid pH and is not even produced *in vitro* when the pH is below 6 (Midtvedt and Norman, 1968; Aries and Hill, 1970). It was suggested therefore (Hill, 1974) that lactulose might greatly reduce the extent of bile acid metabolism to the dehydroxylated products; this has been confirmed recently. Thornton and Heaton (1980) have shown that lactulose therapy causes a reduction in the proportion of deoxycholic acid in bile of humans, and Owen (1982) has shown a similar reduction in the amount of dehydroxylation products in feces, while Kay (1981) obtained similar results with hamsters. Thornton and Heaton (1980) suggested that this result with lactulose is merely a further demonstration of the phenomenon that was reported following bran therapy for gallstones, i.e., that dietary fiber causes a decrease in the rate of production of deoxycholic acid in the human colon (Pomare and Heaton, 1973; Low-Beer and Nutter, 1978), and further suggested that bran might exert its action by the same mechanism.

The metabolism of cholesterol to coprostanol by intestinal bacteria is facile; in normal Western persons about 80% of the total daily amount of fecal neutral steroid has undergone reduction of the 5-6 bond. Lactulose, pectin, and guar, all of which are readily metabolized to fatty acid and CO_2 in the cecum, have little effect on the proportion of cholesterol undergoing reduction in the colon.

5. CONCLUSIONS

There is evidence that dietary fiber may affect the metabolic activity of the gut flora because of either the effects of breakdown products of fiber metabolism on enzyme activity or the effect of fiber on substrate concentration or the physical effect of fiber on bacterial activity.

The overall effect of fiber depends very much on the composition of the fiber itself, because this has a profound effect on the relative importance of these three mechanisms. The effect of the mechanisms individually can be deduced by studying extreme forms of fiber. Thus the effect of breakdown products can be deduced from studies on lactulose or mannitol (both of which are rapidly degraded in the right colon); the effect of fiber on substrate concentration is best illustrated by studies utilizing lignins (which are poorly degraded and which have little effect on ITT), while the physical effects are best illustrated by studies using bran (which is poorly degraded and which, except in large amounts, does not sequester colonic substrates).

ACKNOWLEDGMENTS. The work of my laboratory in this field is financially supported by the Cancer Research Campaign.

REFERENCES

Agostini, L., Down, P. F., Murison, J., and Wrong, O. M., 1972, Faecal ammonia and pH during lactulose administration in man: comparison with other cathartics, *Gut* **13**:859–866.

Aries, V. C., and Hill, M. J., 1970, Degradation of steroids by intestinal bacteria. II. Enzymes catalysing the acid reduction of the 3α-, 7α- and 12α-hydroxyl groups in cholic acid and the dehydroxylation of the 7α-hydroxyl group, *Biochim. Biophys. Acta* **202**:535–543.

Baird, I. M., Walters, R. L., Davies, P. S., Hill, M. J., Drasor, B. S., and Southgate, D. A., 1977, The effect of two dietary fibre supplements on gastrointestinal transit, stool weight and frequency, bacterial flora and faecal-bile acids in normal subjects, *Metabolism* **26**:117–127.

Bircher, J., Haemmerli, U. P., Scollo-Lavizzari, G., and Hoffman, K., 1971, Treatment of chronic portal systemic encephalopathy with lactulose, *Am. J. Med.* **51**:148–159.

Bown, R. L., Gibson, J. A., Sladen, G. E., Hicks, B., and Dawson, A. M., 1974, Effects of lactulose and other laxatives on ileal and colonic pH as measured by a radiotelemetry device, *Gut* **15**:999–1004.

Cummings, J. H., Hill, M. J., Jenkins, D. J., Pearson, J. R., and Wiggins, H. S., 1976, Changes in faecal composition and colonic function due to cereal fiber, *Am. J. Clin. Nutr.* **29**:1468–1473.

Cummings, J. H., Hill, M. J., Jivraj, T., Houston, H., Branch, W. J., and Jenkins, J. H., 1979, The effect of meat protein and dietary fiber on colonic function and metabolism, *Am. J. Clin. Nutr.* **32**:2086–2093.

Eastwood, M. A., and Hamilton, D., 1968, Studies on the adsorption of bile salts to non-absorbed components of the diet, *Biochim. Biophys. Acta* **152**:165–173.

Eastwood, M. A., Kirkpatrick, J. R., Mitchell, W. D., Bone, A., and Hamilton, T., 1973, Effects of dietary supplements of wheat bran and cellulose on faeces and bowel function, *Br. Med. J.* **4**:392–394.

Elkington, S. G., 1970, Lactulose, *Gut* **11**:1043–1048.

Goldin, B. R., and Gorbach, S. L., 1976, The relationship between diet and rat fecal bacterial enzymes implicated in colon cancer, *J. Natl. Cancer Inst.* **57**:371–375.

Hill, M. J., 1974, Steroid nuclear dehydrogenation and colon cancer, *Am. J. Clin. Nutr.* **27**:1475–1480.

Kay, R. M., 1981, Effects of diet on the faecal excretion and bacterial modification of acidic and neutral steroids: Implications for colon carcinogenesis, *Cancer Res.* **41**:3774–3777.

Kay, R. M. and Truswell, A. S., 1977a, Effect of wheat fibre on gastrointestinal function, plasma lipids and steroid excretion in man, *Br. J. Nutr.* **37**:227–234.

Kay, R. M., and Truswell, A. S., 1977b, Effect of citrus pectin on plasma lipids and fecal steroid excretion in man, *Am. J. Clin. Nutr.* **30**:171–175.

Low-Beer, T. S., and Nutter, S., 1978, Colonic bacterial activity, biliary cholesterol saturation and pathogenesis of gallstones, *Lancet* **ii**:1063.

Mastromarino, A., Reddy, B. S., and Wynder, E. L., 1976, Metabolic epidemiology of colon cancer: Enzymatic activity of the fecal flora, *Am. J. Clin. Nutr.* **29**:1455–1460.

Midtvedt, T., and Norman, A., 1968, Parameters in 7α-dehydroxylation of bile acids by anaerobic lactobacilli, *Acta. Pathol. Microbiol. Scand.* **72**:313–329.

Owen, R., 1982, Analysis of metabolic profiles of bile acids in faeces of patients undergoing chenodeoxycholic acid therapy using GLC techniques and GC-MS (in preparation).

Pomare, E. W., and Heaton, K. W., 1973, Alteration of bile salt metabolism by dietary fibre (bran), *Br. Med. J.* **iv**:262.

Stephen, A., and Cummings, J., 1980, Mechanism of action of dietary fibre in the human colon, *Nature* **284**:283–284.

Story, J. A., Kritchevsky, D., and Eastwood, M. A., 1979, Dietary fiber–bile acid interactions, in: *Dietary Fibers: Chemistry and Nutrition* (G. E. Inglett and S. I. Falkehag, eds.), Academic Press, New York, pp. 49–55.

Thompson, M. H., and Hill, M. J., 1980, The effect of dietary fibre on intestinal flora and carcinogenous, in: *Pflanzenfasern-Ballaststoff in der menschlichen Ernährung* (H. Rottka, ed.), Georg Thieme Verlag, Stuttgart, pp. 135–143.

Thornton, J. R. and Heaton, K., 1980, Lactulose and bile: evidence linking colonic bacteria and cholesterol gallstones, *Gut* **21**:906.

Vince, A., and Burridge, S. M., 1980, Ammonia production by intestinal bacteria: The effects of lactose, lactulose and glucose, *J. Med. Microbiol.* **13**:177–192.

Walters, R. L., Baird, I. M., Davies, P. S., Hill, M. J., Drasor, B. S., Southgate, D. A., Green, J., and Morgan, B., 1975, Effects of two types of dietary fibre on faecal steroid and lipid excretion, *Br. Med. J.* **2**:536–538.

5

Digestion and Absorption of Nutrients

D. A. T. SOUTHGATE

1. INTRODUCTION

An increased intake of foods providing plant cell walls (or dietary fiber) almost invariably results in an increase in fecal wet weight, fecal dry matter, and the energy content of the feces. Although this increased fecal loss is to be expected because the diet provides an increased intake of indigestible cell wall material, many workers have argued that there is also evidence for effects on the digestion and absorption of other nutrients (McCance and Walsham, 1948; Widdowson, 1960; Southgate and Durnin, 1970).

More recently it has been suggested that the effects of dietary fiber in reducing the digestibility of energy-yielding components of the diet may be significant in relation to the etiology of obesity (Heaton, 1973).

In this chapter I examine the basis for the assumption that dietary fiber has effects on the digestion and absorption of nutrients.

Since the effects on inorganic nutrients and on carbohydrates are discussed elsewhere in this volume, this chapter is confined to the evidence for effects on the digestion and absorption of protein and fat, with a passing reference to energy.

Before discussing the evidence for the effect of dietary fiber on the digestion and absorption of nutrients, it is necessary to examine critically the concepts of digestibility and availability of nutrients as they have been developed and the methods that have been used to measure these widely used nutritional terms.

D. A. T. SOUTHGATE • Nutrition and Food Quality Division, Agricultural Research Council, Food Research Institute, Norwich NR4 7UA, England.

In the course of this examination I will argue that, while the classical studies of digestibility have been satisfactory in practical situations, where the animal being studied can be regarded as a "black box," the measurement of intake and fecal excretion of nutrients is inadequate if we wish to understand the physiological properties of dietary fiber in the gastrointestinal tract.

2. DIGESTIBILITY AND ITS MEASUREMENT

In classical nutritional studies digestibility is measured by comparison of the intake of a nutrient and its excretion in the feces. Strictly speaking this is *apparent digestibility,* because there is often a finite fecal excretion of many nutrients even when the intake is effectively zero. *True digestibility* (so-called) is calculated from comparisons of intake and fecal excretion in which the latter has been corrected by deducting the excretion on a diet where intake of the nutrient in question is negligible or zero.

Thus *digestibility* in the classical context actually means disappearance during intestinal transit, and the digestibility of an organic nutrient is the sum of digestion and absorption from the small intestine and the proportion degraded or transformed in the large intestine by the intestinal microflora. A true understanding of the physiological effects of dietary fiber depends on the separation of these two components.

At this stage in the argument it is useful to consider why digestibility values derived from this type of study have been so widely used, especially in animal nutrition.

2.1. The Use of Digestibility Studies in Animal Feeding

First, it must be remembered that these values were developed at a time when nutritional research was concerned with establishing that the law of conservation of energy applied in animals.

Second, in the practical context of the feeding of farm animals for growth and production in relation to energy supply, a simple approach was adequate. One could regard the animal as a "black box," and at this level a simple system was needed to predict the digestible energy of the diet. Digestibility trials provided (and still provide) factors that could be applied to intake and which gave a reasonable estimate of the energy that a diet would provide to the animal.

2.2. Limitations of Digestibility Measurements

The "black box" approach has three conceptual limitations. First, as has already been mentioned, it does not distinguish between digestion and absorption in the small intestine, and the effects of the microflora in the large intestine.

Second, measurements of true digestibility assume that there are no interactions between foods in the diet and the gastrointestinal tract. These measurements tacitly assume, for example, that fecal nitrogen losses will be the same irrespective of whether the diet contains nitrogenous materials or not. In a human context there are practical problems in feeding acceptable diets devoid of a nutrient for the length of time required to get reliable values for fecal losses.

Third, it has led to what I believe is an unfounded view that the digestibility relationship (intake minus fecal excretion divided by intake) represents a biological constant for a nutrient in a food.

It is true that this relationship often shows a low variance; nevertheless, in human studies the relationship serves to conceal wide variations in fecal excretion simply because the relationship, intake minus fecal excretion divided by intake, wherever fecal excretion is small will approximate to unity (or 100 if expressed as a percentage).

In the practical context (Section 2.1) this is acceptable, but for studies of the action of dietary fiber I believe that we should focus more of our attention on fecal excretion and its variability.

2.3. The Concept of Availability

Availability can be regarded as an unsatisfactory term because of its wide usage in normal language and it was used by Atwater in the same way as I have described digestibility (Merrill and Watt, 1955). It has I think a place in our discussions in the sense that it was used by McCance and Lawrence (1929) and as defined by a working party of the Group of European Nutritionists (Southgate, 1974). Thus available constituents are those which can be utilized by the human organism, and a constituent may be *unavailable,* either because it cannot be absorbed or, if absorbed, is not metabolized by the normal metabolic processes.

2.4. Separation of Small Intestinal and Large Intestinal Effects

In studies of the effects of dietary fiber on digestion and absorption I therefore believe it is essential to separate the two intestinal phases which influence fecal losses and to be cautious in the interpretation of the results of classical digestibility studies.

The situation that we wish to analyze is illustrated in Fig. 1, which provides a basis for the restricted terminology that I think is necessary.

Thus, *digestion* should be restricted to the hydrolytic and other changes (for example, the formation of fat micelles) which take place in the stomach and small intestine and which are followed by *absorption* of the products of digestion across the mucosa of the small intestine. An effect of dietary fiber on digestion must therefore be a gastric or small intestinal phenomenon.

While digestion and absorption in this strict sense probably continue to

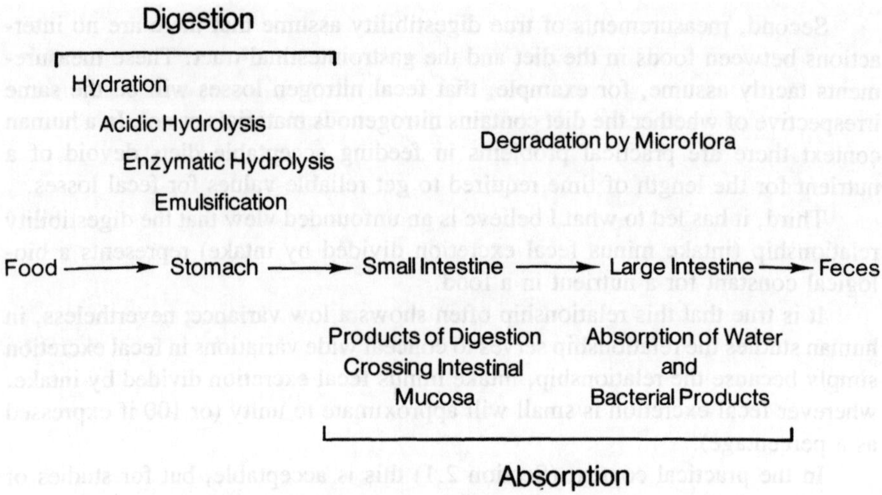

FIGURE 1. Separation of small and large intestinal effects.

some extent in the proximal region of the large bowel, once past the ileocecal valve the activities of the intestinal microflora become quantitatively more important. The microflora are acting on a mixture that includes unabsorbed products of digestion, unabsorbed intestinal secretions, sloughed mucosal material, and the indigestible components of the diet. Fecal excretion represents the product of the activities of the microbial flora on this mixture minus the microbial products that can be absorbed in the large intestine or lost as flatus.

3. EVIDENCE FOR EFFECTS ON DIGESTION AND ABSORPTION

Increasing the intake of dietary fiber is usually associated with an increase in the fecal mass either wet or dry weight. The increased fecal mass is related to the amount and type of dietary fiber ingested (Cummings *et al.*, 1978), but the contribution of dietary fiber itself to the increased mass is often quite small and there is usually an increased excretion of nitrogen and fat (Southgate and Durnin, 1970).

It is possible to interpret this in two ways: that there is an effect of dietary fiber on the digestion and absorption of protein and fats, or alternatively that the protein and lipids associated with the dietary fiber in the diet are intrinsically less digestible. It is more convenient at this point to discuss protein and fat separately.

3.1. Protein

Many studies show that an increased intake of dietary fiber is associated with an increase in fecal nitrogen losses. The results shown in Table I show an effect on mean fecal excretion and the great range observed for the individuals in the groups. In this study the unavailable carbohydrates (dietary fiber) were increased by substitution of foods in the diet (e.g., whole meal for white bread) and thus it is not possible to distinguish between the two possible mechanisms. Thus, if one calculates the expected apparent digestibility of the proteins in the diets eaten by the subjects using the specific factor approach of Merrill and Watt (1955), while one obtains different absolute values, the expected effects of the dietary changes are of the same order. Saunders and Betschart (1980) present analogous data on fecal nitrogen loss as dietary fiber intake is increased and argue that this is evidence that the protein associated with the cell wall material is intrinsically less digestible and effectively part of the dietary fiber. It is of course possible to argue in precisely the reverse sense. McCance and Walsham (1948) found that in man the protein in wholewheat bread was almost completely digested and absorbed and that the nitrogen in the feces was entirely metabolic in origin.

In the conventional digestibility study fecal nitrogen is *unavailable* in the literal sense and the actual form in which the nitrogen is present in the feces is immaterial. If, however, one is trying to deduce the effects of a dietary component on digestion, in the strict sense, and absorption, then the form in which the nitrogen occurs in the feces is of considerable importance.

Studies in which the fecal nitrogens have been fractionated suggest that most of the fecal nitrogen is in the bacterial mass or metabolic, that is, either bacterial products, unabsorbed intestinal secretions, or mucosal cell debris (Stephen and Cummings, 1980a). The bacterial nitrogen is derived from a variety

TABLE I. Effect of Increased Intake of Unavailable Carbohydrates on Fecal Nitrogen Loss[a]

Subjects	Intake of dietary fiber, g/day	Fecal nitrogen, g/day	Range, g/day	Percentage apparent digestibility	
				Found mean ± SE	Calculated
Young men	9.7	1.79	0.94–2.66	89.6 ± 0.77	92.7
	21.5	2.23	1.59–2.79	86.8 ± 0.54	90.3
Young women	6.2	1.03	0.57–1.64	92.1 ± 0.64	92.4
	16.1	1.27	0.69–1.86	90.8 ± 0.50	91.2
	31.9	2.17	1.81–2.56	85.2 ± 0.51	89.3

[a] From Southgate and Durnin (1970).

of sources and the proportion derived from unabsorbed dietary components represents a loss to the body, which, it could be argued, is an effect on the absorption of protein; however, there is also evidence that some bacterial nitrogen is derived from urea secreted into the large bowel, and it has been suggested that some fixation of nitrogen occurs in the large bowel (Oomen and Corden, 1970).

These uncertainties serve to emphasize the point that analyses of fecal excretion are of limited value when making deductions about what are essentially small intestinal events.

3.2. Studies with Isolated Polysaccharides

It may be argued that studies in which the intake of dietary fiber is increased by the supplementation of the diet with an isolated polysaccharide would serve to distinguish between the two views of the action of dietary fiber on protein digestion and absorption.

The results of three typical studies are shown in Table II. The greatest increase in fecal nitrogen was seen with a supplement of pectin (Cummings *et al.*, 1979), which was almost completely degraded during the course of intestinal transit, presumably by the bacterial flora.

At the present time it seems that the evidence for an effect of dietary fiber on the digestion and absorption of protein is equivocal and must await further, more refined, studies, possibly using subjects with ileostomies (Sandberg *et al.*, 1981).

3.3. Lipids

Many studies in which the intake of dietary fiber has been increased show an increased loss of lipid in the feces. This seems to be true for studies where

TABLE II. Effects of Isolated Polysaccharides on Fecal Nitrogen Loss

Supplement	Fecal nitrogen, g/day	Number of subjects	Significance	Reference
Ispaghula (arabinoxylan) 20 g/day	+0.38	4	ns[a]	Greenberg (1976)
Pectin (citrus) 36 g/day	+0.60	5	$p < 0.001$	Cummings *et al.* (1979)
Cellulose (solka floc) 16 g/day	+0.14	7	ns	Slavin and Marlett (1980)

[a] Two subjects showed no change.

TABLE III. Changes in Fecal Fat in Response to Increased Intakes of Dietary Fiber

Subjects	Change in dietary fiber intake, g/day	Change in fecal fat, g/day	Percentage change in fecal fat	Change in percentage apparent digestibility
Young men[a]	11.8 (diet)	+2.27	55.0	−2.8
Young women[a]	9.8 (diet)	+0.68	22.6	−0.4
	25.7 (diet)	+3.23	107.3	−3.6
Young men[b]	36.0 (pectin)	+1.18	80.3	—
Young women[c]	16.0 (solka floc)	+0.60	21.4	−0.9
Men and women[d]	20.0 (ispaghula)	+2.85	40.1	−1.9

[a] Southgate and Durnin (1970).
[b] Cummings *et al.* (1979).
[c] Slavin and Marlett (1980).
[d] Prynne and Southgate (1979).

the diet has undergone qualitative changes (Southgate and Durnin, 1970) or where the increased intake of dietary fiber has been produced by the addition of an isolated polyaccharide (Prynne and Southgate, 1979; Cummings *et al.*, 1979; Slavin and Marlett, 1980). In most studies the increase in fecal excretion was quite high, but the effects on the apparent digestibility values were often barely significant. (Table III).

The argument that the lipids associated with the dietary fiber are intrinsically less digestible has been suggested as the cause of increased fecal fat on bran-enriched diets, but Southgate *et al.* (1976) were unable to detect any changes in fecal lipid that might reflect increased excretion of bran lipids. The pronounced effects of pectin and guar on fecal lipids suggest that some interference with absorption may be taking place. The polysaccharides that form gels or viscous solutions may interfere with emulsification and the formation of fat micelles, possibly by their interaction with bile salts (Story *et al.*, 1979).

4. CONCLUSIONS

At the present time there is circumstantial evidence that increasing the intake of dietary fiber results in an increased loss of fecal nitrogenous material and lipid. The nitrogenous material is largely bacterial and the increase may reflect increased bacterial growth on high-fiber diets (Stephen and Cummings, 1980b).

The increased loss of fecal lipid may be due to an effect of intestinal absorption, but clear evidence is not available. The increased loss in fecal energy may be significant on diets which are very high in dietary fiber but is of questionable significance in relation to the etiology of obesity (Southgate, 1978).

REFERENCES

Cummings, J. H., Southgate, D. A. T., Branch, W., Houston, H., Jenkins, D. J. A., and James, W. P. T., 1978, Colonic response to dietary fibre from carrot, cabbage, apple, bran and guar gum, *Lancet* **i**:5–9.

Cummings, J. H., Southgate, D. A. T., Branch, W., Wiggens, H. S., Houston, H., Jenkins, D. J. A., Jibraj, T., and Hill, M. W., 1979, The digestion of dietary pectin in the human gut, and its effect on calcium absorption and large bowel function, *Br. J. Nutr.* **41**:477–485.

Greenberg, C. J., 1976, Studies on the Fibre in Human Diets and Its Effects on the Absorption and Digestion of Other Nutrients, Ph.D. Thesis, University of Cambridge, Cambridge, England.

Heaton, K. W., 1973, Food fibre as an obstacle to energy intake, *Lancet* **ii**:1418–1421.

McCance, R. A., and Lawrence, R. D., 1929, *The Carbohydrate Content of Foods,* Special Report Series, Medical Research Council, London, No. 135, HMSO, London.

McCance, R. A., and Walsham, C. M., 1948, The digestibility and absorption of the calories, proteins, purines, fat and calcium in wholemeal wheaten bread, *Br. J. Nutr.* **2**:26–41.

Merrill, A. L., and Watt, B. K., 1955, Energy value of foods—basis and derivation, U. S. Department of Agriculture, Handbook 74.

Oomen, H. A. P. C., and Corden, M. W., 1970, Metabolic studies in New Guineans. Nitrogen metabolism in sweet potato eaters, South Pacific Commission Technical Paper No. 163.

Prynne, C. J., and Southgate, D. A. T., 1979, The effects of a supplement of dietary fibre on faecal excretion by human subjects, *Br. J. Nutr.* **41**:495–503.

Sandberg, A. S., Andersson, H., Hallgren, B., Hasselblad K., Isaksson, B., and Hulten, L., 1981, Experimental model for the *in vivo* determination of dietary fibre and its effects on the absorption of nutrients in the small intestine, *Br. J. Nutr.* **45**:283–294.

Saunders, R. M., and Betschart, A. A., 1980, The significance of protein as a component of dietary fibre, *Am. J. Clin. Nutr.* **33**:960–961.

Slavin, J. L., and Marlett, J. A., 1980, Effect of refined cellulose on apparent energy, fat and nitrogen digestibilities, *J. Nutr.* **110**:2020–2026.

Southgate, D. A. T., 1974, *Guidelines for the Preparation of Tables of Food Composition,* S. Karger, Basel.

Southgate, D. A. T., 1978, Has dietary fibre a role in the prevention and treatment of obesity, *Bibleotheca Nutr. Dieta* **26**:70–76.

Southgate, D. A. T., and Durnin, J. V. G. A., 1970, Calorie conversion factors. An experimental re-assessment of the factors used in the calculation of the energy value of human diets, *Br. J. Nutr.* **24**:517–535.

Southgate, D. A. T., Branch, W. J., Hill, M. J., Drasar, B. S., Walters, R. L., Davies, P. S., and McLean Baird, I., 1976, Metabolic responses to dietary supplements of bran, *Metabolism* **25**:1129–1135.

Stephen, A. M., and Cummings, J. H., 1980a, The microbial contribution to human faecal mass, *J. Med. Microbiol.* **13**:45–56.

Stephen, A. M., and Cummings, J. H., 1980b, Mechanism of action of dietary fibre in the human colon, *Nature* **284**:283–284.

Story, J. A., Kritchevsky, D., and Eastwood, M. A., 1979, Dietary fiber–bile acid interactions, in: *Dietary Fibers: Chemistry and Nutrition,* (G. E. Inglett and S. I. Falkehag, eds.), Academic Press, New York, pp. 49–55.

Widdowson, E. M., 1960, Note on the calculation of the calorific value of foods and of diets, in: *The Composition of Foods,* Special Report Series, Medical Research Council, London, No. 297 (R. A. McCance and E. M. Widdowson, eds.), HMSO, London, pp. 153–159.

6

Modification of Intestinal Absorption by Dietary Fiber and Fiber Components

ANTHONY RICHARD LEEDS

1. INTRODUCTION

Some types of dietary fiber have been shown to reduce the magnitude of hyperglycemia and insulinemia after carbohydrate-containing meals in normal man (Jenkins *et al.*, 1977a) and diabetic man (Jenkins *et al.*, 1976). This effect is not confined to the viscous polysaccharides studied by Jenkins *et al.* (1976, 1977a); wheat bran has been shown to have a similar though smaller effect (Jeffreys 1974; Jenkins *et al.*, 1978). Subsequent studies, which have been of two types, have suggested a possible therapeutic role for dietary fiber in the management of diabetes. There have been those in which one form of dietary fiber has been added to several food items in an otherwise low-fiber diet and those in which two diets, one high-fiber and one low-fiber, are compared. Studies of the latter type have in some cases inevitably involved comparison of high-fiber, high-carbohydrate diets with low-fiber, "low"-carbohydrate diets, so that the relative roles of fiber and proportion of energy from carbohydrate are difficult to assess, although the combined effect on diabetic control seems to be an improvement (Simpson *et al.*, 1979a,b). Rivellese (1980) compared low- and high-fiber diets of similar carbohydrate content in diabetic patients and concluded that dietary fiber improved blood glucose control, and Miranda and Horwitz

ANTHONY RICHARD LEEDS • Department of Nutrition, Queen Elizabeth College, University of London, London W8 7AH, England, and Department of General Medicine and Endocrinology, Central Middlesex Hospital, London NW10 7NS, England.

(1978), using low- and high-fiber diets, although with slightly less available carbohydrate in the high-fiber diet (Anderson 1980), demonstrated lower postprandial blood glucose levels in diabetic patients on the high-fiber diets. Others, however, have found high-fiber diets unsatisfactory (Manhire *et al.*, 1981), and the value of dietary fiber has been called into question (Anon, 1981) but stoutly defended (Mann *et al.*, 1981).

The interest in a clinical therapeutic role for dietary fiber in diabetes has also promoted studies into its mechanism of action, in order that ultimately the best material or best combination of foods may be chosen. Information that has become available over the last few years points toward an effect of dietary fiber on the modification of absorption and studies in this area have been undertaken. The amount of dietary fiber in a diet may be only one of many factors which will modify metabolic responses to meals. The degree to which a plant food is

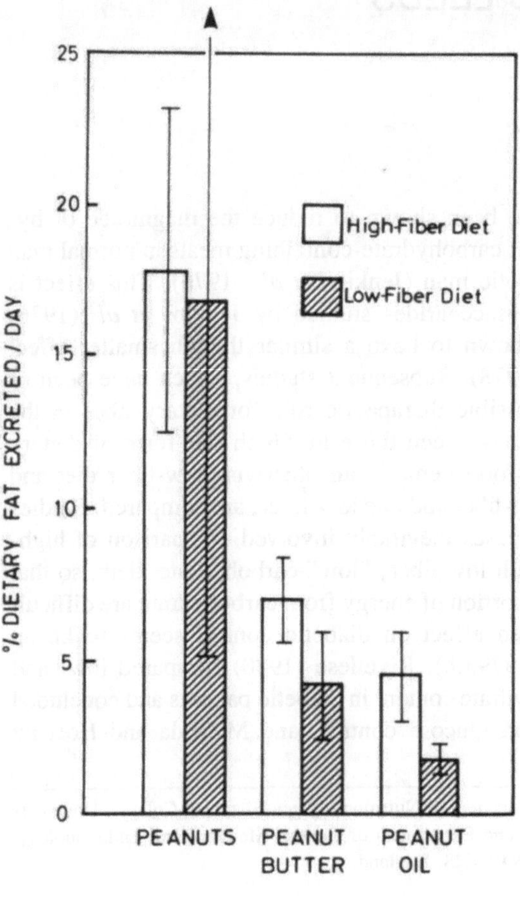

FIGURE 1. Comparison of dietary fecal excretion of fat during six dietary regimens. Values are expressed as mean ± SEM. (Derived from Levine and Silvis, 1980, with permission.)

intact with unbroken cellular architecture may be as important in determining response to the meal as the total amount and type of dietary fiber. Levine and Silvis (1980) studied the effect in ten normal subjects of eating 76 g fat daily as either peanuts, peanut butter, or peanut oil when the subjects were taking a high- or low-fiber diet. Disintegration of the peanut structure resulted in a much greater reduction of fecal fat than reduction of fiber content of the diet (Fig. 1). Collings *et al.* (1981) have shown that the process of cooking starch resulted in greater blood glucose and insulin responses in normal subjects than after raw starch, showing that partial thermal degradation of starch granules is important in determining responses of metabolic variables. Haber *et al.* (1977) have shown that serum insulin responses in normal subjects were greater after apple puree than whole apples, indicating that the destruction of the plant food architecture with no change of fiber content may change glucose and insulin responses. Differences in dietary fiber content and the integrity of food architecture presumably account for different blood glucose and insulin responses to equivalent amounts of carbohydrate in different foods in normal subjects (Crapo *et al.*, 1977; Jenkins *et al.*, 1980b) and diabetic patients (Jenkins *et al.*, 1980a; Crapo *et al.*, 1981).

Dietary fiber is a heterogeneous group of polysaccharides in which actions attributable to one compound may not necessarily be attributable to another. In whatever way dietary fiber affects intestinal absorption of any nutrient, the final effect at the intestinal level is to affect either the rate or completeness of absorption or both.

2. ACUTE EFFECT OF DIETARY FIBER ON FACTORS AFFECTING ABSORPTION (Fig. 2)

2.1. Gastric Filling

Intestinal absorption might be slowed by taking food in small, frequent portions. It has recently been suggested that this would achieve the same effect as some types of dietary fiber, as an aid to diabetic management (Anon., 1981), but in practice small, frequent meals are not likely to achieve high levels of patient compliance. However, high-fiber foods may take longer to eat than low-fiber foods: A meal of wholemeal bread took 45 min to eat, whereas white bread (70% extraction flour) took 34 min (McCance *et al.*, 1953). Haber *et al.* (1977) showed that 60 g of available carbohydrate as 482 g apples, 482 g apple puree, and as 444 ml apple juice took 17.2, 5.9, and 1.5 min to eat, respectively (mean for ten subjects). Thus dietary fiber may affect gastric filling, which might in turn slow intestinal absorption and subsequent glucose and insulin responses. Slower gastric filling might also limit total energy intake.

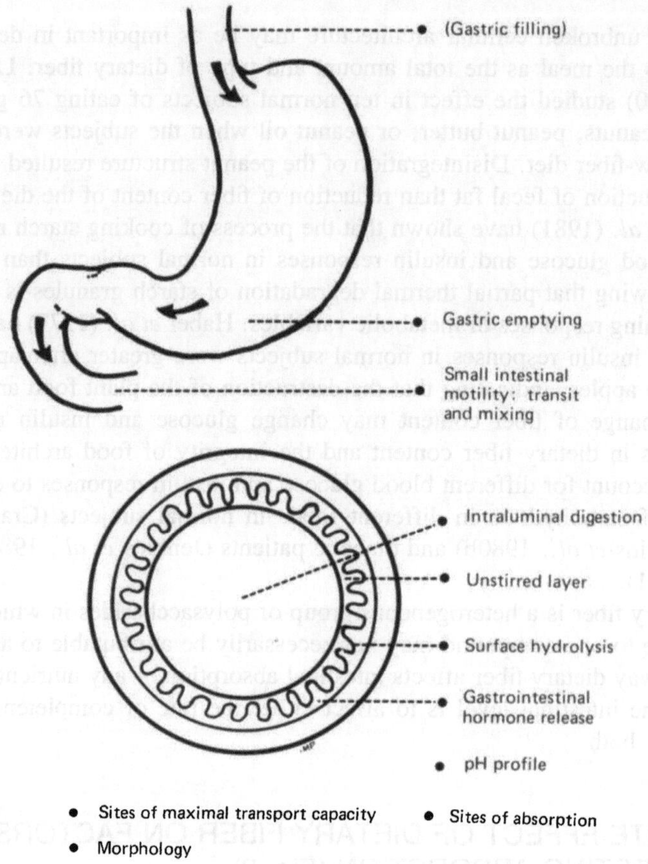

FIGURE 2. Factors affecting intestinal absorption which may be modified by dietary fiber.

2.2. Gastric Emptying

Studies in man of the effects of dietary fiber on gastric emptying are effectively restricted to methods employing radiolabels, since tube aspiration techniques cannot be applied to solid or viscous meals. McCance *et al.* (1953) used barium sulfate as a marker and showed that after white bread (with barium) the marker remained in the stomach for longer than for wholemeal bread. Whether the bread remained associated with the marker is uncertain. Using gamma-emitting labels, Grimes and Goddard (1977) found that after white bread the liquid phase emptied more rapidly than after wholemeal bread. Behavior of the solid phase was not different in the two cases. The viscous polysaccharides, such as guar gum, a galactomannan from the cluster bean, *Cyamopsis tetragonoloba,* and pectin (polygalacturonic acid), appear to slow gastric emptying. Holt *et al.* (1979), studying eight subjects, added 10 g pectin and 16 g guar gum to 400 ml orange juice labeled with ^{113}In and showed that more marker (54% of initial dose) remained in the stomach at 30 min after the meal than remained after the control meal (no guar, no pectin, 32% of initial dose). Using a technique whereby ^{24}Na absorbed from the small gut was detected in the blood in the head by a small external counter, Wilmshurst and Crawley (1980) showed that in seven obese subjects addition of 2 g guar gum to a 200 g low-energy milky drink prolonged gastric transit time from 69 to 112 min (mean values). There was a good direct correlation between mean gastric transit time and time of maximum hunger. In studies on patients with the dumping syndrome addition of pectin to a hypertonic glucose drink so as to achieve a 1000-fold increase of meal viscosity resulted in a prolongation of otherwise abnormally short half-times for gastric emptying (Fig. 3; Leeds *et al.*, 1981). Symptoms provoked by the glucose alone were abolished or reduced in number by addition of pectin. Animal studies, in which more direct measurement of gastric emptying can be undertaken, provide results consistent with those found in man. Hooded Lister rats were killed at varying intervals after orogastric intubation with a glucose-containing meal thickened with one of three types of guar gum (of high, intermediate, and low viscosity), and the gut was clamped into sections, the glucose content of which was analyzed. After the high-viscosity meal, half-emptying of the stomach took longer than after the intermediate- and low-viscosity meals (Fig. 4) (Leeds *et al.*, 1979).

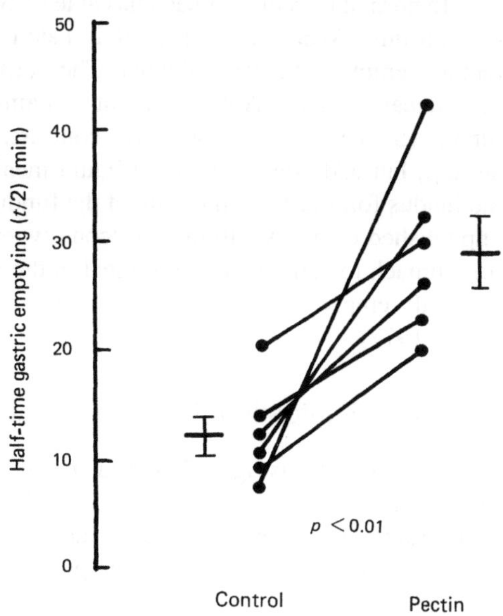

FIGURE 3. Half-time (min) for gastric emptying of hypertonic glucose labeled with ^{113}In in six patients with the dumping syndrome, after control (glucose alone) and test meals (glucose with pectin).

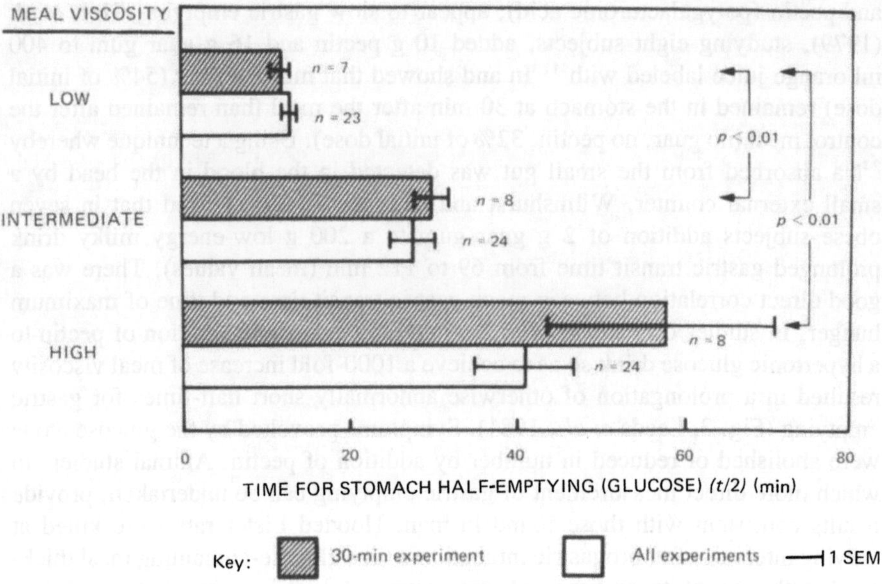

FIGURE 4. Half-time for gastric emptying of glucose in the rat after meals of high, intermediate, and low viscosity.

In normal man the stomach has at least two functional compartments (fundus and antrum). When a solid meal is eaten the fundus fills, while the antrum remains empty for a few minutes. The antrum then fills as the fundus empties in a linear manner. Antral content remains constant during emptying of the fundus (Sheiner *et al.*, 1980). By contrast, liquid meals flow rapidly into the antrum, but addition of pectin to liquid meals results in the "normal" holdup in the fundus followed by emptying of the fundus into the antrum (D. N. L. Ralphs, unpublished data). Addition of viscous types of dietary fiber may simply cause the stomach to handle the liquid meal in the manner of a solid meal. Solid meals generally empty from the stomach more slowly than liquid meals (MacGregor *et al.*, 1977).

2.3. Small Intestinal Motility: Transit and Mixing

The rate of passage of digesta through the small intestine and the degree of mixing are determinants of the completeness and rate of absorption. Very rapid transit would be expected to cause malabsorption of nutrients, especially those not rapidly absorbed from the proximal small gut. In man it is difficult to use any but indirect methods of study. The production of hydrogen by cecal flora from an unabsorbable disaccharide (lactulose) and subsequent expiration of the hydrogen in the breath has been used as a marker for the arrival of the front of a meal at the terminal ileum. Using this technique, Jenkins *et al.* (1978) showed that in six subjects addition of guar gum delayed the appearance of hydrogen in the breath from 1 to $1^1/_2$ hr (Fig. 5). Gum tragacanth and wheat bran appeared to have less pronounced delaying effects and pectin and methyl cellulose no effect. The materials were added at a dosage of 14.5 g to the 400 ml glucose/xylose/lactulose drink, except for wheat bran (41.5 g per 400 ml). Viscosity was measured and correlated well with the mouth-to-cecum transit time, though no account was taken of the shear thinning of guar and pectin. Mouth-to-cecum transit time is, however, composed of gastric filling time (which was standardized in this study), gastric emptying time, small intestinal transit time, and any time delay for substrate/bacteria mixing in the cecum. Unfortunately, it is therefore not possible to come to a conclusion about small intestinal motility. In the same study urine xylose was measured for each of four 2-hr collection periods. Excretion of xylose was reduced in the first 2 hr, but there was a compensatory increase from 2 to 8 hr, suggesting that although absorption was delayed, there was no overall impairment (Fig. 6).

Bond and Levitt (1978) also used breath hydrogen to assess mouth-to-cecum transit time and their findings were consistent with those of Jenkins *et al.* (1978) in that wheat bran added to a lactulose drink did not affect the transit time. Bond and Levitt (1978) also found that when 20 g wheat bran was given on its own,

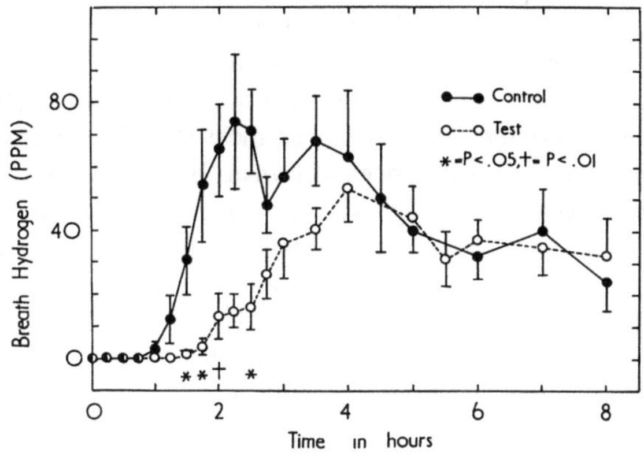

FIGURE 5. Breath hydrogen concentrations (mean ± SEM) in six subjects after meals containing lactulose, with guar gum (test) or without guar gum (control). (With permission of D. J. A. Jenkins.)

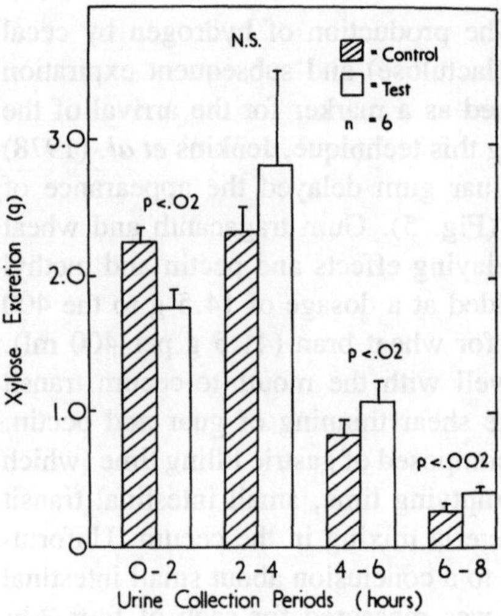

FIGURE 6. Urinary xylose excretion (g/2 hr) over 8 hr in six subjects, after test meals containing xylose, with guar gum (test) or without guar gum (control). (With permission of D. J. A. Jenkins.)

hydrogen was produced, but at a slightly later time than lactulose alone; however, they also showed *in vitro* that wheat bran was not as good a substrate for fecal bacteria as lactulose. Breath hydrogen can also be used as a test for malabsorption and in a preliminary study on young women using appropriate controls to exclude those with gastrointestinal pathology and those not able to produce hydrogen, guar gum given with glucose was shown not to cause malabsorption of the glucose for up to 5 hr after the test meal (Leeds *et al.*, 1978).

As with gastric emptying, more direct measurements can be made in animals. Hooded Lister rats given high-, intermediate-, and low-viscosity guar gum meals, labeled with phenol red by orogastric intubation, were killed at varying intervals up to 2 hr, the gastrointestinal tract was immediately clipped off into sections, and the gut content of marker was measured. In the proximal third of the small gut there were no differences of transit, but values for the middle third of the small gut suggested that high-viscosity meals were moving more slowly (Leeds, unpublished data). However, it has been suggested that the rat is not a suitable model for studying the gastrointestinal effects of dietary fiber and that the pig would be more suitable (Mendeloff, 1978). In a preliminary study in which a ^{103}Ru marker was introduced through a duodenal cannula in a pig given meals (orally) of different composition (with and without guar gum) there appeared to be no difference in the time of appearance of the marker at a terminal ileal cannula (A. G. Low, personal communication). The results of the full study are awaited with interest. Gastrointestinal electromyography studies might also be carried out on a suitable animal model and may help in determining the effects of dietary fiber.

2.4. Intraluminal Digestion

2.4.1. Carbohydrate

In vitro studies on whole foods (Jenkins *et al.*, 1980a) in which 2-g carbohydrate portions of cooked and ground soya beans and lentils and finely crumbed wholemeal bread were mixed with pooled human postprandial jejunal juice and saliva (in a ratio of 3 : 1 by volume jejunal juice : saliva) and placed in closed dialysis bags in stirring water baths containing phosphate buffer showed that glucose, maltose, and oligosaccharides were released from the food very much more slowly from the soya beans than from the bread. The release of glucose, maltose, and oligosaccharides from the lentils took place at an intermediate rate. These results may have been due to different rates of enzymic cleavage of the starch or to trapping of glucose, maltose, and oligosaccharides in the interstices of the food. To account for the latter possibility these studies were then run by placing 3 mM glucose or maltose without digestive enzymes into the dialysis bags and observing the transfer of glucose or maltose through the membrane into the external compartment. In these studies glucose and maltose transfer to the outside at 3 hr was 40%, 65%, and 100%, compared with food-free controls, for soya beans, lentils, and bread, respectively. The proportion of food actually digested by 3 hr was then calculated as 6%, 15%, and 27% for soya beans, lentils, and bread, respectively. Differences of the quantity and type of dietary fiber of these foods might have accounted for different rates of digestion, although the different architecture of the food (modified by cooking and grinding) may have played a part in modifying digestion, as well as "trapping" glucose, maltose, and oligosaccharides and making them unavailable for dialysis to the external compartment. The results of *in vitro* studies were consistent with blood glucose responses to the same foods fed to normal volunteers (soya beans and lentils provoked smaller rises of blood glucose than bread), suggesting that rates of digestion *in vivo* may differ from food to food.

More direct methods of study are again required. Sambrook (1979a) studied carbohydrate absorption in large white pigs fitted with reentrant cannulas in the duodenum, midjejunum, or terminal ileum and fed the animals three diets: a barley, wheat offal, and whitefish meal diet; a starch, sucrose, maize oil, cellulose (20 g/kg), and groundnut meal diet; and a starch, sucrose, maize oil, cellulose (30 g/kg), and casein diet. Total carbohydrate, total reducing substances (in a supernatant and in a hydrolysate of the total digesta), and glucose were measured in samples from the cannulas and also in feces. Table I shows values for 24 hr output : intake of glucose in the partially hydrolyzed digesta (this excluded cellulose–glucose). There was considerable net absorption of glucose anterior to the duodenal cannula and nearly 100% absorption before the terminal ileum. The glucose contents of the supernatants from all sites were low, suggesting that

TABLE I. Mean 24-hr Output : Intake for Glucose in Pigs with Intestinal Reentrant Cannulas and Pigs without Cannulas[a]

	Barley, wheatmeal, fishmeal diet	Starch, sucrose, groundnut meal diet	Starch, sucrose, casein diet
Digesta from pigs cannulated in:			
Duodenum	0.799[b]	0.502[e]	0.807[b]
Jejunum	0.488[c]	0.378[c]	0.466[d]
Ileum	0.039[c]	0.010[f]	0.024[c]
Feces from pigs without cannulas	0.006[b]	0.004[b]	0.001[b]

[a] Derived from Sambrook (1979a), with permission.
[b] $N = 6$. [c] $N = 5$. [d] $N = 4$. [e] $N = 2$. [f] $N = 1$.

once released by digestion, absorption of glucose was rapid. The differences between semisynthetic diets (those containing starch, sucrose, and maize oil) and diets of normal foods, in terms of food architecture, seemed to make no difference to the rate of absorption of glucose. Some indication that digestion of carbohydrate was slowed after the barley/wheat offal/fishmeal diet can be seen in Table II, where total carbohydrate figures are given. Total carbohydrate was determined from the equation, total carbohydrate = dry matter − (nitrogen × 6.25) + ash + total lipid, so it would include dietary fiber. The higher ratios at the ileum of pigs on the barley/wheat offal/fishmeal diet, which suggest less complete digestion and absorption of carbohydrate, may be partly accounted for by dietary fiber.

In man direct studies might be carried out by intubation techniques or studies

TABLE II. Mean 24-hr Output : Intake for Total Carbohydrate in Pigs with Intestinal Reentrant Cannulas and Pigs without Cannulas[a]

	Barley, wheatmeal, fishmeal diet	Starch, sucrose, groundnut meal diet	Starch, sucrose, casein diet
Digesta from pigs cannulated in:			
Duodenum	0.896[b]	0.825[e]	0.874[b]
Jejunum	0.605[c]	0.613[d]	0.684[d]
Ileum	0.254[c]	0.147[e]	0.059[b]
Feces from pigs without cannulas	0.177[b]	0.103[b]	0.025[b]

[a] Derived from Sambrook (1979a), with permission.
[b] $N = 6$. [c] $N = 5$. [d] $N = 4$. [e] $N = 2$.

could be carried out in patients with ileostomies, although there would be some doubt about whether results from such patients truly reflected digestion in intact man.

2.4.2. Fat

The effect of different types of dietary fiber on fecal fat outputs has been studied (Losowsky, 1978). In most instances dietary fiber causes an increase in daily fecal fat output, in some cases of small magnitude and in others, large. Fecal fats are, however, only an indirect reflection of events in the small gut. The source of the increased fecal fat is uncertain and more direct methods of study are needed. Sambrook (1979b) has also looked at the flow of total lipid in cannulated pigs, using diets and techniques described in his previous work (Sambrook, 1979a). There was a substantial secretion of total lipid into the proximal small intestine after all diets, the major absorption taking place between the midjejunal and terminal ileal cannulas. There was no major difference between the three diets used. The use of pigs with reentrant cannulas, perhaps together with pulsed intravenous injection of a radiolabel for incorporation into endogenous lipid, might help elucidate the effects of dietary fiber on fat digestion in pigs.

2.4.3. Protein

Changes in the rates of digestion of proteins and the completeness of absorption of protein may have important implications for the use of high-fiber diets for diabetics. Fecal nitrogen in man is probably more of a reflection of events in the colon than in the small bowel (Stephen and Cummings, 1979). Techniques for studying the flow of nitrogen and amino acids in cannulated pigs are well developed (Low, 1979a,b) and could easily be applied to the question of the effects of dietary fiber on protein digestion. (Schneeman considers other factors that would affect intraluminal digestion in Chapter 7, and Southgate considers digestion in Chapter 5.)

2.5. The Unstirred Layer

Johnson and Gee (1980) studied the uptake of glucose by everted rat jejunum from 28 mM glucose containing 0, 0.1, 0.25, and 0.5 g/dl guar gum. The 0.5 g/dl medium had a viscosity 100 times that of the medium free of guar gum. There was an inverse relationship between viscosity and glucose uptake, and with carboxymethylcellulose the same effect was demonstrated; the authors suggested that the gum probably increased the thickness of the unstirred layer. Elsenhans *et al.* (1980), using everted rat jejunal and ileal rings, showed that the uptakes of α-methyl-D-glucoside, D-galactose, L-leucine, and L-phenylalanine were inhibited by guar gum, pectin, gum tragacanth, carubin, and carrageenan in relation to the viscosity of the solution. They also showed that inhibition by gums was reversible by washing the tissues, reincubating the tissue in medium without gums, and demonstrating a return of uptake to control values. The inhibition of uptake by tissues was dependent on the shaking rate—increased shaking increased tissue uptake in the presence of gums as well as in their absence. Detailed studies of transport parameters were performed for L-leucine with guar gum. There were no significant differences of V_{max}, but K_m values were greater in the presence of guar gum. The possibility that substrates might bind to the polysaccharides was excluded by binding studies in which substrates were mixed with the polysaccharides and filtered; recovery of substrates was virtually 100% in all cases. The authors concluded that inhibition of intestinal transport *in vitro* by carbohydrate gelling agents is due to an enlargement of the unstirred layer resistance. One of the consequences of this conclusion is that the effects of dietary fiber on the motility of the small intestine (especially movements resulting in mixing of contents, rather than propulsion) need further investigation.

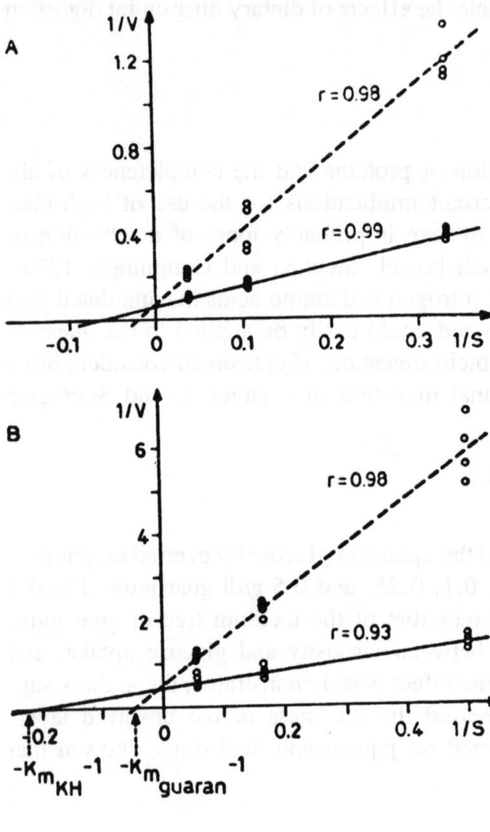

FIGURE 7. Kinetic analysis of the inhibition of surface hydrolysis of (A) maltose and (B) Phe-Gly by guar gum. The 3-min incubations at 37°C were performed in Krebs–Henseleit phosphate buffer with 0.5% guar (- - -) and without guar (—) using a shaking rate of 120 cycles/min. Units: 1/*S* (1/mM), 1/*V* (min · g protein/mM). (From Elsenhans *et al.*, 1981, with permission.)

Increased mixing might overcome any effect of increasing unstirred layer thickness.

2.6. Surface Hydrolysis

Elsenhans *et al.* (1981) have also studied the effects of guar gum on intestinal surface hydrolysis of D-maltose and L-phenylalanylglycine. Using everted segments of rat jejunum, they found that guar gum (5 g/liter) inhibited production of the products of hydrolysis in a competitive manner (Fig. 7). Increasing the shaking rate did not affect the maximal hydrolytic capacity, but the Michaelis constants for hydrolysis were reduced (Table III). Hydrolysis in mucosal homogenates was also studied with and without guar gum; there were no significant differences of hydrolytic rates. These results point toward an effect of guar gum in increasing unstirred layer thickness, in causing an apparent competitive inhibition of surface hydrolysis, but not actually competing for substrate-binding sites on the enzymes.

2.7. Gastrointestinal Hormone Release

Although the reduction of postprandial plasma insulin levels by incorporation of guar gum into meals may be due to slower glucose absorption, another possibility would be that the release of gastrointestinal hormones which stimulate insulin release might be modified by guar gum in the gut contents. Gastric inhibitory peptide (or glucose-dependent insulinotropic polypeptide, GIP) and gut glucagonlike immunoreactivity (GLI) are able to stimulate insulin release (Marks and Turner, 1977). Five healthy volunteers and six diabetics were given a mixed test meal on two occasions by Morgan *et al.* (1979). On one occasion the test meal contained guar gum. Plasma GIP levels were lower after the guar-containing meal in both volunteers and diabetics, but GLI levels were not significantly different. Another group of volunteers and diabetics took a carbohydrate meal with and without guar gum and in these the reduction of plasma GIP levels after the guar-containing meals was more pronounced than after guar in the mixed meal study. GLI levels were lower at some points in normal volunteers

TABLE III. Effect of Guar Gum on the Kinetic Parameters of Brush Border Hydrolysis of D-Maltose[a]

	K_m, mM		V_{max}, mM/min·g	
Shaking rate, cycles/min	Buffer	Buffer + 0.5% guar	Buffer	Buffer + 0.5% guar
120	14.9	52.8	15.9	15.7
240	9.3	25.2	16.3	16.3

[a] Derived from Elsenhans *et al.* (1981), with permission.

after the guar-containing carbohydrate meal. Plasma GIP and GLI levels were also lowered by pectin in 11 postgastric surgery patients after 50 g oral glucose taken on two occasions, one with pectin (Fig. 8) (Jenkins *et al.*, 1980c).

The results of these two studies point toward a modification of GIP release by two types of dietary fiber, guar gum and pectin. Carbohydrates and fat are the most important stimuli of GIP release, which occurs from the proximal and mid-small gut (Sarson, 1978). Presumably factors discussed above (slowing of gastric emptying and digestion, increased thickness of the unstirred layer) operate to decrease the stimulus for GIP secretion. The reduction of GIP levels may partly account for the lower postprandial serum insulin levels occurring after meals containing guar gum and pectin. Nutrients in the lumen of the gut are also responsible for stimulating the release of GLI. The GLI responses to meals are high in diseases such as celiac disease, where nutrients pass further down the small gut than is normal. The lowering of GLI levels by dietary fiber points

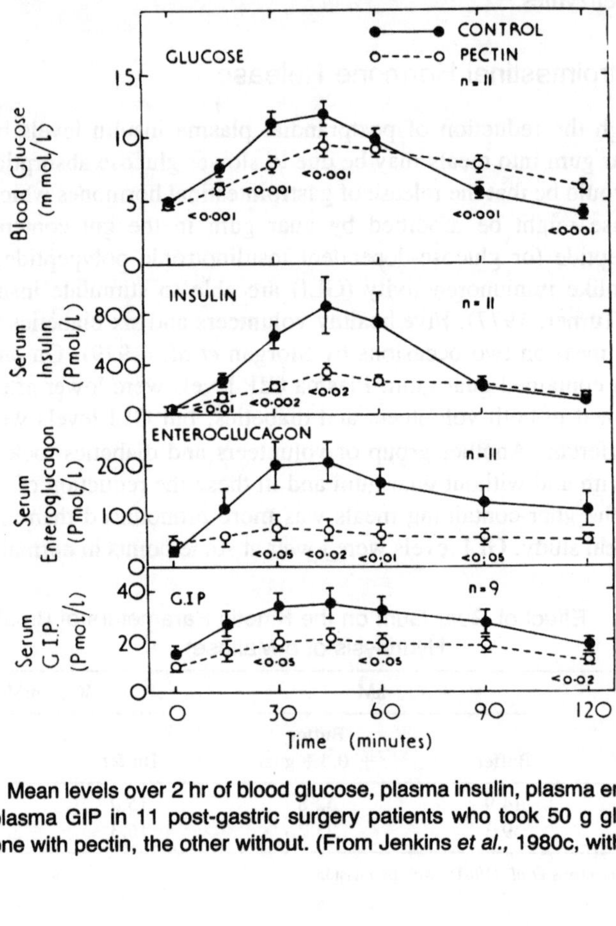

FIGURE 8. Mean levels over 2 hr of blood glucose, plasma insulin, plasma enteroglucagon (GLI), and plasma GIP in 11 post-gastric surgery patients who took 50 g glucose on two occasions, one with pectin, the other without. (From Jenkins *et al.*, 1980c, with permission.)

to either a local decrease of the stimulus for secretion, as with GIP, or, in patients after gastric surgery, an improvement by the fiber in proximal absorption as a consequence of the slowing of otherwise abnormally fast transit through the upper gut (Jenkins *et al.*, 1980c).

2.8. The pH Profile

Dietary fiber of some types may alter pH levels or rates of change of pH in response to meals in all parts of the gut. This might be expected to alter enzyme function and possibly the release of gastrointestinal hormones. Other aspects of food architecture may have similar effects; for example, after a maize meal duodenal ulcer patients had higher gastric pH than after a cornflour meal (Lennard-Jones *et al.*, 1968). The viscous types of dietary fiber might be expected to have a "dampening effect" on pH changes simply by slowing mixing of gut contents and intestinal secretions; however, this would not happen if the dietary fiber affected small intestinal motility in such a way as to increase mixing. In the rather artificial situation where a purified acidic polysaccharide such as pectin is taken with a meal, intestinal pH might be lower. The traditional use of apples in treating infant diarrhoea and the use of coarse pectin in treating scouring in swine may relate to the effects of pH changes on bacterial growth.

2.9. Sites of Absorption

If some types of dietary fiber slow nutrient absorption, then sites of absorption would shift distally. This would have no acute implications, other than slowing of absorption rates, unless there were qualitative differences in the different sites of absorption. Wu *et al.* (1980) showed that in rats chylomicrons secreted by the distal small gut were larger and contained relatively larger proportions of triglyceride and apoprotein than those secreted from the proximal small gut. The results suggested a reduction of the utilization of phospholipid for incorporation into chylomicrons by the distal small gut. If dietary fiber were to shift fat absorption distally, chylomicrons might be larger and this might affect subsequent utilization of the triglyceride.

3. CHRONIC EFFECTS OF DIETARY FIBER

3.1. Sites of Maximal Transport Capacity

Chronic feeding of diets high in dietary fiber may result in delivery of nutrients to parts of the small gut more distal than would be the case after low-fiber diets. This might induce transport at distal sites and deinduce transport at proximal sites. One effect of this would be to improve glucose tolerance when tested with glucose alone. Such an improvement of glucose tolerance after 6 months of wheat bran (24 g/day) has been demonstrated by Brodribb and Humphreys (1976) in 37 patients with diverticular disease. However, Jenkins *et al.* (1977b) demonstrated no improvement of glucose tolerance after 6 weeks of pectin (36 g/day) in three normal volunteers.

In rats Schwartz and Levine (1980) have shown that feeding pectin or cellulose for 5 weeks resulted in impairment of absorption of glucose by the proximal jejunum as measured by a perfusion technique; however, it is possible that some of the impairment of absorption may have been due to residual fiber overlying the mucosa. Use of a rat mucosal ring technique has produced some evidence for a reduction of maximal transport capacity for glucose in the proximal small gut after feeding of guar gum-containing diets, but none for an increase of transport capacity in the more distal small gut (J. Daley, unpublished data).

3.2. Morphology

There is evidence that glucose absorption in the jejunum is directly related to the absorptive mucosal area (Anon., 1978). Morphological changes caused by dietary fiber would therefore be expected to affect absorptive capacity. Tasman-Jones *et al.* (1978) have shown that rats fed a fiber-free or cellulose-containing diet maintained an immature villous pattern, while rats fed a diet with pectin developed mucosal ridges. Schwartz and Levine (1980) showed some reduction of jejunal mucosal thickness but no change of crypt/villus ratio in rats after feeding pectin for 5 weeks, and Cassidy *et al.* (1981) showed that feeding pectin or alfalfa (15 g/100 g diet) to rats for 6 weeks resulted in about 30% of jejunal intestinal villi with structural deviations (swelling with microvillar disarray), whereas feeding of bran and cellulose (also 15 g/100 g diet) resulted in only 5–7% of villi with structural deviations (the same as after the control chow). The degree of morphological damage caused by different fibers correlated with their ability to bind bile salts.

4. IMPLICATIONS OF CHANGES OF ABSORPTION BY DIETARY FIBER FOR DIABETES MELLITUS AND OBESITY

Changes of rates of absorption and the completeness of absorption as may be caused by dietary fiber may have considerable advantages for diabetics. Apart from the immediate advantages of decreased hyperglycemia and glycosuria, there may be secondary metabolic changes, such as a decrease of hepatic triglyceride synthesis, which might prevent hypertriglyceridemia. Lower insulin levels throughout the day may result in changes of tissue sensitivity to insulin, and atherogenic processes in the arterial intima may be slowed. A small reduction in the completeness of absorption need not necessarily be harmful—a slight reduction of net energy input may be advantageous, especially for obese diabetics.

With respect to obesity, reduction of postprandial blood glucose and insulin levels would be expected to decrease satiety rather than increase it, but prolonged distension of the stomach and upper small intestine by dietary fiber may provide a strong signal for satiety.

5. SUMMARY

The effects of dietary fiber on intestinal absorption seem to be mainly due to a slowing of all processes involved. Gastric emptying and intestinal digestion are probably slowed and movement of nutrients to the absorptive mucosal surface is slowed. Some types of dietary fiber modify intestinal morphology and this, too, may impair absorption. Chronic feeding of fiber may alter the sites of transport of nutrients.

Further studies, especially in man or a suitable animal model, are needed.

ACKNOWLEDGMENTS. I am grateful to Mrs. I. M. Prentice and Mike Ethrington for preparation of figures and plates, and to Mrs. J. Glass and Mr. G. Snow for preparation of the typescript.

REFERENCES

Anderson, J. W., 1980, Dietary fiber and diabetes, in: *Medical Aspects of Dietary Fiber* (G. A. Spiller and R. M. Kay, eds.), Plenum Press, New York, pp. 193–221.

Anon., 1978, Intestinal surface area and sugar malabsorption, *Nutr. Rev.* **36:**293–295.

Anon., 1981, High fibre diets and diabetes, *Lancet* **i:**423–424.

Bond, J. M., and Levitt, M. D., 1978, Effect of dietary fiber on intestinal gas production and small bowel transit time in man, *Am. J. Clin. Nutr.* **31:**S169–S174.

Brodribb, A. J. M., and Humphreys, D. M., 1976, Diverticular disease: three studies. Part III. Metabolic effect of bran in patients with diverticular disease, *Br. Med. J.* **i:**428–430.

Cassidy, M. M., Lightfoot, F. G., Gran, L. E., Story, J. A., Kritchevsky, D., and Vahouny, G. V., 1981, Effect of chronic intake of dietary fiber on the ultrastructural topography of rat jejunum and colon: A scanning electron microscopy study, *Am. J. Clin. Nutr.* **34:**218–228.

Collings, P., Williams, C., and MacDonald, I., 1981, Effects of cooking on serum glucose and insulin responses to starch, *Br. Med. J.* **282:**1032.

Crapo, P. A., Reaven, G., and Olefsky, J., 1977, Postprandial plasma-glucose and insulin responses to different complex carbohydrates, *Diabetes* **26:**1178–1183.

Crapo, P. A., Insel, J., Sperling, M., and Kolterman, D., 1981, Comparison of serum glucose, insulin and glucagon responses to different types of complex carbohydrate in non-insulin-dependent diabetic patients, *Am. J. Clin. Nutr.* **34:**184–190.

Elsenhans, B., Sufke, U., Blume, R., and Caspary, W. F., 1980, The influence of carbohydrate gelling agents on rat intestinal transport of monosaccharides and neutral amino acids *in vitro*, *Clin. Sci.* **59**:373–380.

Elsenhans, B., Sufke, U., Blume, R., and Caspary, W. F., 1981, *In vitro* inhibition of rat intestinal surface hydrolysis of disaccharides and dipeptides by Guaran, *Digestion* **21**:98–103.

Grimes, D. S., and Goddard, J., 1977, Gastric emptying of wholemeal and white bread, *Gut* **18**:725–729.

Haber, G. B., Heaton, K. W., Murphy, D., and Burroughs, L., 1977, Depletion and disruption of dietary fibre. Effects on satiety, plasma glucose and serum insulin, *Lancet* **2**:679–682.

Holt, S., Heading, R. C., Carter, D. C., Prescott, L. F., and Tothill, P., 1979, Effect of gel fibre on gastric emptying and absorption of glucose and paracetamol, *Lancet* **1**:636–639.

Jeffreys, D. B., 1974, The effect of dietary fibre on the response to orally administered glucose, *Proc. Nutr. Soc.* **33**:11A.

Jenkins, D. J. A., Leeds, A. R., Gassull, M. A., Wolever, T. M. S., Goff, D. V., Alberti, K. G. M. M., and Hockaday, T. D. R., 1976, Unabsorbable carbohydrates and diabetes: Decreased postprandial hyperglucaemia, *Lancet* **2**:172–174.

Jenkins, D. J. A., Leeds, A. R., Gassull, M. A., Cochet, B., and Alberti, K. G. M. M., 1977a, Decrease in postprandial insulin and glucose concentrations by guar and pectin, *Ann. Intern. Med.* **86**:20–23.

Jenkins, D. J. A., Leeds, A. R., Houston, H., Hinks, L., and Alberti, K. G. M. M., 1977b, Carbohydrate tolerance in man after six weeks of pectin administration, *Proc. Nutr. Soc.* **36**:62A.

Jenkins, D. J. A., Wolever, T. M. S., Leeds, A. R., Gassull, M. A., Haisman, P., Dilawari, J., Goff, D. V., Metz, G. L., and Alberti, K. G. M. M., 1978, Dietary fibres, fibre analogues and glucose tolerance, importance of viscosity, *Br. Med. J.* **1**:1392–1394.

Jenkins, D. J. A., Wolever, T. M. S., Taylor, R. H., Ghafari, H., Jenkins, A. L., Barker, H., and Jenkins, M. J. A., 1980a, Rate of digestion of foods and postprandial glycaemia in normal and diabetic subjects, *Br. Med. J.* **281**:14–17.

Jenkins, D. J. A., Wolever, T. M. S., Taylor, R. H., Barker, H. M., and Fielden, H., 1980b, Exceptionally low blood glucose response to dried beans: Comparison with other carbohydrate foods, *Br. Med. J.* **281**:578–580.

Jenkins, D. J. A., Bloom, S. R., Albuquerque, R. H., Leeds, A. R., Sarson, D. L., Metz, G. L., and Alberti, K. G. M. M., 1980c, Pectin and complications after gastric surgery: Normalisation of postprandial glucose and endocrine responses, *Gut* **21**:574–579.

Johnson, I. T., and Gee, J. M., 1980, Inhibitory effect of guar gum on the intestinal absorption of glucose *in vitro*, *Proc. Nutr. Soc.* **39**:52A.

Leeds, A. R., Bolster, N., and Truswell, A. S., 1978, Guar gum and glucose absorption: Absence of evidence for malabsorption, *Proc. Nutr. Soc.* **37**:88A.

Leeds, A. R., Bolster, N. R., Andrews, R., and Truswell, A. S., 1979, Meal viscosity, gastric emptying and glucose absorption in the rat, *Proc. Nutr. Soc.* **38**:44A.

Leeds, A. R., Ralphs, D. N. L., Ebied, F., Metz, G., and Dilawari, J. B., 1981, Pectin and the dumping syndrome: Reduction of symptoms and plasma volume changes, *Lancet* **1**:1075–1078.

Lennard-Jones, J. E., Fletcher, J., and Shaw, D. G., 1968, Effect of different foods on the acidity of the gastric contents in patients with duodenal ulcer. Part III. Effect of altering the proportions of protein and carbohydrate, *Gut* **9**:177–182.

Levine, A. S., and Silvis, S. E., 1980, Absorption of whole peanuts, peanut oil and peanut butter, *N. Engl. J. Med.* **303**:917–918.

Losowsky, M. S., 1978, Effects of dietary fibre on intestinal absorption, in: *Dietary Fibre: Current Development of Importance to Health* (K.W. Heaton, ed.), John Libbey and Co., London, pp. 129–139.

Low, A. G., 1979a, Studies on digestion and absorption in the intestines of growing pigs. 5. Measurements of the flow of nitrogen, *Br. J. Nutr.* **41:**137–146.

Low, A. G., 1979b, Studies on digestion and absorption in the intestines of growing pigs. 6. Measurements of the flow of amino acid, *Br. J. Nutr.* **41:**147–156.

MacGregor, I. L., Martin, P., and Meyer, J. H., 1977, Gastric emptying of solid food in normal man and after subtotal gastrectomy and truncal vagotomy with pyloroplasty, *Gastroenterology* **72:**206–211.

Manhire, A., Henry, C. L., Hartog, M., and Heaton, K. W., 1981, Unrefined carbohydrate and dietary fibre in treatment of diabetes mellitus, *J. Hum. Nutr.* **35:**99–101.

Mann, J. I., Kinmouth, A. L., Todd, E., Angus, R. M., Simpson, H. C. R., and Hockaday, T. D. R., 1981, High fibre diets and diabetes, *Lancet* **i:**731–732.

Marks, V., and Turner, D. S., 1977, The gastrointestinal hormones with particular reference to their role in the regulation of insulin secretion, *Essays Biochem.* **3:**109–152.

McCance, R. A., Prior, K. M., and Widdowson, E. M., 1953, A radiological study of the rate of passage of brown and white bread through the digestive tract of man, *Br. J. Nutr.* **7:**98–104.

Mendeloff, A. I., 1978, Workshop III—Fiber and the gastrointestinal tract. Summary and recommendations, *Am. J. Clin. Nutr.* **31:**5145–5147.

Miranda, P. M., and Horwitz, D. L., 1978, High fiber diets in the treatment of diabetes mellitus, *Ann. Intern. Med.* **88:**482.

Morgan, L. M., Goulder, I. J., Tsiolakis, D., Marks, V., and Alberti, K. G. M. M., 1979, The effect of unabsorbable carbohydrate on gut hormones, *Diabetologia* **17:**85–89.

Rivellese, A., Riccardi, G., Giacco, A., Pacioni, D., Genovese, S., Mattioli, P., and Mancini, M., 1980, Effect of dietary fibre on glucose control and serum lipoproteins in diabetic patients, *Lancet* **2:**447–450.

Sambrook, I. E., 1979a, Studies on digestion and absorption in the intestines of growing pigs. 7. Measurements of the flow of total carbohydrates, total reducing substances and glucose, *Br. J. Nutr.* **42:**267–277.

Sambrook, I. E., 1979b, Studies on digestion and absorption in the intestines of growing pigs. 8. Measurements of the flow of total lipid, acid-detergent fibre and volatile fatty acids, *Br. J. Nutr.* **42:**279–287.

Sarson, D. L., 1978, Gastric inhibitory polypeptide (GIP), *J. Clin. Pathol.* **33**(Suppl. 8)**:**31–37.

Schwartz, S. E., and Levine, G. D., 1980, Effects of dietary fibre on intestinal glucose absorption and glucose tolerance in rats, *Gastroenterology* **79:**833–836.

Sheiner, H. J., Quinlan, M. F., and Thompson, I. J., 1980, Gastric motility and emptying in normal and post-vagotomy subjects, *Gut* **21:**753–759.

Simpson, R. W., Mann, J. I., Eaton, J., Moore, R. A., Carter, R., and Hockaday, T. D. R., 1979a, Improved glucose control in maturity onset diabetes treated with high carbohydrate-modified fat diet, *Br. Med. J.* **i:**1753–1756.

Simpson, R. W., Mann, J. I., and Eaton, J., 1979b, High carbohydrate diets and insulin dependent diabetics, *Br. Med. J.* **ii:**523–525.

Stephen, A. M., and Cummings, J. H., 1979, The influence of dietary fibre on faecal nitrogen excretion in man, *Proc. Nutr. Soc.* **38:**141A.

Tasman-Jones, C., Jones, A. L., and Owery, R. L., 1978, Jejunal morphological consequences of dietary fibre in rats, *Gastroenterology* **74:**1102.

Wilmshurst, P., and Crawley, J. C. W., 1980, The measurement of gastric transit time in obese subjects using ^{24}Na and the effects of energy content and guar gum on gastric emptying and satiety, *Br. J. Nutr.* **44:**1–6.

Wu, A. L., Clark, S. B., and Holt, P. R., 1980, Composition of lymph chylomicrons from proximal or distal rat small intestine, *Am. J. Clin. Nutr.* **33:**582–589.

7

Pancreatic and Digestive Function

BARBARA OLDS SCHNEEMAN

1. INTRODUCTION

The gastrointestinal (GI) tract influences nutrient metabolism through several routes. First, enzymes and other cofactors necessary for digestion are secreted into or are present in the small intestine. These factors are essential to break down the complex mixture of food to available nutrients, which then are absorbed by mechanisms in the small intestinal cells. Second, the gut is capable of adapting to diet composition. Adaptation to diet may reflect the GI tract's function in homeostasis. Third, the GI tract has its own nutrient needs. The small intestine maintains a high rate of cell renewal and several GI organs actively synthesize secretory proteins. Because of its metabolic activity, the nutrient needs of this organ system are high. The small intestine is in an advantageous position in that it can utilize available nutrients directly from the lumen.

The GI tract is the first organ system to come in contact with food and must convert the complexity of food to nutrients the rest of the body recognizes and can utilize. Hence, the influence of diet composition on the functioning of the GI tract could have several effects on metabolism. A review of the numerous studies that have investigated the metabolic responses to dietary fiber suggests that its primary effects may be on the functioning of the GI tract. Data have been gathered by balance studies, measuring apparent digestibility, fecal excretion, or plasma and tissue responses (Kelsay, 1978). The effects vary by fiber source but, in general, suggest that digestion and absorption may occur more

BARBARA OLDS SCHNEEMAN • Departments of Nutrition and Food Science and Technology, University of California, Davis, California 95616.

slowly in subjects fed high-fiber diets. An investigation of the intraluminal effects of dietary fiber in the small intestine could prove useful in understanding basic mechanisms by which a nondigestible dietary component can influence metabolism.

Dietary fiber could cause functional changes in the gastrointestinal tract through three general mechanisms. The first possibility is that fiber sources could change secretion in the gastrointestinal tract; hence the digestive process could be altered by the availability of enzymes or bile acids secreted into the gut. As a component of this, certain fiber sources may interact with enzymes or bile acids in the small intestine, limiting their activity. Second, the composition and physical characteristics of the small intestinal contents can be altered by fiber sources. Differences in viscosity, pH, and quantity of the chyme will affect the interaction of enzymes and substrates and their mobility. Finally, because of fiber's physical attributes, it could cause changes in the morphology of the small intestine. Hence, structural changes in the gastrointestinal tract could be associated with the functional changes caused by fiber in the gut.

2. AVAILABILITY OF ENZYMES AND BILE ACIDS

Initially, the interaction of pancreatic enzyme activity with various sources of dietary fiber was examined *in vitro* (Fig. 1) (Dunaif and Schneeman, 1981). Pure human pancreatic juice was incubated for 5 min at 37°C with 5.0% (w/v) of alfalfa, oat bran, wheat bran, cellulose, xylan, or 2.5% pectin. The supernatant was analyzed for amylase and lipase activity and for trypsin and chymotrypsin activity after activation by enterokinase. The control sample, which does not

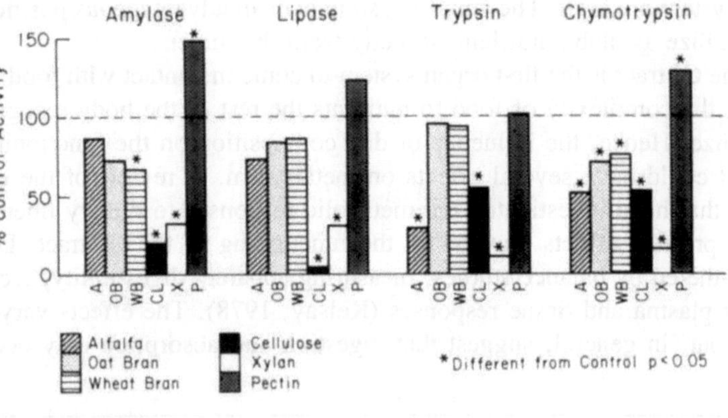

FIGURE 1. Percentage of human pancreatic enzyme activity remaining after a 5-min incubation with one of the fiber sources. Control (no fiber) is shown as 100% activity.

contain a fiber source, is shown in the figure as 100% activity. The fiber-treated samples are shown as percentages of control activity. This study has indicated that potentially an interaction between fiber source and human pancreatic enzymes does exist. For example, cellulose and xylan significantly depressed the activity of each enzyme measured, alfalfa depressed the proteolytic enzymes' activities, and, in contrast to the other fiber sources, pectin tended to elevate activity. Interestingly, chymotrypsin activity was affected significantly by every fiber treatment. This study with human enzymes supports our earlier observations using a variety of commercial enzyme sources and rat pancreatic juice (Schneeman, 1978). The mechanism by which enzyme activity is altered is not known. For some fiber sources it could be due to enzyme inhibitors, such as are known to exist in alfalfa (Liener and Kakade, 1980), and for others perhaps a nonspecific binding. The important point is that an interaction is possible and the results suggest that the effect of dietary fiber on the availability of digestive enzyme activity *in vivo* should be examined.

Our approach has been to use an animal model, the rat. Since we are primarily interested in small bowel function, not large bowel, the rat is a useful model. Some of the *in vitro* interactions observed with fibers and human pancreatic enzymes have been observed with rat pancreatic juice, suggesting that experiments with the rat will be useful. The general approach has been to adapt rats to either a fiber-free diet or one containing the fiber source to be studied. After several weeks, groups of animals are killed in a fed or unfed state. The fed groups are examined for the response in the small intestine to a meal, whereas the unfed groups are examined for long-term changes in the GI tract. The data in Tables I–III are a compilation of four separate studies (Schneeman and Gallaher, 1980; Sheard and Schneeman, 1980; Forman and Schneeman, 1980;

TABLE I. Pancreatic Enzyme Activity[a] in the Intestine

	Total units × 10^2				
	Amylase	Lipase	Trypsin	Chymotrypsin	Reference
Fiber-free	12.2	3.4	2.74	2.15	
5% Pectin	18.5[b]	7.4[b]	6.57[b]	2.98[b]	Forman and Schneeman (1980)
20% Cellulose	17.29	3.79	3.11	1.49	Schneeman and Gallaher (1980)
5% Wheat bran	8.44	4.92	1.81	2.15	Sheard and Schneeman (1980)
20% Wheat bran	16.47	5.90[b]	2.72	1.73	Schneeman *et al.* (1982)

[a] Adjusted based on fiber-free control for each experiment.
[b] Significantly different from fiber-free ($p < 0.05$).

TABLE II. Enzyme Activity[a] in the Pancreas

	Units/pancreas × 10^3				
	Amylase	Lipase	Trypsin	Chymotrypsin	Reference
Fiber-free	8.3	2.83	0.82	0.87	
5% Pectin	9.85	2.95	0.79	1.04	Forman and Schneeman (1980)
20% Cellulose	8.46	2.83	1.02	0.83	Schneeman and Gallaher (1980)
5% Wheat bran	18.5[b]	15.13[b]	1.16[b]	1.15[b]	Sheard and Schneeman (1980)
20% Wheat bran	14.8[b]	3.34	1.48[b]	1.00	Schneeman *et al.* (1982)

[a] Adjusted based on fiber-free control for each experiment.
[b] Significantly different from fiber-free ($p < 0.05$).

Schneeman *et al.*, 1982). The data from the fiber-free group for each study have been pooled and the values for the fiber-treated groups adjusted based on each experiment's fiber-free group. However, because each experiment was performed under different conditions, each experimental group can only be compared to the fiber-free control.

Total amylase, lipase, trypsin, and chymotrypsin activities measured in the small intestine are similar among the fiber-free, 20% cellulose, and 5% wheat bran groups. Total lipase activity was elevated in the 20% wheat bran group. Activity of all four enzymes was significantly elevated in rats fed the 5% pectin diet (Table I). Pectin lead to an elevation of enzyme activity *in vitro* (Fig. 1). In humans the concentration of enzyme in a duodenal aspirate tended to be elevated after receiving a pectin-containing test meal; however, the total enzyme activity in the duodenum of humans is not known (Sommer and Kasper, 1980). The *in vivo* and *in vitro* interactions of pectin with enzyme activity seem parallel—a tendency to increase activity. In contrast, the *in vitro* interactions of wheat bran and cellulose tended to decrease enzyme activity, yet total measureable enzyme levels in the intestine are similar between the fiber-free and wheat

TABLE III. Protein in the Intestinal Contents

	Amount,[a] mg	Reference
Fiber-free	31.5	
20% Cellulose	44.8[b]	Schneeman and Gallaher (1980)
5% Wheat bran	27.4	Sheard and Schneeman (1980)
20% Wheat bran	53.4[b]	Schneeman *et al.* (1982)

[a] Adjusted based on fiber-free control for each experiment.
[b] Significantly different from fiber-free control ($p < 0.05$).

bran or cellulose groups. An important point to realize is that decreases in digestibility caused by the fiber sources tested cannot be accounted for by lower levels of total enzyme activity. The levels of protein in the intestinal contents of the rats fed 20% cellulose or wheat bran (Table III) would suggest that digestion of protein is slower. The fiber-treated groups tend to have more protein remaining in the intestine. Although total activity is similar, the enzymes and substrates may be more dilute with the addition of nondigestible material, or the fiber may interfere with the interaction of enzyme and substrate by other mechanisms.

The level of pancreatic enzyme activity in the small intestine represents a dynamic state reflecting secretion and degradation of the enzymes. Enhancement of enzyme levels in the intestine could reflect either an increase in secretion or a decrease in degradation. It has been demonstrated that the presence of nondigested protein in the intestine will slow the breakdown of pancreatic enzymes (Percival and Schneeman, 1979; Snook and Meyer, 1964). Pectin or other nondigested materials could help stabilize enzyme activity within the intestinal contents.

The enzyme levels in the intestinal contents provide some indication of the pancreatic response to a meal. The enzyme activity within the pancreatic tissue was also measured to assess adaptation to the diet since adaptation is often correlated with stimulation of the pancreas (Schneeman *et al.*, 1977).

Table II contains data for the pancreatic levels of these four enzymes. Pectin and cellulose did not cause any changes relative to the fiber-free group. Hence it is unlikely that these fibers had a pronounced stimulatory effect, since the adaptive response of the pancreas to diet is related, in part, to its stimulatory ability (Schneeman *et al.*, 1977). In contrast, wheat bran caused an elevation of enzyme activity and may have stimulated pancreatic secretion (Sheard and Schneeman, 1980). Adaptation of pancreatic enzyme composition to the digestible components of the diet, protein, fat, or carbohydrate, is well documented (Abdeljlil *et al.*, 1963; Snook, 1971; Forman and Schneeman, 1980). However, this is the first observation that changes in the nondigestible component of the diet can lead to such changes. Mechanisms involving stimulation of pancreatic secretion or perhaps changes in the absorption of nutrients or in gut hormone release could be involved (Corring, 1977). Sommer and Kasper (1980) implied that certain fiber sources suppressed pancreatic secretion. However, their studies were done with total diversion of pancreatic juice from the intestine. Diversion of pancreatic juice in the rat causes a hypersecretion of enzymes (Green *et al.*, 1973) during which the pancreas is unable to respond to normal stimuli (Schneeman *et al.*, 1977).

Dietary fiber, depending on the type, may also be able to alter bile acid secretion. Portman measured bile acid output from bile cannulated rats 15 days after they had adapted to various dietary regimes (Table IV). Total bile acid

TABLE IV. Bile Acid Secretion

	Bile acids, mg/kg BW/24 hr			
	Cholic acid[a]	Dihydroxy cholanic acid[a]	Total bile acids[a]	Cholic half-life,[b] days
Purina Chow	126.8[c]	30.8	157.6[c]	2.00[c]
Semisynthetic	37.1	38.6	75.7	4.17
Semisynthetic + 20% Celluflour	76.0[c]	26.4	102.4[c]	1.44[c]

[a] Portman *et al.* (1955).
[b] Portman and Murphy (1958).
[c] Significantly different from semisynthetic group.

output was lower in rats fed a low-fiber, semisynthetic diet compared to the output of the groups given laboratory chow or a cellulose supplement (Portman *et al.*, 1955). The half-life of cholic acid was shorter in animals consuming high-fiber diets as well (Portman and Murphy, 1958). With an acute infusion of cellulose into rats with bile duct cannulae we did not see a change in bile acid secretion, suggesting that a long-term adaptation is involved (Schneeman, 1979). Following a meal about 80% of the bile acid pool in a rat is in the small intestine (Uchida *et al.*, 1978). Rats fed cellulose-containing diets have elevated bile acid levels in the small intestine relative to rats fed low-fiber rations (Schneeman and Gallaher, 1980; Boyd and Eastwood, 1968). However, this effect does not seem to be generally true for all fiber sources (Sheard and Schneeman, 1980; Forman and Schneeman, 1980). In measuring bile acid levels in the small intestine, it is not clear if changes reflect shifts in the pool size, increased secretion, or decreased absorption. Other chapters discuss the importance of bile acids and the composition of bile to other aspects of metabolism.

Clearly a potential interaction exists between dietary fibers and digestive function; however, the nature of the interaction varies by the fiber source.

3. CHANGES IN COMPOSITION OF THE INTESTINAL CONTENTS

One of the more striking effects of dietary fiber on the small intestinal contents is the increase in the dry weight of material. Dry weight of contents was about 40% greater in rats fed pectin, cellulose, or wheat bran than in the fiber-free groups (Table V). In another study, we demonstrated that both wet and dry weight of the intestine are heavier at 1, 2, or 4 hr after consumption of the meal (Table VI) (Forman and Schneeman, 1982). This elevation of the

TABLE V. Calculated Dry Weight of Small Intestinal Contents

	Amount,[a] mg	Reference
Fiber-free	171	
5% Pectin	295[b]	Forman and Schneeman (1980)
20% Cellulose	262[b]	Schneeman and Gallaher (1980)
20% Wheat bran	287[b]	Schneeman *et al.* (1982)

[a] Adjusted based on fiber-free control for each experiment.
[b] Significantly different from fiber-free ($p < 0.05$).

volume and quantity of material in the small intestine could be important. For digestion and absorption to proceed, enzymes and substrates must be able to diffuse together, and the end products of digestion must diffuse to the mucosal surface for final hydrolysis and absorption. The greater bulk of material in the gut could slow this process and hence account for, in part, the apparent slower absorption of nutrients. In addition, this increase in material could dilute enzymes and absorbable compounds in the gut.

The viscosity of the intestinal contents can be changed by the addition of fiber. High methoxy-pectin and guar gum increased the viscosity of human duodenal fluid around tenfold or more; wheat bran increased the viscosity threefold (I. Ihse, G. Isaksson, and I. Lundquist, personal communication, 1981). The mobility of enzymes, substrates, and digestion products will be slower in a high-viscosity solution. Jenkins *et al.* (1978) have demonstrated that the reduction of blood glucose concentration during a glucose tolerance test by guar gum is abolished if hydrolyzed, nonviscous guar is used.

Less is known concerning other potential changes in the composition of intestinal contents. For example, alteration in the pH of the chyme could occur and would influence enzyme activity as well as hormone release from the upper intestine (Grossman, 1967).

TABLE VI. Weight of Small Intestine Plus Contents[a]

	One hour	Two hour	Four hour
Wet weight, g			
Fiber-free	4.82	4.94	4.16
5% Pectin	6.33	6.99	5.74
Dry weight, g			
Fiber-free	1.26	1.24	1.12
5% Pectin	1.49	1.61	1.43

[a] Forman and Schneeman (1982). In all cases pectin group is heavier ($p < 0.05$).

4. MORPHOLOGY OF THE SMALL INTESTINE

Several studies have implied that some sources of dietary fiber may cause morphological changes in the structure of the small intestine which are related to the functional changes that have been observed. Previous studies have suggested that mucosal protein levels are lower in rats fed wheat bran or cellulose and mucosal peptidase activity is lower in rats fed cellulose (Schneeman and Gallaher, 1980; Sheard and Schneeman, 1980).

In a study by Brown *et al.* (1979), alkaline phosphatase and peptidase were lower in pectin-fed rats (Table VII). In their study the intestine was divided into five segments, three of which were used for biochemical measurements. The number in the significance column indicates the number of segments in which a significant difference was observed. The lower enzyme values in the fiber-fed mucosae could imply increases in cell sloughing, perhaps indicative of differences in cell turnover, that the cells on the villi are less mature with lower enzyme content, or reflect adaptive changes in the composition of brush border enzymes. In the study of Brown *et al.* (1979) (Table VII) mucosal DNA levels had not changed, indicating that cell numbers were not different, but that the enzyme content per cell had changed.

Other indirect evidence that the morphology of the small intestine can be altered by feeding fiber is evident from absorption studies. The absorption of labeled cholesterol into lymph is slower in rats that have been pretreated with fiber-containing diets (Vahouny *et al.*, 1980). Likewise, glucose absorption in an isolated intestinal segment from rats that had been pretreated with fiber-containing diets is less (Schwartz and Levine, 1980). In both the thoracic-duct cannulated experiment and the isolated intestinal segment experiments, the fiber sources were not present in the perfusion mixtures, yet absorption was slower. These studies suggest that alterations in the intestinal mucosa could contribute to slower absorption.

We examined sections of duodenum and ileum from rats that consumed

TABLE VII. Small Intestine Biochemistry in Rats Fed Pectin[a]

	Basal	Pectin	Significance[b]
Mucosal DNA, μg/mm	16.8	18.5	ns
Mucosal protein, μg/mm	285	362	ns
Alkaline phosphatase,[c] units/mg DNA	66.6	22.2	2/3
Leucyl β-naphthylamidase,[c] units/mg DNA	131.0	76.1	1/3

[a] Brown *et al.* (1979).
[b] Number of intestinal segments in which a significant difference was observed. ns, Not significant.
[c] Pooled values for three out of five segments.

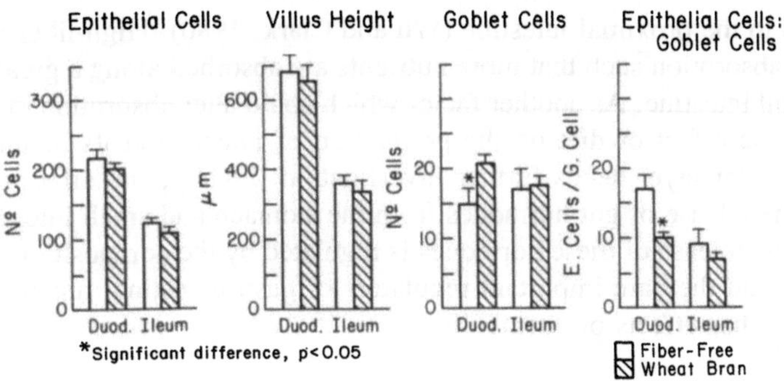

FIGURE 2. Epithelial cells per villus, villus height, and goblet cells in the duodenum or ileum of rats fed a fiber-free diet or a 20% wheat bran diet.

either a fiber-free or 20% wheat bran diet (Fig. 2). No differences in the villus height or the number of epithelial cells per villus were observed. However, the number of goblet cells per villus was greater in the wheat bran group. Consequently the ratio of epithelial cells per goblet cell was reduced in this group. Because of the greater percentage of goblet cells, animals fed wheat bran could potentially produce more mucus in the duodenum. Experimental evidence suggests that the mucus layer is an important component of the unstirred water layer (Nimmerfall and Rosenthaler, 1980). This layer provides the rate-limiting barrier for absorption in the small intestine, particularly for lipid-soluble compounds (Thomson, 1978). An examination of mucus production in the small intestine, particularly in the duodenum and upper jejunum, could provide insights to explain the slower absorption of nutrients that seems to occur with high-fiber diets.

5. CONCLUSIONS

The studies that have been discussed have begun to expose the potential intraluminal effects of dietary fiber during its transit through the small intestine. These intraluminal interactions can provide a basic understanding of how a nondigestible dietary component can alter metabolism.

To expand our understanding in this regard, research is needed in several areas. We need to determine if *in vivo* rates of digestion have been slowed by the presence of dietary fiber. Evidence suggests that the rate of digestion could be slowed by fiber-containing diets. Such an effect would slow the rate of absorption from the intestine. The increase in small intestinal contents could slow the digestive, absorptive processes. It is known that the site of absorption from the small intestine will influence nutrient metabolism. For example, larger chylomicrons, which can be cleared more rapidly, are made in the distal intestine compared to the proximal intestine (Wu and Clark, 1980). High-fiber diets may influence absorption such that more nutrients are absorbed along a greater length of the small intestine. As another factor which could alter absorption in the small intestine, the effect of diet on the production of mucus and its relation to the unstirred water layer needs further investigation. Finally, the effect of dietary fiber on the release of gut hormones from the stomach and small intestine is not known. The release of these hormones is regulated by the composition of chyme in the gut and they are important regulators of gastrointestinal function as well as having other effects on metabolism.

ACKNOWLEDGMENTS. This work has been supported in part by NIH grant AM 20446 and Nutrition Foundation grant 542. The author is grateful for the valuable contributions that have been made by Dan Gallaher, L. P. Forman, Diane Richter, and George Dunaif.

REFERENCES

Abdeljlil, B. A., Visani, A. M., and Desnuelle, P., 1963, Adaptation of the rat exocrine pancreas to the composition of the diet, *Biochem. Biophys. Res. Commun.* **10:**112–116.

Boyd, G. S., and Eastwood, M. A., 1968, Studies on the quantitative distribution of bile salts along the rat small intestine under varying dietary regimes, *Biochim. Biophys. Acta* **52:**159–164.

Brown, R. C., Kelleher, J., and Losowsky, M. S., 1979, The effect of pectin on the structure and function of the rat small intestine, *Br. J. Nutr.* **42:**357–365.

Corring, T., 1977, Possible role of hydrolysis products of the dietary components in the mechanisms of the exocrine pancreatic adaptation of the diet, *World Rev. Nutr.* **27:**132–144.

Dunaif, G., and Schneeman, B. O., 1981, The effect of dietary fiber on human pancreatic enzyme activity *in vitro, Am. J. Clin. Nutr.* **34:**1034–1035.

Forman, L. P., and Schneeman, B. O., 1980, Effects of dietary pectin and fat on the small intestinal contents and exocrine pancreas in rats, *J. Nutr.* **110:**1992–1999.

Forman, L. P., and Schneeman, B. O., 1982, Dietary pectin's effect on starch utilization in the rat, *J. Nutr.* (in press).

Green, G. M., Olds, B. A., Matthews, G., and Lyman, R. L. 1973, Protein as a regulator of pancreatic enzyme secretion in the rat, *Proc. Soc. Exp. Biol. Med.* **142:**1162–1167.

Grossman, M. I., 1967, in: *Handbook of Physiology, Section 6: Alimentary Canal,* Vol. w (C. F. Code, ed.), American Physiological Society, Washington, D.C., pp. 835–863.

Jenkins, D. J. A., Wolever, T. M. S., Leeds, A. R., Gassull, M. A., Haisman, P., Dilaware, J., Goff, D. V., Metz, G. L., and Alberti, K. G. M. M., 1978, Dietary fibres, fibre analogues, and glucose tolerance: Importance of viscosity, *Br. Med. J.* **1:**1392–1394.

Kelsay, J. L., 1978, A review of research on effects of fiber intake in man, *Am. J. Clin. Nutr.* **31:**142–159.

Liener, I. E., and Kakade, M. L., 1980, Protease inhibitors, in: *Toxic Constituents of Plant Foodstuff,* 2nd ed. (I. E. Liener, ed.), Academic Press, New York, pp. 7–71.

Nimmerfall, F., and Rosenthaler, J., 1980, Significance of the goblet-cell mucin layer, the outermost barrier to passage through the gut wall, *Biochem. Biophys. Res. Commun.* **94:**960–966.

Percival, S. S., and Schneeman, B. O., 1979, Long term pancreatic response to feeding heat damaged casein in rats, *J. Nutr.* **109:**1609–1614.

Portman, O. W., and Murphy, P., 1958, Excretion of bile acids and β-hydroxysterols by rats, *Arch. Biochem. Biophys.* **76**:367–376.

Portman, O. W., Mann, G. V., and Wysocki, A. P., 1955, Bile acid excretion by the rat: Nutritional effects, *Arch. Biochem. Biophys.* **59**:224–232.

Schneeman, B. O., 1978, Effect of plant fiber on lipase, trypsin and chymotrypsin activity, *J. Food Sci.* **43**:634–635.

Schneeman, B. O., 1979, Acute pancreatic and biliary response to protein, cellulose and pectin, *Nutr. Rep. Int.* **20**:45–48.

Schneeman, B. O., and Gallaher, D., 1980, Changes in small intestinal digestive enzyme activity and bile acids with dietary cellulose in rats, *J. Nutr.* **110**:584–590.

Schneeman, B. O., Jacobs, L. R., and Richter, D., 1982, Response to dietary wheat bran in the exocrine pancreas and intestine of rats, *J. Nutr.* (in press).

Schneeman, B. O., Chang, I., Smith, L. B., and Lyman, R. L., 1977, The effect of dietary amino acids, casein, and soybean trypsin inhibitor on pancreatic protein secretion in the rat, *J. Nutr.* **107**:281–288.

Schwartz, S. E., and Levine, G. D., 1980, Effects of dietary fiber on intestinal glucose absorption and glucose tolerance in rats, *Gastroenterology* **79**:833–836.

Sheard, N. F., and Schneeman, B. O., 1980, Wheat bran's effect on digestive enzyme activity and bile acid levels in rats, *J. Food Sci.* **45**:1645–1648.

Snook, J. T., 1971, Dietary regulation of pancreatic enzymes in the rat with emphasis on carbohydrate, *Am. J. Physiol.* **221**:1383–1387.

Snook, J. T., and Meyer, J. H., 1964, Factors influencing the significance of endogenous nitrogen to the non-ruminant, in: *The Role of the Gastrointestinal Tract in Protein Metabolism* (H. N. Munro, ed.), F. A. Davis, Philadelphia, pp. 97–110.

Sommer, H., and Kasper, H., 1980, The effect of dietary fiber on the pancreatic excretory function, *Hepato-Gastroenterology* **27**:477–483.

Thomson, A. B. R., 1978, Intestinal absorption of lipids: Influence of the unstirred water layer and bile acid micelle, in: *Disturbances in Lipid and Lipoprotein Metabolism* (J. M. Dietschy, A. M. Gotto, Jr., and J. A. Ontko, eds.), American Physiological Society, Bethesda, pp. 29–55.

Uchida, K., Okuno, I., Takase, H., Nomura, Y., and Kadowaki, M., 1978, Distribution of bile acids in rats, *Lipids* **13**:42–48.

Vahouny, G. V., Roy, T., Gallo, L. L., Story, J. A., Kritchevsky, D., and Cassidy, M., 1980, Dietary fibers. III. Effects of chronic intake on cholesterol absorption and metabolism in the rat, *Am. J. Clin. Nutr.* **33**:2182–2191.

Wu, A. L., Clark, S. B., and Holt, P. R., 1980, Composition of lymph chylomicrons from proximal or distal rat small intestine, *Am. J. Clin. Nutr.* **33**:582–589.

8

Interactions of Dietary Fiber and Nutrients

JUAN M. MUNOZ

There is some evidence that the metabolic effects of dietary fiber may depend not only on its physical chemical properties (i.e., water-holding capacity, bile absorption, cation exchange) but also on its interaction with other nutrients, and perhaps with other gastrointestinal factors as well. Realizing that fiber's influence on metabolism may be due not only to its physical properties but to other factors as well may help explain the seemingly contradictory reports we see in the literature. It has recently been shown, for example (Levine and Silvis, 1980), that the ability of dietary fiber to bind fat depends on the physical form of fat present in the diet, the binding being directly related to the degree of refinement of fat. When the content of fat in the diet was in the form of whole peanuts, the addition of 20 g of crude fiber had no effect on the daily fecal excretion of fat; however, when the content of fat was in the form of peanut oil, the addition of 20 g of crude fiber to the diet significantly increased the fecal excretion of fat. Macleod and Blacklock (1979) have reported that the ability of dietary fiber to bind calcium is diminished or abolished by the simultaneous administration of high-carbohydrate diets, and Sandstead *et al.* (1979) and Munoz *et al.* (1982) have observed that the content or level of protein in the diet may also influence the metabolic effects of dietary fiber.

This chapter will summarize our previous observations of how different protein levels in the diet influence the effects of dietary fiber on carbohydrate metabolism and on the gastrointestinal absorption of trace elements.

At the USDA Human Nutrition Laboratory in Grand Forks, North Dakota, various sources of dietary fiber (Table I) and their possible effects on *metabolism*

JUAN M. MUNOZ • Fargo Clinic, Fargo, North Dakota 58123.

TABLE I. Sources of Dietary Fiber and Their Main Fraction and Their Effects on Lipid Acid Carbohydrate Metabolism[a]

	SWW	HRSW	CB	SH	Apples	Carrots
Main fraction	Hemicellulose	Hemicellulose	Hemicellulose	Cellulose and hemicellulose	Pectin	Pectin
Lowering effect on:						
Triglyceride	Yes	Yes	Yes	Yes	Yes	Yes
Cholesterol	No	Yes	Yes	Yes	Yes	Yes
Glucose tolerance test	No	No	Yes	Yes	Yes	Yes

[a]SWW, Soft white winter wheat; HRSW, hard red spring wheat; CB, corn bran; SH, soy hulls.

have been studied (Munoz *et al.*, 1979a,b). Munoz *et al.* (1982) found that all fiber sources, with the exception of the wheat sources, were associated with improved glucose tolerance tests. In this experiment the composition of the control diet was typical of a mixed Western diet: 15% of energy as protein (or about 120–140 g), 40% fat, and 45% carbohydrate. The addition of modest amounts of dietary fiber (26 g of either corn bran, soybean hulls, apples, or carrots) to this diet was associated with a significant improvement in the oral glucose tolerance test (Fig. 1A); however, when the control diet had a lower level of protein—8% of energy as protein (about 60 g), 40% fat, and 52% carbohydrate—these effects of dietary fiber on the glucose tolerance test *were not observed* (Fig. 1B). These results are difficult to explain. Schteingart *et al.* (1978) have recently reported that protein-restricted diets decrease basal plasma insulin levels in obese people and this was accompanied by a fall in plasma

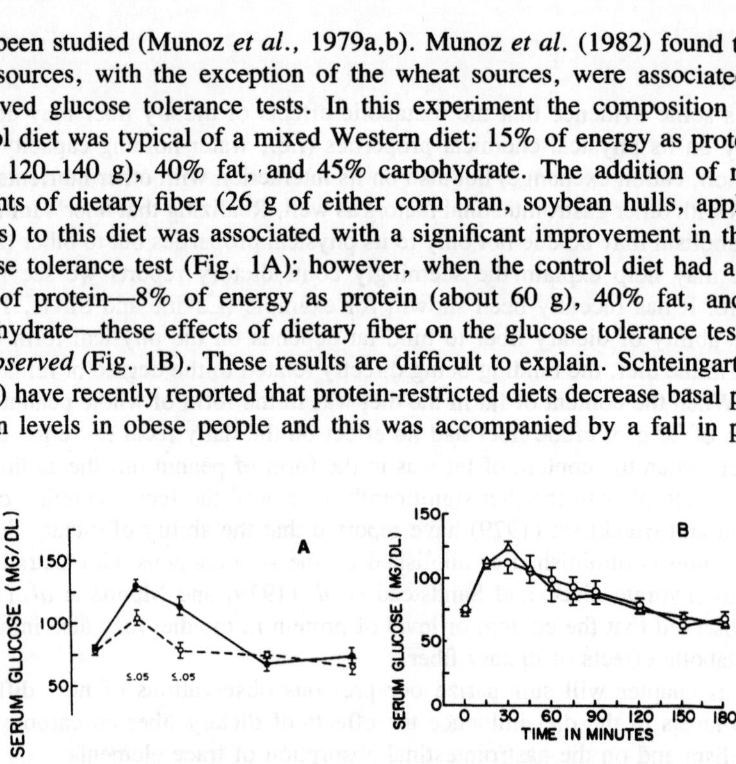

FIGURE 1. Effects of dietary fiber and two levels of protein on the oral glucose tolerance test of normal men. (A) Daily intake of protein, 120–140 g. Solid circles, control, low-fiber diet. Open circles, control diet supplemented with 26 g of dietary fiber. (B) Daily intake of protein, 60 g. Open circles, control, low-fiber diet. Open triangles, control diet supplemented with 26 g of dietary fiber. For details see Munoz *et al.* (1979).

glucose, suggesting an increase in insulin sensitivity. It is obvious that the interactions of protein and dietary fiber need further evaluation.

In our study the oral glucose load was given alone, not with the fiber source; thus the results indicate that the effects of dietary fiber on carbohydrate metabolism may not be due solely to its physical chemical properties of binding carbohydrates and slowing the gastrointestinal absorption of sugars, as has been suggested by other investigators, who have shown that the simultaneous oral administration of a source of dietary fiber with glucose (Crapo *et al.*, 1978; Kanter *et al.*, 1980; Goulder *et al.*, 1978; Jenkins *et al.*, 1976, 1977, 1978a, 1980) or D-xylose (Jenkins *et al.*, 1978b) are associated with lower postprandial glucose and D-xylose concentrations. Several studies have suggested that the effects of dietary fiber on carbohydrate metabolism are mediated by changes in the enteral-pancreatic axis. Jenkins *et al.* (1977) have reported that the decrease in postprandial glucose concentrations observed with high-fiber diets was associated with lower insulin levels, and Munoz *et al.* (1979a) observed that the improvement of the glucose tolerance tests observed with high-fiber diets was associated with no changes in the insulin levels. These studies have suggested that the plasma glucose-lowering effects of high-fiber diets may be the consequence of an increased total glucose utilization, or increased peripheral insulin sensitivity. Munoz *et al.* (1979a) and Miranda and Horwitz (1978) observed lower glucagon levels and Morgan *et al.* (1979) reported lower GIP levels associated with the consumption of high-fiber diets. All these studies support the concept that dietary fiber somehow affects the enteral–pancreatic axis.

Sandstead *et al.* (1978) reported that modest additions of dietary fiber (26 g) to a controlled diet (15% protein, 40% fat, 45% carbohydrate) did not interfere with intestinal absorption of copper, zinc, calcium, and phosphorus. This was more evident when the amount of protein in the diet was decreased from between 120 and 140 g to 60 g (Sandstead *et al.*, 1979), as shown in Table II. The lower protein diet was associated with a more positive balance for zinc, copper, calcium, and phosphorus, resulting in lower daily requirements for these elements.

TABLE II. Influence of Protein and Dietary Fiber on the Daily Requirements of Zinc and Calcium

	Daily requirements, mg/day	
Diet	Zinc	Calcium
High protein		
Low fiber	12.5	841
High fiber	12.8	985
Low protein		
Low fiber	7.0	685
High fiber	7.4	736

This finding suggests that the negative effects of dietary fiber on the gastrointestinal availability of nutrients reported by some investigators may depend on (1) the amount of dietary fiber, and (2) the amount of protein in the diet.

In summary, the metabolic effects of dietary fiber may not only depend on its physical chemical properties, but also on its interactions with other nutrients, such as carbohydrates and protein, and perhaps with other gastrointestinal factors as well.

REFERENCES

Crapo, P. A., Reaven, G., Olefsky, J., and Alto, P., 1978, Plasma glucose and insulin responses to orally administered simple and complex carbohydrates, *Ann. Intern. Med.* **88:**482.

Goulder, T. J., Alberti, K. G. M. M., and Jenkins, D. A., 1978, Effect of added fiber on the glucose and metabolic response to a mixed meal in normal and diabetic subjects, *Diabetes Care* **1**(6):351–355.

Jenkins, D. J. A., Goff, D. V., Leeds, A. R., Alberti, K. G. M. M., Wolever, T. M. S., Gassull, M. A., and Hockaday, T. D. R., 1976, Unabsorbable carbohydrates and diabetes; decreased post-prandial hyperglycemia, *Lancet* **2:**172–174.

Jenkins, D. J. A., Leeds, A. R., Gassull, M. A., Cochet, B., and Alberti, K. G. M. M., 1977, Decrease in postprandial insulin and glucose concentrations by guar and pectin, *Ann. Intern. Med.* **86:**20–23.

Jenkins, D. J. A., Wolever, T. M. S., Nineham, R., Taylor, T., Metz, G. L., Bacon, S., and Hockaday, T. D. R., 1978a, Guar crispbread in the diabetic diet, *Br. Med. J.* **2:**1744–1746.

Jenkins, D. J. A., Wolever, T. M. S., Leeds, A. R., Gassull, M. A., Haisman, P., Dilawari, J., Goff, D. V., Metz, G. L., and Alberti, K. G. M. M., 1978b, Dietary fibres, fibre analogues, and glucose tolerance: Importance of viscosity, *Br. Med. J.* **1:**1392–1394.

Jenkins, D. J. A., Wolever, T. M. S., Bacon, S., Nineham, R., Lees, R., Rowden, R., Love, M., and Hockaday, T. D. R., 1980, Diabetic diets: High carbohydrate combined with high fiber, *Am. J. Clin. Nutr.* **33:**1729–1733.

Kanter, Y., Eitan, N., Brook, G., and Barzilai, D., 1980, Improved glucose tolerance and insulin response in obese and diabetic patients on a fiber-enriched diet, *Isr. J. Med. Sci.* **16:**1–6.

Levine, A. S., and Silvis, S. E., 1980, Absorption of whole peanuts, peanut oil, and peanut butter, *N. Engl. J. Med.* **303:**917–918.

Macleod, M. A., and Blacklock, N. J., 1979, The influence of glucose and crude fibre (wheat bran) on the rate of intestinal Ca absorption. The influence of glucose and wheat bran on calcium absorption, *J. R. Naval Med. Ser. Winter* **65:**143–146.

Miranda, P. M., and Horwitz, D. L., 1978, High-fiber diets in the treatment of diabetes mellitus, *Ann. Intern. Med.* **88:**482–486.

Morgan, L. M., Goulder, T. J., Tsiolakis, D., Marks, V., and Alberti, K. G. M. M., 1979, The effect of unabsorbable carbohydrate on gut hormones, *Diabetologia* **17:**85–89.

Munoz, J. M., Sandstead, H. H., Jacob, R. A., Johnson, L., and Mako, M. E., 1979a, Effects of dietary fiber on glucose tolerance of normal men, *Diabetes* **28**(5):496–502.

Munoz, J. M., Sandstead, H. H., Jacob, R. A., Logan, G. M., Reck, S. J., Klevay, L. M., Dintzis, F. R., Inglett, G. E., and Shuey, W. C., 1979b, Effects of some cereal brans and textured vegetable protein on plasma lipds, *Am. J. Clin. Nutr.* **32:**580–592.

Munoz, J. M., Sandstead, H. H., and Jacob, R. A., 1982, Effects of protein and dietary fiber on glucose tolerance test of normal men (in preparation).

Sandstead, H. H., Munoz, J. M., Jacob, R. A., Klevay, L. M., Reck, S. J., Logan, G. M., Jr., Dintzis, F. R., Inglett, G. E., and Shuey, W. C., 1978, Influence of dietary fiber on trace element balance, *Am. J. Clin. Nutr.* **31**:S180–S184.

Sandstead, H. H., Klevay, L. M., Jacob, R. A., Munoz, J. M., Logan, G. M., Jr., and Reck, S. J., 1979, Effects of dietary fiber and protein level on mineral element metabolism, in: *Dietary Fibers: Chemistry and Nutrition* (G. E. Inglett and S. I. Falkehag, eds.), Academic Press, New York, pp. 147–156.

Schteingart, D. E., McKenzie, A. K., Victoria, R. S., and Tsao, H. S., 1978, Suppression of insulin secretion by protein deprivation in obesity, in: *Treatment of Early Diabetes* (R. A. Camerini-Davalos and B. Hanover, eds.), Plenum Press, New York, pp. 125–132.

9

Effects of Fiber on Mineral and Vitamin Bioavailability

JUNE L. KELSAY

1. INTRODUCTION

Along with positive effects of increased fiber in the diet, there have been reports of decreased bioavailability of nutrients. Some nutrients may be bound by fiber and thus rendered unavailable for absorption by the body.

A number of reports indicate that fiber in the diet may result in decreased or negative mineral balances in human subjects. A review of the effects of fiber on iron, zinc, and copper balances in human subjects has been published recently (Kelsay, 1981), and these minerals will not be discussed here. In that review it was reported that in most studies increased fiber in the diet did not affect iron balance. The results in the studies reviewed may have been at least partially due to high levels of iron intake. However, fiber decreased zinc and copper balances in some studies.

The next section of this chapter deals with the effects of fiber on calcium, magnesium, and phosphorus balances of human subjects. In Section 3, studies on the effects of fiber on vitamin bioavailability are reviewed. The effects of fiber on vitamin A, thiamine, nicotinic acid, vitamin B6, and vitamin B12 have been investigated in rats. The effects of fiber on vitamin A and carotene, riboflavin, nicotinic acid, vitamin B6, folic acid, and pantothenic acid have been studied in human subjects.

JUNE L. KELSAY • Carbohydrate Nutrition Laboratory, Beltsville Human Nutrition Research Center, Human Nutrition, Science and Education Administration, U.S. Department of Agriculture, Beltsville, Maryland 20705.

TABLE I. Effect of Fiber on Calcium, Magnesium, and Phosphorus Balances

Investigators	Fiber source	Effect of fiber[a]
McCance and Widdowson (1942a)	Brown bread (40–50% of kcal)	Ca, Mg, and P absorptions decreased
McCance and Widdowson (1942b)	Brown bread and dephytinized brown bread (40–50% of kcal)	Ca, Mg, and P absorptions lowest with brown bread not dephytinized
Walker *et al.* (1948)	Brown bread (1 lb/day)	Ca and Mg balances negative, but improved with time; P not affected
Cullumbine *et al.* (1950)	Unpolished rice (27 oz cooked per day)	Ca and Mg balances decreased, but improved with time; P retentions improved
Reinhold *et al.* (1973)	Unleavened wholemeal bread (350 g/day)	Ca balances negative; P not affected
Reinhold *et al.* (1976)	Leavened wholemeal bread (60% of kcal)	Ca, Mg, and P balances negative
Cummings *et al.* (1979a)	31 g Wheat fiber	Ca balances negative
Ismail-Beigi *et al.* (1977)	10 g Cellulose in 150 g apple compote	Ca and Mg balances negative; P not generally affected
Cummings *et al.* (1979b)	36 g Pectin	Ca balances not affected
Drews *et al.* (1979)	14.2 g Cellulose, hemicellulose, or pectin	Mg balance lowered by hemicellulose; other fibers had no effect
Slavin and Marlett (1980)	16 g Cellulose	Ca and Mg balances decreased
Kelsay *et al.* (1979a,b)	24 g NDF in fruits and vegetables	Ca and Mg balances negative; P not affected
Kelsay *et al.* (1981)	10, 18, or 25 g NDF in fruits and vegetables	Ca and Mg balances not affected

[a] If the mineral is not mentioned in this column, it was not determined.

2. EFFECTS OF FIBER IN THE DIET ON CALCIUM, MAGNESIUM, AND PHOSPHORUS BALANCES OF HUMAN SUBJECTS

Results of human studies in which calcium, magnesium, and phosphorus balances were determined when fiber was increased in the diet are tabulated in Table I. Sources of fiber included: (1) cereal products, which contain cellulose, hemicellulose, and lignin; (2) commercial preparations of pectin, cellulose, or hemicellulose; and (3) fruits and vegetables, which contain pectin, cellulose, hemicellulose, and lignin.

2.1. Cereal Products

In a study of ten subjects on a controlled diet for 14–28 days, McCance and Widdowson (1942a) compared brown and white breads which furnished 40–50% of the kcal. Calcium, magnesium, and phosphorus were less completely absorbed when the brown bread was consumed. In a further study on six subjects the same investigators (McCance and Widdowson, 1942b) fed brown bread, dephytinized brown bread, and white bread for 3 weeks at a level of 40–50% of the kcal. Both sources of brown bread decreased calcium, magnesium, and phosphorus absorptions, but the decrease was greatest when brown bread containing phytate was fed.

Walker *et al.* (1948) added 1 lb of brown or white bread to an experimental diet fed to three subjects. Two subjects consuming the brown bread had negative calcium and magnesium balances for the first 4 weeks, but had positive balances at 8 weeks. There was no effect on phosphorus balance.

Cullumbine *et al.* (1950) replaced 27 oz of polished rice per day with an equal amount of unpolished rice in the diets of 12 subjects for 3 weeks. Calcium and magnesium balances were lower on the unpolished rice diets. The unpolished rice resulted in greater phosphorus retentions, although there were greater excretions in urine and feces. The intake of phosphorus was greater on the unpolished rice diet due to the phytate phosphorus. In three subjects fed the unpolished rice diet for 18 weeks, retention of calcium and magnesium improved with time.

Reinhold *et al.* (1973) studied three subjects on a diet containing 350 g of white bread for 16 days and 350 g of unleavened wholemeal wheat bread for 32 days. Reinhold *et al.* (1976) also studied the effects of substituting leavened wholemeal wheat bread for white bread. These investigators had previously reported that phytate was destroyed by leaven. Two subjects were studied for 20 days each on the white bread and wholemeal bread diets. In both studies wholemeal bread resulted in negative calcium and magnesium balances; phosphorus balances were not affected in the first study, but were negative on wholemeal bread in the second study.

Cummings *et al.* (1979a) added 31 g of wheat fiber to the diets of four subjects consuming a high-protein diet. The subjects consumed the high-protein diet for 3 weeks before and 3 weeks after addition of fiber to the diet. Calcium balances were negative when fiber was added to the diet, even though calcium intakes were 1300 mg/day on the diet with fiber added and 960 mg/day without the added fiber.

2.2. Commercial Preparations

Ismail-Bengei *et al.* (1977) added 10 g of cellulose in 150 g of apple compote to a low-fiber diet for 20 days, and to a high-fiber, high-phosphorus diet for 14 days. Three subjects were studied on the low-fiber diet and two subjects on the high-fiber diet and results were compared with those found when the cellulose was not added. With the cellulose and low-fiber diet, negative balances of calcium and magnesium resulted, but phosphorus balances were not affected. With the high-fiber diet, previously negative balances of calcium and magnesium became more negative when the cellulose was added.

Cummings *et al.* (1979b) studied the effect of pectin on calcium balance. Five subjects were fed a controlled diet for 3 weeks and the controlled diet with 36 g of pectin per day for 6 weeks. Pectin had no effect on calcium balance.

Drews *et al.* (1979) added 14.2 g of cellulose, hemicellulose, or pectin to a basal diet fed eight adolescent males for 4 days each. Hemicellulose significantly lowered magnesium balance, but the other fibers had no effect.

Slavin and Marlett (1980) fed 16 g of refined cellulose per day to seven women consuming a high-protein diet for one month and compared mineral balances with those obtained when they consumed the same diet for one month before cellulose addition. When cellulose was added, both calcium and magnesium balances became negative, but only calcium balance was significantly lower.

2.3. Fruits and Vegetables

In a study conducted in our laboratory in cooperation with the Department of Food, Nutrition, and Institution Administration of the University of Maryland, we fed 12 men a low-fiber diet containing 4 g of neutral detergent fiber (NDF) per day and a diet with fruits and vegetables containing 24 g of NDF per day for 26 days each. The diet containing fruits and vegetables resulted in negative balances for calcium and magnesium (Kelsay *et al.*, 1979a). Phosphorus balances were not affected (Kelsay *et al.*, 1979b).

In a second study, we fed a low-fiber diet containing 2 g of NDF and three diets containing 10, 18, or 25 g of NDF from fruits and vegetables (Kelsay *et al.*, 1981). Fiber content of the diets in the second study was determined using the American Association of Cereal Chemists' modification of the NDF method (1976), which resulted in lower values than those obtained with the original method used for analysis of samples from the first study. The diet found to contain 18 g of NDF in this study is comparable to the diet found to contain 24 g of NDF in the first study. Twelve subjects were fed each of the four diets for 21 days. Determination of mineral balances in the second study revealed that the diets containing fiber had no effect on calcium and magnesium balances. This difference in results between the two studies was attributed to the presence of spinach, which contains oxalic acid, in the first study. Spinach was replaced with cauliflower, which is low in oxalic acid, in the second study.

In a third study (Kelsay and Prather, 1981) the separate and combined effects of fiber and oxalic acid on mineral balances were investigated. Three diets were fed to 12 subjects for 4 weeks each. Diet 1 was a low-fiber diet which included spinach every other day. Diet 2 contained fiber in fruits and vegetables, including spinach every other day, and was the same as the higher fiber diet fed in our first study. Diet 3 was the same as diet 2 except that cauliflower replaced the spinach. During the fourth week, calcium balance was negative on diet 2 and was significantly lower on diet 1. The calcium balance on diet 3 was lower than on diet 1 and higher than on diet 2, but was not significantly different from either. Magnesium balance was negative on diet 2 and significantly lower than on diets 1 and 3. Results of this third study indicate that the combined effect of oxalic acid and fiber can lead to decreased calcium and magnesium balances.

2.4. Summary

Although in many of the studies reviewed here it appears that increasing fiber in the diet caused decreased calcium and magnesium balances, other complicating factors may be involved.

In assessing the effects of fiber on mineral balances the following factors should be considered: kind of fiber, level of fiber intake, level of mineral intake, length of study period, presence of other mineral-binding agents, such as phytic acid and oxalic acid, and possibly the level of protein intake.

Pectin was reported to have no effect on calcium balance in one study or on magnesium balance in another study. In two studies cellulose decreased calcium and magnesium balances, and in a third it had no effect. Hemicellulose was reported to decrease magnesium balance in one study.

The levels of fiber intake in some of the studies were greater than in others. The higher level of fiber intake in some of the studies may have been responsible for decreased mineral balances.

It seems logical that with higher levels of intake of the mineral in question, decreased availability due to binding would be less important. However, in one study a higher calcium intake did not prevent negative calcium balances when fiber was added to the diet.

Length of time allowed for adaptation to an experimental diet is very important. Most of the studies cited here were conducted for 2–4 weeks. However, subjects may become adapted to decreased availability of minerals over a longer period of time.

Results of studies in which cereal products were fed are difficult to evaluate because the relative and combined effects of fiber and phytic acid on mineral binding have not been completely resolved. Also, the presence of oxalic acid in diets containing fiber in fruits and vegetables complicated evaluation of results in other studies.

In two of the studies cited here the level of protein intake was high and may have influenced results.

3. EFFECTS OF FIBER ON VITAMIN BIOAVAILABILITY

There is not too much information on effects of fiber on availability of some vitamins to human subjects, and care must always be exercised in extrapolating results of animal studies to humans. The varied procedures used in evaluating vitamin responses also make comparison of results difficult.

3.1. Vitamin A and Carotene

Methylcellulose was given daily by stomach tube for 28 days along with vitamin A to vitamin A-depleted rats. Methylcellulose did not decrease the effectiveness of vitamin A as evaluated by growth response (Ellingson and Massengale, 1952).

Gronowska-Senger *et al.* (1979) fed rats 5–20% fiber in the diet as a mixture of methylcellulose, agar, pectin, and wheat grain with 30 IU of retinyl acetate. With 10 or 20% fiber, vitamin A in liver was decreased and retinyl acetate in serum increased. In a second study 30, 90, or 300 IU of retinyl acetate and 10% fiber were fed. Vitamin A in liver and blood serum was decreased when 300 IU of the vitamin was fed with 10% fiber.

Phillips and Brien (1970) found that 3% pectin added to ground cube diet or semipurified casein–sucrose diet did not affect carotene or vitamin A absorption as determined by vitamin A liver stores of rats.

Barnard and Heaton (1973) found that in 14 healthy human subjects addition of lignin to a vitamin A-containing test meal had no effect on serum rise of vitamin A.

Carotene appears to be bound to plant cell walls, rendering it largely unavailable in human subjects. Excretion of carotene from vegetables was less (Leonhardi, 1947; Kreula and Virtanen, 1939) and carotene in blood greater (Van Zeben and Hendriks, 1948) when vegetables were finely chopped or homogenized so that cell walls were disrupted. Rao and Rao (1970) suggested that

TABLE II. IU Vitamin A Activity[a]

	Low fiber diet plus carotene	Diet with fruits and vegetables
Total vitamin A intake (calc.)[b]	18,677 ± 498	15,438 ± 421
Vitamin A activity in feces		
α-Carotene	623 ± 43	1,068 ± 49
β-Carotene	4,770 ± 342	11,213 ± 567
Total carotene	5,392 ± 382	12,283 ± 604

[a] Intakes and excretions of 12 men consuming a low-fiber diet and a diet containing fruits and vegetables. Values given are mean ± SEM.
[b] Adams (1975).

differences in excretion of carotenes may be related to fiber content of foods. Results of other human studies also show that more carotene from vegetable sources than from pure carotene in oil was excreted in feces (Van Eekelen and Pannevis, 1938; Hume and Krebs, 1949; Roels *et al.*, 1958).

We determined carotene excretion of 12 subjects who had consumed a low-fiber diet and a diet containing fruits and vegetables. Since the diet containing fruits and vegetables was high in vitamin A activity, we added a mixture of 17% α-carotene and 83% β-carotene to the low-fiber diet in an attempt to make the two diets more comparable in this respect. The calculated vitamin A activities (Adams, 1975) of the low-fiber diet plus added carotene and the diet containing fruits and vegetables were approximately 18,000 and 15,000 IU per day, respectively. A greater amount of vitamin A activity was excreted on the diet containing fruits and vegetables than on the low-fiber diet with crystalline carotene added (Table II).

3.2. Thiamine, Riboflavin, and Nicotinic Acid

Methylcellulose given daily by stomach tube for 28 days along with thiamine hydrochloride to thiamine-depleted or normal rats did not affect growth response (Ellingson and Massengale, 1952).

Holman (1954) carried out studies on the nutritive value of bread and on the effect of variations in the extraction rate of flour on undernourished children. The results provided some indication that children eating wholemeal bread excreted less riboflavin and metabolites of nicotinic acid in their urine than did children on enriched white bread. This observation led to the suggestion that these two vitamins in whole-wheat flour might not be freely absorbed by children.

On the other hand, Roe *et al.* (1978) reported that dietary fiber promoted riboflavin absorption. Twelve men were fed diets containing either coarse bran, fine bran, cellulose, or cabbage, followed by a low-fiber diet. A load dose of riboflavin was given with breakfast and riboflavin content of urine samples was measured at intervals for 8 hr. Each of the dietary fiber sources increased urinary riboflavin excretion.

When nicotinic acid was identified as a factor in pellagra prevention and treatment, there were a number of reports which indicated that nicotinic acid was present in maize, but that it was in a "bound" form. Therefore, the nicotinic acid was not available to microorganisms, rats, and pigs (Carpenter *et al.*, 1960; Chaudhuri and Kodicek, 1960; Harper *et al.*, 1958; Kodicek, 1960; Kodicek and Wilson, 1959; Kodicek *et al.*, 1956, 1959; Pearson *et al.*, 1957). Nicotinic acid could be liberated by alkaline hydrolysis or treatment with lime water and was effective in promoting weight gain of animals that had developed a deficiency due to consumption of maize containing the "bound" form. Das and Guha (1960) and Kodicek and Wilson (1960) isolated bound nicotinic acid from cereal brans, and later Mason and Kodicek (1973) reported that the binding agent was probably cellulose or hemicellulose.

Goldsmith *et al.* (1956) found that lime treatment of corn, to release bound nicotinic acid, failed to influence the production of pellagra symptoms in human subjects as had been shown in studies on animals. Pellagra was induced more rapidly with diets containing Guatemalan whole corn than with diets containing unenriched milled corn obtained in the U.S.

3.3. Vitamin B6 and Vitamin B12

Gregory (1980) determined biologically available vitamin B6 in nonfat dry milk and a fortified rice base cereal product fed to rats at a 5% level. Results of microbiological and high-performance liquid chromatographic methods for total B6 were compared with rat bioassay results. Total bioavailability was found for vitamin B6 in nonfat dry milk, but bioavailability of the vitamin B6 in the cereal product was low.

Miller *et al.* (1979) studied the effect of wheat bran on bioavailability of vitamin B6 in ten men. A basal diet with 15 g of wheat bran was fed for 18 days. Bran increased fecal B6 and decreased 4-pyridoxic acid excretion in urine in eight men. There was no effect on urinary B6. Bran also lowered plasma B6 in eight men and pyridoxal phosphate in four men.

Leklem *et al.* (1980) compared the bioavailability of vitamin B6 from soybeans and from beef in eight men studied for 15 days. Soybeans and beef each provided 60% of vitamin B6 in the diet. In feces, 7% more B6 was excreted when soy was fed than when beef was fed. Six percent less B6 was excreted in urine as 4-pyridoxic acid when soy was fed. There was no effect on urinary or blood B6.

Cullen and Oace (1978) investigated the rate of B12 depletion in rats on a fiber-free diet and diets containing 10–50% cellulose or 5–30% pectin. After

10 weeks rats were tested for vitamin B12 depletion by measurement of methylmalonic acid (MMA) excretion. This metabolite is increased in urine when there is a block in the propionate metabolic pathway due to lack of vitamin B12. Rats fed diets of 40–50% cellulose excreted about twice as much MMA as did rats in the fiber-free group. Rats consuming 5–15% pectin had a 9- to 21-fold increase in MMA excretion as compared with the fiber-free group. Rats fed cellulose or pectin also had significantly greater fecal excretions of a radioactive B12 injection given after several weeks of B12 depletion than did rats in the fiber-free group.

Cullen and Oace (1980) further investigated the impact of pectin and six other dietary fibers on B12-deprived rats. Cellulose, lignin, alginate, and wheat bran fed at the 5% level had only a slight negative effect on B12 status. These fibers are poorly digested by intestinal bacteria. However, xylan, guar gum, and pectin, all digestible by intestinal bacteria, increased MMA excretion and decreased propionate oxidation rate and hemoglobin to a significantly greater extent than did the other fibers.

3.4. Folic Acid and Pantothenic Acid

Luther *et al.* (1965) found that nonfilterable dietary residues were capable of binding significant amounts of synthetic folic acid and two of its formyl derivatives. These investigators postulated that adsorption of folate compounds to insoluble dietary residues could be a factor affecting human folate nutrition.

Russell *et al.* (1976) fed four subjects breakfasts containing white bread, wholemeal bread, or cellulose, plus labeled folic acid. Sequential folate adsorption tests, based on percentage of radiolabel recovered in urine after 24 hr, were carried out. They reported no interference of folate absorption either by bread or cellulose. *In vitro* studies also showed no formation of insoluble complexes between bread fiber and folic acid.

Babu and Srikantia (1976) studied the availability of folic acid from different foods in ten subjects. All foods were consumed as a single meal and 24-hr urine collections were made. Availabilities of folate from egg and liver were 72 and 70%, respectively, from bengal gram and green gram 70%, and from spinach 63%. Thus fiber content of the food appeared not to affect folate availability.

Colman *et al.* (1975) investigated the absorption of folic acid from fortified maize meal, polished rice, and whole wheat bread as compared with an aqueous solution of the vitamin. Seven subjects were given each test material and blood was taken before and at 1 and 2 hr after feeding. The increase in serum folate 1 and 2 hr after feeding was 56%, 54%, and 29% for maize, rice, and bread, as compared with the increase after the vitamin alone.

Trimbo *et al.* (1979) supplemented diets of 16 adolescents with 20 g of wheat bran, 20 g of corn bran, 10 or 20 g cellulose, or 10 or 20 g hemicellulose for 6 days. There were no definite trends in blood serum pantothenic acid levels due to these fiber sources.

3.5. Summary

Methylcellulose or pectin did not affect utilization of vitamin A in rats, but a mixture of fibers did appear to decrease liver stores of vitamin A. Lignin given with a test meal did not affect serum vitamin A of human subjects. Carotene appears to be bound to plant cell walls, rendering it less available than pure carotene.

Methylcellulose had no effect on growth of thiamine-depleted or normal rats. Riboflavin and nicotinic acid may be less available from wholemeal bread than from enriched white bread. Several fiber sources improved absorption of a load dose of riboflavin in human subjects. Nicotinic acid is bound to fiber in cereal brans and is not available unless released.

Lower bioavailability of vitamin B6 from cereal than from nonfat dry milk was reported for rats. Vitamin B6 was less available from soybeans than from beef and the addition of wheat bran to a diet decreased B6 availability in human subjects. Fibers digestible by intestinal bacteria intensified symptoms of B12 depletion in rats.

In two studies fiber did not affect urinary folate excretion, but in another study fiber decreased the rise in serum folate as compared to that of the vitamin alone.

4. CONCLUSIONS

Calcium and magnesium balances were reported to be decreased by increasing fiber in the diet in most studies. Phosphorus was not generally affected. Results of studies on effect of fiber on mineral availability may be affected by kind and amount of fiber in the diet, level of mineral intake, length of study period, the présence of phytic acid or oxalic acid, and possibly the level of protein intake.

It is difficult to make a general statement about the effects of fiber on availability of vitamins, due to the paucity of studies on human subjects and the variety of methods employed in evaluating vitamin response. Some vitamins, such as carotene and nicotinic acid, and possibly others, may not be readily available because they occur bound to plant cell walls. It is not clear whether fiber can alter availability of vitamins that do not occur bound by fiber in the plant cell walls.

REFERENCES

Adams, C. F., 1975, Nutritive Value of American Foods in Common Units, Agriculture Handbook No. 456, Agriculture Research Service, U.S. Department of Agriculture, U.S. Government Printing Office, Washington, D.C.

American Association of Cereal Chemists, 1976, Approved Methods of the AACC, Method 32-20, approved October 1977, American Association of Cereal Chemists, St. Paul, Minnesota.

Babu, S., and Srikantia, S. G., 1976, Availability of folates from some foods, *Am. J. Clin. Nutr.* **29**:376–379.

Barnard, D. L., and Heaton, D.W., 1973, Bile acids and vitamin A absorption in man: The effects of two bile acid-binding agents, cholestyramine and lignin, *Gut* **14**:316–318.

Carpenter, K. J., Kodicek, E., and Wilson, P. W., 1960, The availability of bound nicotinic acid to the rat. 3. The effect of boiling maize in water, *Br. J. Nutr.* **14**:25–34.

Chaudhuri, D. K., and Kodicek, E., 1960, The availability of bound nicotinic acid to the rat. 4. The effect of treating wheat, rice and barley brans and a purified preparation of bound nicotinic acid with sodium hydroxide. *Br. J. Nutr.* **14**:35–42.

Colman, N., Green, R., and Metz, J., 1975, Prevention of folate deficiency by food fortification. II. Absorption of folic acid from fortified staple foods, *Am. J. Clin. Nutr.* **28**:459–464.

Cullen, R. W., and Oace, S. M., 1978, Methylmalonic acid and vitamin B12 excretion of rats consuming diets varying in cellulose and pectin, *J. Nutr.* **108**:640–647.

Cullen, R. W., and Oace, S. M., 1980, Impact on B12 status of pectin and six dietary fibers in rats, *Fed. Proc.* **39**:785.

Cullumbine, H., Basnayake, V., Lemottee, J., and Wickramanayake, T. W., 1950, Mineral metabolism on rice diets, *Br. J. Nutr.* **4**:101–111.

Cummings, J. H., Hill, M. J., Jivraj, T., Houston, H., Branch, W. J., and Jenkins, D. J. A., 1979a, The effect of meat protein and dietary fiber on colonic function and metabolism. 1. Changes in bowel habit, bile acid excretion, and calcium absorption, *Am. J. Clin. Nutr.* **32**:2086–2093.

Cummings, J. H., Southgate, D. A. T., Branch, W. J., Wiggins, H. S., Houston, H., Jenkins, D. J. A., Jivraj, T., and Hill, M. J., 1979b, The digestion of pectin in the human gut and its effect on calcium absorption and large bowel function, *Br. J. Nutr.* **41**:477–485.

Das, M. L., and Guha, B. C., 1960, Isolation and chemical characteristics of bound niacin (niacinogen) in cereal grains, *J. Biol. Chem.* **235**:2971–2976.

Drews, L. M., Kies, C., and Fox, H. M., 1979, Effect of dietary fiber on copper, zinc, and magnesium utilization by adolescent boys, *Am. J. Clin. Nutr.* **32**:1893–1897.

Ellingson, R. C., and Massengale, O. N., 1952, Effect of methylcellulose on growth response of rats to low vitamin intakes, *Proc. Soc. Exp. Biol. Med.* **79**:92–94.

Goldsmith, G. A., Gibbens, J., Unglaub, W. G., and Miller, O. N., 1956, Studies of niacin requirement in man. III. Comparative effects of diets containing lime-treated and untreated corn in the production of experimental pellagra, *Am. J. Clin. Nutr.* **4**:151–160.

Gregory, J. F., III, 1980, Bioavailability of vitamin B6 in nonfat dry milk and a fortified rice breakfast cereal product, *J. Food Sci.* **45**:84–86.

Gronowska-Senger, A., Chudy, D., and Smaczny, E., 1979, The content of fibre in the diet and the utilization of vitamin A by the organism, *Roczniki Panstwowego Zakladu Higieny* **30**:553–558 (Poland); cited in *Nutr. Abs. Rev.* **50**:805, Abs. No. 7418 (1980).

Harper, A. E., Punekar, B. D., and Elvehjem, C. A., 1958, Effect of alkali treatment on the availability of niacin and amino acids in maize, *J. Nutr.* **66**:163–172.

Holman, W. I. M., 1954, Studies on the Nutritive Value of Bread and on the Effect of Variations in the Extraction Rate of Flour on the Growth of Undernourished Children, Medical Research Council Special Report Series No. 287, HMSO, London, pp. 92–118.

Hume, E. M., and Krebs, H. A., 1949, Vitamin A Requirements of Human Adults. An Experimental Study of Vitamin A Deprivation in Man, Medical Research Council Special Report Series No. 264, HMSO, London, 145 pp.

Ismail-Beigei, F., Reinhold, J. G., Faradji, B., and Abadi, P., 1977, Effects of cellulose added to diets of low and high fiber content upon the metabolism of calcium, magnesium, zinc and phosphorus by man, *J. Nutr.* **107**:510–518.

Kelsay, J. L., 1981, Effect of diet fiber level on bowel function and trace mineral balances of human subjects, *Cereal Chem.* **58**:2–5.

Kelsay, J. L., and Prather, E. S., 1981, Effect of fiber and oxalic acid on mineral balances of adult human subjects, *Fed. Proc.* **40**:854.

Kelsay, J. L., Behall, K. M., and Prather, E. S., 1979a, Effect of fiber from fruits and vegetables on metabolic responses of human subjects. II. Calcium, magnesium, iron, and silicon balances, *Am. J. Clin. Nutr.* **32**:1876–1880.

Kelsay, J. L., Jacob, R. A., and Prather, E. S., 1979b, Effect of fiber from fruits and vegetables on metabolic responses of human subjects. III. Zinc, copper, and phosphorus balances, *Am. J. Clin. Nutr.* **32**:2307–2311.

Kelsay, J. L., Clark, W. M., Herbst, B. J., and Prather, E. S., 1981, Nutrient utilization by human subjects consuming fruits and vegetables as sources of fiber, *J. Agr. Food Chem.* **29**:461–465.

Kodicek, E., 1960, The availability of bound nicotinic acid to the rat. 2. The effect of treating maize and other materials with sodium hydroxide, *Br. J. Nutr.* **14**:13–24.

Kodicek, E., and Wilson, P. W., 1959, The availability of bound nicotinic acid to the rat. 1. The effect of lime-water treatment of maize and subsequent baking into tortilla, *Br. J. Nutr.* **13**:418–430.

Kodicek, E., and Wilson, P. W., 1960, The isolation of niacytin, the bound form of nicotinic acid, *Biochem. J.* **76**:27p.

Kodicek, E., Braude, R., Kon, S. K., and Mitchell, K. G., 1956, The effect of alkaline hydrolysis of maize on the availability of its nicotinic acid to the pig, *Br. J. Nutr.* **10**:51–67.

Kodicek, E., Braude, R., Kon, S. K., and Mitchell, K. G., 1959, The availability to pigs of nicotinic acid in tortilla baked with maize treated with lime-water, *Br. J. Nutr.* **13**:363–384.

Kreula, M., and Virtanen, A. I., 1939, Absorption of carotene from carrots in humans, *Upsala Lakareforen. Forh.* **45**:357–362.

Leklem, J. E., Schultz, T. D., and Miller, L. T., 1980, Comparative bioavailability of vitamin B6 from soybeans and beef, *Fed. Proc.* **39**:558.

Leonhardi, G., 1947, Uber die Resorption von Karotin aus Gemüsen beim Menschen. 1. Mitteilung, *Z. Gesamte Inn. Med.* **2**:376–384; cited in *Chem. Abstr.* **42**:Abs. No. 3473e (1948).

Luther, L., Santini, R., Brewster, C., Perez-Santiago, E., and Butterworth, C. E., Jr., 1965, Folate binding by insoluble components of American and Puerto Rican diets, *Ala. J. Med. Sci.* **2**:389–393.

Mason, J. B., and Kodicek, E., 1973, The chemical nature of the bound nicotinic acid of wheat bran: Studies of partial hydrolysis products, *Cereal Chem.* **50**:637–646.

McCance, R. A., and Widdowson, E. M., 1942a, Mineral metabolism of healthy adults on white and brown bread dietaries, *J. Physiol.* **101**:44–85.

McCance, R. A., and Widdowson, E. M., 1942b, Mineral metabolism on dephytinized bread, *J. Physiol.* **101**:304–313.

Miller, L. T., Lindberg, A. S., Whanger, P., and Leklem, J. E., 1979, Effect of wheat bran on the bioavailability of vitamin B6 and the excretion of selenium in men, *Fed. Proc.* **38**:767.

Pearson, W. N., Stempfel, S. J., Valenzuela, J. S., Utley, M. H., and Darby, W. J., 1957, The influence of cooked vs. raw maize on the growth of rats receiving a 9% casein ration, *J. Nutr.* **62**:445–463.

Phillips, W. E. J., and Brien, R. L., 1970, Effect of pectin, a hypocholesterolemic polysaccharide, on vitamin A utilization in the rat, *J. Nutr.* **100**:289–292.

Rao, C. N., and Rao, B. S. N., 1970, Absorption of dietary carotenes in human subjects, *Am. J. Clin. Nutr.* **23:**105–109.

Reinhold, J. G., Nasr, K., Lahimgarzadeh, A., and Hedayati, H., 1973, Effects of purified phytate and phytate-rich bread upon metabolism of zinc, calcium, phosphorus and nitrogen in man, *Lancet* **1:**283–288.

Reinhold, J. G., Faradji, B., Abadi, P., and Ismail-Beigi, F., 1976, Decreased absorption of calcium, magnesium, zinc and phosphorus by humans due to increased fiber and phosphorus consumption as wheat bread, *J. Nutr.* **106:**493–503.

Roe, D. A., Wrick, K., McLain, D., and Van Soest, P., 1978, Effects of dietary fiber sources on riboflavin absorption, *Fed. Proc.* **37:**756.

Roels, O. A., Trout, M., and Dujacquier, 1958, Carotene balances on boys in Ruanda where vitamin A deficiency is prevalent, *J. Nutr.* **65:**115–127.

Russell, R. M., Ismail-Beigei, F., and Reinhold, J. G., 1976, Folate content of Iranian breads and the effect of their fiber content on the intestinal absorption of folic acid, *Am. J. Clin. Nutr.* **29:**799–802.

Slavin, J. L., and Marlett, J. A., 1980, Influence of refined cellulose on human bowel function and calcium and magnesium balance, *Am. J. Clin. Nutr.* **33:**1932–1939.

Trimbo, S., Kathman, J., Kies, C., and Fox, H. M., 1979, Pantothenic acid nutritional status of adolescent humans as affected by dietary fiber and bran, *Fed. Proc.* **38:**556.

Van Eekelen, M., and Pannevis, W., 1938, Absorption of carotenoids from the human intestine, *Nature* **141:**203–204.

Van Zeben, W., and Hendricks, T. F., 1948, The absorption of carotene from cooked carrots, *Int. Z. Vitaminforsch.* **19:**265–266.

Walker, A. R. P., Fox, F. W., and Irving, J. T., 1948, Studies in human mineral metabolism. 1. The effect of bread rich in phytate phosphorus on the metabolism of certain mineral salts with special reference to calcium, *Biochem. J.* **42:**452–462.

10

Effects of Phytic Acid on Mineral Availability

N. T. DAVIES

1. HISTORICAL PERSPECTIVE

During research on experimental rickets in young dogs in the early 1920s Mellanby observed that diets poor in vitamin D and rich in cereals were highly effective in affecting mineralization of bones and teeth. In a series of classical studies it was shown that the rachitogenic effects of cereals depended on cereal type, degree of extraction, ripeness, and extent of germination (reviewed by Gontzea and Sutzescu, 1968). Bruce and Callow (1934) finally demonstrated that the agent responsible was phytic acid (myoinositol 1,2,3,4,5,6-hexakis dihydrogen phosphate), a phosphorus storage compound ubiquitously distributed in the plant kingdom and found in all cereals, many legumes and nuts, and a few fruits, tubers, and roots. The rickets-producing effects of phytate are to this day not fully understood but appear to be due in part to the poor availability of phytate phosphorus and partly to the ability of phytate to bind Ca to form an insoluble complex from which Ca is unavailable for absorption in subjects whose vitamin D status is marginal.

Numerous balance studies have demonstrated that consumption of phytate-rich foods similarly impair Ca absorption in man and these observations led directly to the fortification of UK flours with Ca in 1943.

Recently, the possible rachitogenic properties of dietary phytate has again been of interest to human nutritionists following the recent finding of a high incidence of rickets in some Asian immigrant communities in the UK. In a series of elegant studies, Dunnigan and his associates in Glasgow have demonstrated

N. T. DAVIES • Rowett Research Institute, Bucksburn, Aberdeen AB2 9SB, Scotland.

a clear association between the consumption of phytate-rich, high-extraction-rate wheat flour in the form of chapattis and the incidence of rickets (Dunnigan, 1977). However, it must be pointed out that the precise and direct role, if any, of phytic acid in the etiology of "Asian rickets" has yet to be established.

Following demonstrations of its effects on Ca metabolism, dietary phytate was shown to adversely affect Fe utilization in both experimental animals and man (Fuhr and Steenback, 1943; Widdowson and McCance, 1942). These latter authors showed in balance trials in man that subjects consuming phytate-rich brown bread made from 92% extracted flour absorbed only 1% of dietary Fe compared with 7.3–12.5% when low-phytate white bread was consumed. Since these initial reports numerous studies, including some using isotopic techniques, have shown that the addition of sodium phytate to white bread markedly impaired Fe availability, indicating that poor utilization of this element from high-extraction-rate cereal products is due at least in part to their high contents of phytic acid (reviewed by Gontzea and Sutzescu, 1968).

2. THE EFFECT OF DIETARY PHYTATE ON TRACE ELEMENT AVAILABILITY

The demonstration that dietary phytate may adversely affect the utilization of Zn arose from practical problems encountered in pig and chick production. During the late 1950s the increased incidence of a skin disease in pigs (parakeratosis, or dry skin dermatitis) seemed to be associated with increased usage of corn and soya products as sources of dietary protein. A similar association of diet with productive losses was also demonstrated in the poultry industry. At first sight it seemed unlikely that these effects resulted from nutritional Zn deficiency, since the Zn content of these diets was as high as or higher than those containing animal products as the major source of dietary protein. However, O'Dell and Savage (1957) showed that the Zn requirement for poultry was higher when cereal–soya diets were offered than when casein was used as the protein source. During the 1960s these observations were confirmed and extended by many workers who showed that phytic was the agent responsible (O'Dell and Savage, 1960; Oberleas *et al.*, 1961; Likuski and Forbes, 1965).

Early studies on the effects of dietary phytate on Zn availability showed that the severity of the induced Zn deficiency was dependent upon the dietary Ca content. Thus if dietary Ca contents were low, much higher concentrations of phytate were necessary before Zn availability was affected (Likuski and Forbes, 1965). The significance of this observation was not clarified until Oberleas *et al.* (1966) demonstrated *in vitro* that in the pH range found within the intestinal tract the amount of Zn precipitated by phytate was considerably greater if Ca were present. It has been suggested therefore that the mechanism by which phytate impairs Zn utilization is due to its ability to complex Zn and Ca within the intestinal lumen to form an insoluble Zn–Ca–phytate complex from which Zn is unavailable for absorption.

The exact quantitative relationships between dietary concentrations of Zn, Ca, and phytate which give rise to impaired Zn absorption have yet to be fully investigated. Recently we investigated the proposal that a reliable indication of Zn availability from phytate-rich foods might be obtained from the phytate and zinc contents expressed as the molar ratio phytate : Zn (Davies and Olpin, 1979). In this study diets were formulated to contain phytate : Zn values of 0 : 1 to 40 : 1. Two series of diets were prepared. The first series had the same Zn contents of 18 mg Zn/kg and the amount of phytate varied. This level of Zn is 50% above the requirement of rats fed the basal diet. In the second series the phytate contents were the same (0.74 g/100 g) and identical phytate : Zn values to the first series were achieved by varying the amount of added $ZnSO_4 \cdot 7H_2O$. Young rats were offered their respective diets for 42 days.

In the first 14-day period of the experiment the Ca content was 0.6 g Ca/100 g, after which it was increased to 1.2 g Ca/100 g for the remaining 28 days of the experiment. During the first period when the dietary Ca contents were lower, average daily weight gains were significantly impaired in rats consuming diets containing phytate : Zn ratios of 25 : 1 and above. However, when the higher level of Ca was fed, significantly impaired growth and reductions in plasma Zn concentrations were observed at ratios of 15 : 1. In the first series of diets when the Zn content was constant, plasma Zn was significantly reduced at a ratio as low as 10 : 1. In general there was no significant difference in the responses of rats receiving the same phytate : zinc ratios in the two series of diets. Furthermore, when rats were offered diets in which soya-based textured vegetable protein was the protein source their growth rates could be accurately predicted from the phytate : Zn ratios of their diets. These findings clearly indicated that for soya-based proteins phytate : Zn ratios of contents are an accurate means of predicting Zn availability and suggest that in these products phytate is the sole factor responsible for impaired Zn absorption.

In a somewhat similar yet complementary study Morris and Ellis (1980) have shown that if rats consume diets containing only marginally adequate Zn contents (10 mg Zn/kg) and high Ca contents (1.75 g/100 g) ratios of 9 : 1 were effective in reducing growth rates.

The possible effects of dietary phytate on the availability of trace elements and minerals other than Zn, Fe, and Ca have yet to be systematically investigated. However in one report (Davies and Nightingale, 1975) the inclusion of phytate (1 g/100 g) in Ca-rich diets consumed by young rats resulted in significantly reduced whole-body retention of Mn and Cu as well as the expected effects on Zn and Fe. High dietary contents of phytic acid have also been shown to affect Mg availability in rats (Sathe and Krishnamurthy, 1953), indicating perhaps that under the appropriate circumstances this agent may promote a nutritional deficiency of any of these essential elements, depending upon which first becomes limiting in the diet.

3. DIETARY PHYTATE AND TRACE ELEMENT NUTRITION IN MAN

The significance of dietary phytic acid in the nutrition of man has yet to be adequately investigated. The observation that in some subjects impaired Ca balances due to dietary phytate were only transient and that adaptation occurred (Walker, 1951) has led to the tendency to dismiss phytate in food as unimportant. However, there has been renewed interest in the nutritional importance of phytate since the finding that the high incidence of clinically overt Zn deficiency in some sections of the populations of Iran and Egypt may be associated with the consumption of large amounts of phytate-rich cereal staples (reviewed by Reinhold *et al.*, 1976, and Davies, 1978).

Despite this increasing interest in phytic acid as a nutritionally deleterious agent the amount of information concerning its concentration in foodstuffs is sadly lacking. In the most recently compiled Food Composition Tables published in the UK (Paul and Southgate, 1978) the phytate content of foodstuffs is contained in a single table containing data for less than 60 food items. Furthermore, the table is identical to those in the previous two editions of similar publications! Even less information has been published of the phytic acid contents of "total diets" and thus it is difficult to assess the nutritional significance of dietary phytate on the trace element status of the populations of either the developed or developing world.

My own assessment of the phytate intakes by the population of the UK based on scant published analytical data and my own unpublished findings suggests it is in the range 600–800 mg daily. This estimate is lent support by a recent study of the phytate content of some representative UK total diets which indicated an average daily consumption of 806 mg/day (D. H. Buss, personal communication). In this latter study, 70% of the phytate intake was derived from cereal products, 20% from fruit, and the remainder from vegetables and nuts. Using these estimates and assuming a mean daily zinc intake of 11 mg/day (Davies, 1977), one finds that the phytate : Zn ratio of a "typical" UK diet is in the range of 6–8. Recently, Harland and Peterson (1978) reported that the phytate : Zn ratio of a typical US hospital diet was 6.

If the relationships between Ca and phytate : Zn ratios and Zn availability demonstrated in rats are applicable to man, then clearly dietary phytate is probably of minor significance in affecting the Zn status of most of the populations of the Western world. Furthermore, it is unlikely, except in a few exceptional cases, that the dietary intakes of Ca are sufficiently high as to promote a significant phytate-induced impairment of Zn availability. When Ca intakes are expressed in terms of energy intakes, a reasonable value for a Western type diet would be 1 g Ca/2300 kcal (US recommended dietary allowances, RDA). When the Ca contents of the diets used in my own rat studies (0.6 and 1.2 g Ca/100 g) are expressed in similar units this corresponds to Ca to energy ratios of 3.45 and 6.9 g Ca/2300 kcal. It would be surprising if there are many individuals who consume between 3 and 7 g Ca/day.

Notwithstanding these arguments, there may still be some sections of the population for whom high intakes of dietary phytate and low Zn intakes may be of significance. In a recent study of the nutritional habits of a group of lacto-ovo-vegetarians the phytate : Zn ratios of their self-selected diets ranged from 9.7 to 244 and some Ca intakes were up to 3 times the US RDA (Harland and Peterson, 1978). Furthermore, the finding by Reinhold *et al.* (1973) that the addition of purified sodium phytate to a diet which raised the phytate : zinc ratio from 2–3 to just 10–13 resulted in decreased Zn balances in three human subjects may indicate that man may be more susceptible to phytate-induced disturbances in Zn utilization than other species.

In the developing world, where many populations consume little or no protein from animal sources and exist on mixtures of various cereals and legumes, high intakes of phytate may similarly be of some significance. In a recently completed study on the trace element and phytate content of various diets of rural communities in the Gambia, both Zn and Cu contents were low and phytate levels high, with phytate : zinc ratios ranging from 25 to 40 (N. T. Davies, unpublished).

4. THE RELATIVE IMPORTANCE OF PHYTATE AND FIBER IN LIMITING ZINC AVAILABILITY

The possibility that consumption of a fiber-rich diet could lead to Zn and other mineral deficiency states in human populations was first recognized and studied in detail by Reinhold and his co-workers in Iran (reviewed by Reinhold *et al.*, 1976).

Initially, Reinhold considered that the high incidence of Zn deficiency in rural communities in Iran was associated with the high phytate content of tanok, an unleavened bread made from high-extraction-rate wheat flour, which was their diet staple. However the results of extensive studies, both *in vitro* and *in vivo,* have led him to suggest that the fiber component of village wholemeal bread is of greater importance in limiting Zn (and also Ca, Fe, and P) availability. When ^{65}Zn-labeled flour was used to prepare leavened and unleavened bread, the leavening process resulted in a 3.7-fold increase in Zn solubility, whereas the phytate content was reduced by a mere 20% (reviewed by Davies, 1978). In studies of ^{65}Zn binding by the various fractions of wheat, all components, including the indigestible fiber fraction, bound considerable amounts of zinc. A comparison of Zn binding by tanok and phytate-free tanok (prepared either by extraction with hydrochloric acid or by treatment with phytase) showed no difference in the quantity of Zn bound, suggesting that, in tanok, fiber rather than phytate is the major determinant of Zn availability.

Balance studies with human subjects support this proposal. Thus, Reinhold and co-workers (1973) measured zinc balance and plasma zinc concentration in three subjects maintained on a diet rich in a low-fiber, low-phytate bread (bazari), or a similar diet supplemented with sodium phytate, and on a diet rich in tanok. The addition of phytate to bazari reduced Zn balance and caused a lowering of plasma Zn concentration. However, considerably greater reductions in Zn balance were seen when the naturally phytate-rich tanok was eaten, and two subjects were in negative Zn balance. Similar effects were observed on Ca and P balance. Since the phytate content of the tanok and phytate-supplemented bazari was the same, it seems probably that the greater effect of tanok on Zn availability was due to its higher fiber content.

In a more recent study, Reinhold has demonstrated directly the effects of fiber on Zn, P, and Ca balances (Reinhold *et al.*, 1976). Subjects were maintained on a low-fiber diet in which 50% of the energy intake was provided by low-fiber, low-phytate bazari. After 14 days they consumed the same diet but in addition they received 10 g of cellulose as filter paper dispersed in 150 g of apple compote. Fecal excretion and balance of Zn, Ca, and P were monitored throughout the experiment. On the "high-fiber" diet there was increased fecal excretion of Zn, Ca, and P, and both subjects exhibited negative zinc and phosphorus balances.

These studies strongly support the proposal that it is the complexing action of fiber rather than that of phytate which accounts for the high incidence of mineral-deficiency diseases in the villages of Iran.

In contrast to these findings, studies in experimental animals have failed to demonstrate impaired Zn utilization due to dietary fiber. In one study (Davies *et al.*, 1977) rats were fed diets containing either wheat bran (containing both phytate and fiber), extracted bran fiber, or phytate. Those that received bran or phytate showed impaired Zn utilization as assessed from decreased growth rates, whereas those receiving fiber grew at the same rate as control rats receiving no supplements. In subsequent studies (Davies, 1978) dietary supplements of individual components of dietary fiber to rat diets, namely cellulose, hemicellulose, and pectin, similarly failed to influence zinc availability.

In the last decade there have been a number of studies of the effects of fiber on Zn balances in human subjects, some of which are summarized in Table I. The fiber intakes are taken directly from the cited papers or estimated from the

TABLE I. The Effects of Fiber Phytate and Dietary Protein on Zinc Balances

Study	Fiber	Fiber type	g/day	Phytate,[a] g/day	Zinc, mg/day	Phytate/ zinc	Protein, g/day	Percent animal	Daily zinc balances[b]
1. Reinhold *et al.* (1973)	Low	Mixed	17	0.7	25–30	2–3	80–90	65–70	+ + + +
	Low + phytate	Mixed	17	2.5	25–30	8–10	80–90	65–70	+
	High	Mixed + wheat	40	2.5–3.4	25–30	8–14	80–90	25–30	− −
	Mixed	Mixed	20	0.5	25–30	2–3	80–90	25–30	+ + + +
2. Reinhold *et al.* (1976)	Low	Mixed	23	0.5	18–20	2–3	90–100	26–30	+
	High	Mixed + wheat	34	1.9	18–20	9–10	90–100	26–30	− −
3. Ismail-Beigi *et al.* (1977)	Low	Mixed	23	0.5	12–14	3–4	90–100	26–30	−
	High	Mixed + cellulose	35	0.5	12–14	3–4	90–100	26–30	− −
4. Sandstead *et al.* (1978)	Low	Mixed	12	0.8	10–15	5–8	120	70	+ + +
	High	Soft wheat bran	23	1.3	14–16	8–10	120	70	+ +
	High	Corn bran	28	0.8	12–16	5–8	120	70	+ + +
5. Drews *et al.* (1979)	Low	Mixed	7	0.7	11	6–8	<50	50	+
	High	Mixed + hemicellulose	21	0.7	11	6–8	<50	50	− − −
	High	Mixed + cellulose	21	0.7	11	6–8	<50	50	+
	High	Mixed + pectin	21	0.7	11	6–8	<50	50	+
6. Kies *et al.* (1979)	Low	Mixed	12	0.9–1.5	11	8–15	<50	0	+
	High	Mixed + hemicellulose	22	0.9–1.5	11	8–15	<50	0	+
	High	Mixed + hemicellulose	32	0.9–1.5	11	8–15	<50	0	− −
7. Kelsay *et al.* (1979)	Low	Mixed	5	0.8	13	6	95	70–80	+ + +
	High	Fruit and vegetables	24	1.2	13	10	95	70–80	−

[a] With the exception of studies 1–4, these have been estimated from phytate contents of foods from a number of published and unpublished sources.
[b] +, + +, + + +, + + + +, Positive balances 0–1, 1–2, 2–3, >5 mg. −, − −, − − −, Negative balances 0–1, 1–2, 2–3 mg.

dietary details presented and published values of food dietary fiber contents. Similarly, the estimates of daily phytate intakes in the studies of Reinhold *et al.* (1973, 1976) and Ismail-Beigi *et al.* (1977) are those given by the authors, whereas the remainder either have been assessed from the dietary details presented or a value of 0.8 g/day has been assumed (see above).

Space will not allow a detailed discussion of all of these studies, but no clear pattern can be seen. Fiber intakes as low as 20–25 g/day have resulted in negative Zn balances in some studies (studies 5–7), whereas not in others (studies 1, 3, 4, and 6). Similarly, in one investigation (study 3), the addition of 10 g of cellulose resulted in negative Zn balances, while it was without significant effect in another (study 5).

Even within individual investigations, apparently contradictory results have been obtained. In the study of Sandstead *et al.* (1978) (study 4) a dietary supplement of 26 g/day of soft white wheat apparently reduced Zn balances in some subjects, while the same quantity of corn bran was without effect even though it contained more fiber. It may perhaps be significant that the wheat bran contained phytic acid whereas the corn bran did not.

Various explanations may account for these conflicting findings, such as differences in sources of fiber and its composition, and/or differences in dietary Ca, Zn, and phytate contents. One other variable which might be involved is the amount or source of dietary protein. Thus, in the studies of Drews *et al.* (1979) and Kies *et al.* (1979) (studies 5 and 6) and the Iranian investigations (studies 1–3), in which increased fiber consumption was associated with markedly impaired Zn balances, most of the dietary protein was derived from plant sources. However, in the study of Sandstead *et al.* (1978) (study 4), in which the diets supplied very much greater quantities of protein (120 g/day), 70% of which was of animal origin, dietary supplementation with high levels of dietary fiber produced no significant effect on Zn balances.

Support for the proposal of a beneficial effect of increased intakes of animal protein on Zn availability from fiber and phytate-rich food is given by the report of Sandström *et al.* (1980). In this study, Zn absorption from a meal of wholemeal or white bread was measured directly in human subjects using an extrinsic ^{65}Zn label and whole-body counting. Zinc absorption from white bread was 13% compared with only 8% from wholemeal bread. However, protein in the forms of eggs, milk, cheese, beef, and casein added either singly or as mixtures significantly improved the absorption from wholemeal bread such that at the highest protein level tested (26 g/meal) it exceeded that from white bread.

In a recent study in rats we at the Rowett Institute have investigated a possible effect of increased levels of dietary protein on Zn availability from diets rich in fiber and phytate (Davies and McKenzie, unpublished).

Groups of rats were offered diets containing either 10%, 20%, or 30%

protein. At each protein level diets were offered which contained either egg albumen as the sole protein source, or egg albumen together with either wholemeal bread (40 g/100 g diet), dephytinized bread fiber (prepared by neutral detergent extraction), sodium phytate, or a mixture of phytate and fiber in similar proportions to those supplied by the bread diets. The phytate : Zn ratio of the bread, phytate, and phytate + fiber diets was 20 : 1 and the fiber content of diets containing bread and fiber was 3.6 g/100 g. After 28 days the rats were killed and whole-body retention of Zn determined by carcass analysis (Davies and Nightingale, 1975). The results are presented in Table II. At each level of

TABLE II. The Effect of Wholemeal Bread, Bread Fiber, and Phytate on Percent Retention of Dietary Zinc by Young Rats[a]

		Percent		
10%	Control	38.8 ± 2.2	B	**
	Phytate	26.1 ± 1.21	P	***
	Fiber	49.8 ± 2.5	F	ns
	Phytate + fiber	19.7 ± 2.68	P × F	***
	Bread	28.9 ± 1.76		
20%	Control	65.7 ± 2.02	B	***
	Phytate	32.1 ± 2.21	P	***
	Fiber	62.7 ± 1.04	F	ns
	Phytate + fiber	38.4 ± 1.27	P × F	*
	Bread	35.5 ± 2.57		
30%	Control	63.0 ± 1.56	B	***
	Phytate	38.6 ± 2.20	P	***
	Fiber	65.9 ± 1.77	F	ns+
	Phytate + fiber	46.3 ± 4.77	P × F	ns
	Bread	41.0 ± 3.19		

Joint analysis of variance of combined results

B	***	F	*
P	***	P × F	ns

Interaction with protein: Overall protein effect PL ***, PQ *

B × PL	*	F × PL	ns
B × PQ	***	F × PQ	ns
P × PL	ns+	F × P × PL	**
P × PQ	**	F × P × PQ	**

[a] *, **, ***, $p < 0.05$, $p < 0.01$, $p < 0.001$. ns, Not significant; ns+, $p < 0.1$. B, P, F, effects due to bread, phytate, and fiber, respectively. P × F, interaction between phytate and fiber effects. PL, PQ, linear and quadratic interactions with protein.

protein, phytate and bread significantly reduced Zn retention, whereas fiber alone was without effect. However, at the lowest level of dietary protein (10%) there was a significant interaction between fiber and phytate, which indicated that phytate was more effective in reducing Zn retention in the presence of fiber than when fiber was absent. However, at the higher levels of dietary protein the interaction was either not evident or reversed, i.e., the presence of fiber reduced the phytate-induced impairment of Zn retention. When these results were combined and subjected to an overall joint analysis of variance, the effects of bread and phytate were both highly significant. Surprisingly, the effect of fiber overall was to significantly improve Zn retention.

When the results were analyzed for interactions between protein level and effects of bread, phytate, and fiber it was evident that increasing the dietary protein level significantly improved Zn retention from the phytate and bread diets. Furthermore, the interaction between protein level and the phytate–fiber interaction was significant.

5. SUMMARY

It has been known for many years that dietary phytic acid can impair the absorption of iron and calcium in both man and experimental animals. It has also been shown in pigs, rats, and chicks that high dietary intakes of phytic acid can induce a nutritional deficiency of zinc, particularly if the diets also contain a high calcium content. The exact quantitative relationships between the dietary contents of zinc, calcium, and phytate that will promote decreased zinc availability are not known. However, recent studies with rats indicate that if the phytate and zinc contents of a diet expressed as a molar ratio exceed 15–25, the impairment of zinc availability is sufficiently severe as to cause growth failure. A survey of the phytate and zinc contents of "Western-type" food indicates a ratio of phytate : zinc contents in "typical diets" of 6–8. However, some sections of Western communities, such as vegetarians, and many populations in the developing world may be consuming diets with phytate : zinc ratios in excess of 30 : 1, the significance of which is not known.

Studies in man suggest that when phytate- and fiber-rich foods are consumed fiber may be of greater significance in reducing zinc availability. However, from a survey of recently published investigations in man on the effect of dietary fiber on zinc balances, it is clear that other dietary variables are involved. One of these, the amount of dietary protein of animal origin, has recently been studied in both man and rats. Both studies indicate that zinc availability from fiber- and phytate-rich wholemeal bread is improved by increased intakes of animal protein.

6. CONCLUSIONS

The results of this study clearly do not resolve whether in man dietary fiber or phytate is more effective in reducing Zn availability. However, when taken alongside the study of ^{65}Zn absorption from wholemeal bread (Sandström *et al.*, 1980), they suggest that people who habitually consume diets containing low levels of animal protein and high quantities of fiber and phytate may be the subjects at most risk of suffering impairment of Zn status.

Finally, the results of the many studies described here serve yet again to emphasize the importance of a *balanced* diet for optimal nutritional health.

REFERENCES

Bruce, H. M., and Callow, R. K., 1934, Cereals and rickets. The role of inositolhexaphosphoric acid, *Biochem. J.* **28:**517–528.

Davies, N. T., 1977, Trace elements: Factors affecting zinc availability, in: *Proc. Symp. Child Nutrition in Relation to Mental and Physical Development,* Kellogg, London, pp. 21–34.

Davies, N. T., 1978, The effects of dietary fibre on mineral availability, *J. Plant Foods* **3:**113–212.

Davies, N. T., and Nightingale, R., 1975, The effects of phytate on intestinal absorption and excretion of zinc, and whole body retention of Zn, copper, iron and manganese in rats, *Br. J. Nutr.* **34:**243–258.

Davies, N. T., and Olpin, S. E., 1979, Studies on the phytate : zinc molar contents in diets as a determinant of Zn availability to young rats, *Br. J. Nutr.* **41:**591–603.

Davies, N. T., Hristic, V., and Flett, A., 1977, Phytate rather than fibre in bran as the major determinant of zinc availability to young rats, *Nutr. Rep. Int.* **15:**207–214.

Drews, L. M., Kies, C., and Fox, H. M., 1979, Effect of dietary fiber on copper, zinc and magnesium utilization by adolescent boys, *Am. J. Clin. Nutr.* **32:**1893–1897.

Dunnigan, M. G., 1977, Asian rickets and osteomalacia in Britain, in: *Proc. Symp. Child Nutrition in Relation to Mental and Physical Development,* Kellogg, London, pp. 43–70.

Fuhr, I., and Steenback, H., 1943, The effect of dietary calcium, phosphorus and Vitamin D on the utilization of iron. I. The effect of phytate acid on the availability of iron. II. The effect of Vitamin D on the body iron and hemoglobin production. III. The relation of rickets to anemia, *J. Biol. Chem.* **147:**59–74.

Gontzea, I., and Sutzescu, P., 1968, *Natural Antinutritive Substances in Foodstuffs and Forages,* S. Karger, New York.

Harland, B. F., and Petersen, M., 1978, Nutritional status of lacto-ovo vegetarian Trappist Monks, *J. Am. Dietet. Assoc.* **72:**259–264.

Ismail-Beigi, F., Reinhold, J. G., Faraji, B., and Abadi, P., 1977, The effects of cellulose added to diets of low and high fibre content upon the metabolism of calcium, magnesium, zinc and phosphorus by man, *J. Nutr.* **107:**510–518.

Kelsay, J. L., Behall, K. M., and Prather, E. S., 1979, Effect of fiber from fruits and vegetables on metabolic responses of human subjects. III. Zinc, copper and phosphorus balances, *Am. J. Clin. Nutr.* **32:**2307–2311.

Kies, C., Fox, H. M., and Beshgetoor, D., 1979, Effect of various levels of dietary hemicellulose on zinc nutritional status of men, *Cereal Chem.* **56:**133–136.

Likuski, H. J. A., and Forbes, R. M., 1965, Mineral utilization in the rat. IV. Effects of calcium and phytic acid on the utilization of dietary zinc, *J. Nutr.* **85**:230–234.

Morris, E. R., and Ellis, R., 1980, Bioavailability to rats of iron and zinc in wheat bran: Response to low-phytate bran and effect of the phytate/zinc molar ratio, *J. Nutr.* **110**:2000–2011.

Oberleas, D., Muhrer, M. E., O'Dell, B. L., and Kuntner, L. D., 1961, Effects of phytic acid on zinc availability in rats and swine, *J. Anim. Sci.* **20**:945–949.

Oberleas, D., Muhrer, M. E., and O'Dell, B. L., 1966, Dietary metal-complexing agents and zinc availability in the rat, *J. Nutr.* **90**:56–62.

O'Dell, B. L., and Savage, J. E., 1957, *Fed. Proc.* **16**:394 (abstract).

O'Dell, B. L., and Savage, J. E., 1960, Effect of phytic acid on zinc availability, *Proc. Soc. Exp. Biol. Med.* **103**:304–306.

Paul, A., and Southgate, D. A. T., 1978, *Widdowson and McCance's Food Composition Tables*, HMSO, London.

Reinhold, J. G., Nasr, K., Lahimgarzadeh, A., and Hedayafi, H., 1973, Effects of purified phytate and phytate-rich bread upon metabolism of zinc, calcium, phosphorus, and nitrogen in man, *Lancet* **i**:238–288.

Reinhold, J. G., Faradji, B., Abadi, P., and Ismail-Beigi, F., 1976, Decreased absorption of calcium, magnesium, zinc and phosphorus by humans due to increased fiber and phosphorus consumption as wheat bread, *J. Nutr.* **106**:493–503.

Sandstead, H. H., Munoz, J. M., Jacob, R. A., Klevay, L. M., Reck, S. J., Logan, G. M., Dintzis, F. R., Inglett, G. E., and Shuey, W. C., 1978, Influence of dietary fiber on trace element balance, *Am. J. Clin. Nutr.* **31**:S180–S184.

Sandström, B., Arvidsson, B., Lederblad, A., and Björn-Rasmussen, E., 1980, Zinc absorption from composition meals. I. The significance of wheat extraction rate, zinc, calcium and protein content in meals based on bread, *Am. J. Clin. Nutr.* **33**:739–745.

Sathe, V., and Krishnamurthy, K., 1953, Phytic acid and absorption of iron, *Indian J. Med. Res.* **41**:453–461.

Walker, A. R. P., 1951, Cereals, phytic acid and calcification, *Lancet* **ii**:244–248.

Widdowson, E. M., and McCance, R. A., 1942, Iron exchanges of adults on white and brown bread diets, *Nature* **148**:219–220.

11

Dietary Fiber–Iron Interactions

Fiber-Modified Uptakes of Iron by Segments of Rat Intestine

JOHN G. REINHOLD, PEDRO M. GARCÍA L., LUIS ARIAS-AMADO, and PEDRO GARZÓN

1. INTRODUCTION

Nutritionally important bivalent metals combine with dietary fiber to form complexes that may decrease availability for absorption. Bremner (1970) found such complexes of manganese and zinc to be present in rye grass and digesta of sheep. Both metals were liberated by peptic digestion and treatment with fungal cellulase. The ability of fiber-rich Iranian wholemeal flat breads and of cellulose to bind firmly zinc, iron, and calcium *in vitro* was described by Reinhold and associates (Reinhold *et al.*, 1975; Ismail-Beigi *et al.*, 1977). Removal of phytate, previously thought to be the agent binding these metals, by acid extraction or action of phytase left the binding capability unimpaired. The binding of each of the metals was pH dependent. Thompson and Weber (1979) and Reilly (1979) confirmed the pH dependence of the binding of zinc and iron (and of copper) by dietary fibers from various sources and showed that the effect of pH was reversible. Eastwood and Kay (1979) listed cation exchange as an important property of dietary fiber that modifies excretion of minerals in feces.

JOHN G. REINHOLD, PEDRO M. GARCÍA L., and LUIS ARIAS-AMADO • Centro de Investigación en Ciencias de la Alimentación, Facultad de Medicina Veterinaria y Zootecnia, Universidad de Guadalajara, Guadalajara, Mexico. PEDRO GARZÓN • Laboratorio del Bioquímica, División de Biología del Desarrollo, Unidad de Investigaciones Biomédicas de Occidente, Instituto Mexicano del Seguro Social, Guadalajara, Mexico.

TABLE I. Measurement of Iron Binding by Fiber

Fiber	20 mg
Glucose–saline solution	10 ml
Equilibration on rotator	20 min
pH adjustment	
Equilibration on rotator (24–26°C)	30 min
Centrifugation	
Measurement of iron in solution[a]	
Bound iron = (iron in a control solution) − (iron in supernatant)	

[a] With addition of ascorbate to overcome interference by, e.g., citrate, phytate.

More information is needed regarding factors that determine metal–fiber interactions and their nutritional importance. Because iron deficiency, manifested by anemia, is prevalent in all regions of Mexico despite adequate intakes of iron (Perez Hidalgo *et al.*, 1971), and because fiber intakes of the affected population tend to be moderately high, there exists the possibility that binding of iron by fiber contributes to the iron deficiency. Consequently, we have investigated certain factors that affect binding of iron by fiber.

2. PREVIOUS STUDIES

A review of our previous studies will serve as an introduction to more recent work. The techniques used for the study of iron binding *in vitro* are described in Table I, and the iron-containing glucose–saline used in previous and present studies is described in Table II (Reinhold *et al.*, 1981).

TABLE II. Glucose–Saline–Iron Solution

	mmol/liter
Sodium chloride	128.0
Potassium chloride	4.0
D-Glucose	28.0
Sodium acetate	1.0
Imidazole	0.5
Hydrochloric acid	0.2
Ascorbate	0.028
Iron II (0.725 μg/ml)	0.0129
Iron-59 (tissue studies)	q.s.[a]
pH	6.45 ± 0.05

[a] Quantity sufficient.

2.1. Iron Concentration and Binding

The quantity of iron that combines with neutral detergent fiber (NDF), acid-detergent fiber (ADF), or cellulose differs but is directly proportional to iron concentration up to 1.0 μg iron/ml. At higher concentrations, instability becomes a problem, despite the presence of glucose, imidazole, and ascorbate as stabilizers. Iron concentrations were measured by means of bathophenanthroline sulfonate.

2.2. Estimation of Iron Bound by Fiber

A plot of bound iron/free iron against bound iron indicated that NDF of wheat bound approximately 0.38 μg iron/mg fiber. Maize NDF bound somewhat less, about 0.30 μg/mg. These values are equivalent to 0.38 and 0.30 mg/g, respectively. ADF of wheat and maize did not differ, each binding about 0.24 μg iron/mg. Binding by cellulose (finely divided filter paper or cotton) was nearly the same as that by ADF. These measurements were made at pH 6.45 ± 0.05 and about 25°C. The NDF preparations examined were made from the standard wheat bran distributed by the American Association of Cereal Chemists (St. Paul, Minnesota), or from maize bran obtained from Productos de Maiz, S.A. de C.V., Guadalajara, Mexico.

2.3. Effect of pH on Binding of Iron by Fiber

Binding of iron by NDF or by enzymatic preparations of cereal fiber was strongly dependent upon pH. Binding approached a maximum in the region of pH 7.0. It decreased as the pH declined, first gradually, then abruptly between pH 6.0 and 5.0. At the latter pH, the proportion of iron bound was one-third or less of that bound at pH 7.0. However, complete release of iron from fiber occurred only in the region of pH 1.0. The pH–iron binding relationship was reversible. Iron binding by ADF and cellulose was also pH dependent, but much less so than was NDF. Iron hydroxide formation was negligible.

2.4. Inhibition of Binding

Various substances interfered with binding of iron by fiber. Ascorbate, citrate, cysteine, EDTA, and phytate were among the more potent inhibitors. Active interference with binding occurred at inhibitor concentrations as low as 100 μM/liter. Inhibition by amino acids varied. Besides strongly active cysteine, the dicarboxylic acids and histidine were moderately inhibitory, but most amino acids were inactive. Some of the strong inhibitors of the binding of iron by dietary fiber will be recognized as substances that have been used as adjuvants to promote iron absorption, e.g., ascorbate and EDTA.

With this summary of earlier work, we turn to a description of recent studies dealing with the stability of fiber-bound iron in the presence of segments of the intestine. The objective was to discover whether the absorptive capabilities of the intestine enabled it to liberate and absorb iron combined with dietary fiber. We also wished to learn if a separation of iron from combination with fiber would be facilitated by inhibitors of iron binding *in vitro* with consequent improvement of iron absorption. Uptake by segments of rat intestine has been found to be a sensitive method for estimating the availability of zinc (Oberleas *et al.*, 1966; Sahagian *et al.*, 1966; Reinhold *et al.*, 1974). We have adapted the same technique to the measurement of the availability of iron.

3. METHODS

Segments of intestine were obtained from mature male rats weighing from 194 to 369 g. Light anesthesia was induced by ether or chloroform and the rat killed by dislocation of the cerebral vertebrae. The small intestine was removed, flushed with cold glucose–saline solution (5.0 g D-glucose, 7.5 g sodium chloride, and 0.3 g potassium chloride per liter of water), opened by slitting lengthwise, and sliced into segments of about 22 mm length. The first 50 mm proximal to the pylorus was rejected. Adjacent segments were placed in paired flasks of 25 ml capacity containing 6.0 ml of the solution described in Table II. One flask of each pair contained 30 mg of the fiber undergoing study in finely divided form. Oxygen was passed through the flasks for 15 sec. They were placed in a reciprocating bath and shaken for 20 min at 38°C at the rate of about 25 excursions/min. The segments were removed, washed in two portions of glucose–saline solution, drained on paper toweling, and weighed to the nearest 10 mg. They were liquefied by digestion in 3 ml of concentrated sulfuric acid with the aid of heat. Then 2 ml was placed in cuvettes and the emission of gamma radiation was measured using an Autowall Gamma Counter (Picker Corporation, Northford, Conn.). The results are expressed as CPM/g. Three pairs of segments from proximal, medial, and distal portions of the small intestine were exposed to each treatment. The mean of the three was used for comparison. Results shown in the figures as means ± SE are drived from four to eight experiments unless otherwise indicated. Samples of cecum and colon were examined by a similar technique.

In some experiments, the intestine was everted, ligated with double ligatures, and cut between ligatures so that only mucosa was exposed to the bathing solution.

The pH of the solutions was measured at the beginning and end of an experiment to take into account changes brought about by contact with the tissue.

Acidified solutions became much less acid during incubation. Initial and final pH values were averaged.

3.1. Preparation of Fiber

NDF was prepared by the method of Robertson and Van Soest (1977), which includes treatment with bacterial amylase (Sigma Chemical Co., Type IIIA). Ten-gram batches of bran were treated at a time with reagent additions scaled up accordingly.

Dietary fiber was also prepared by treatment of bran with digestive enzymes, using several procedures. In each, treatment with bacterial amylase at boiling temperatures preceded digestion with peptidases, including pepsin, 0.5% (Sigma Chemical Co.) in 0.1 M hydrochloric acid, and after filtration and washing with water, 250 mg of Pronase Type V (Sigma Chemical Co.) in 300 ml of water with incubation in an oscillating bath at 38°C for a minimum of 4 hr. Pepsin–hydrochloric acid was omitted from some preparation procedures.

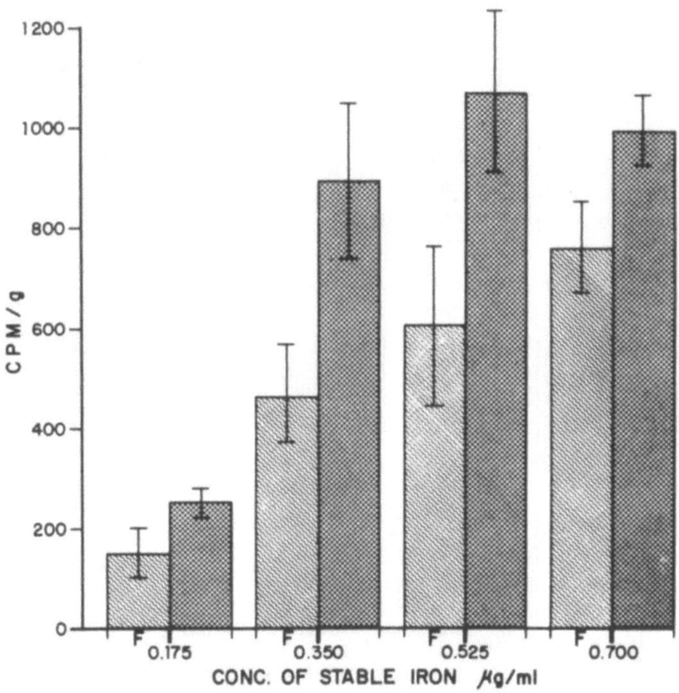

FIGURE 1. Uptake of iron-59 by segments of small intestine at several iron concentrations. Note the decreased uptake of iron in the presence of fiber at all concentrations of iron. F, fiber present.

Nitrogen concentrations of wheat NDF ranged from 0.81 to 0.99%; those of maize NDF from 0.55 to 0.70%. In the enzymatic preparations, nitrogen varied beteen 0.53 and 0.73% for wheat fiber and 0.13 and 0.59% for maize fiber.

4. RESULTS

4.1. Iron Concentration and Intestinal Uptake

The effect of iron concentration on the uptake of iron by segments of intestine in the presence and absence of dietary fiber is shown in Fig. 1. Uptake increases with concentration of iron until the latter reaches approximately 0.5 μg/ml. Beyond this concentration no further increase in uptake occurred. It will

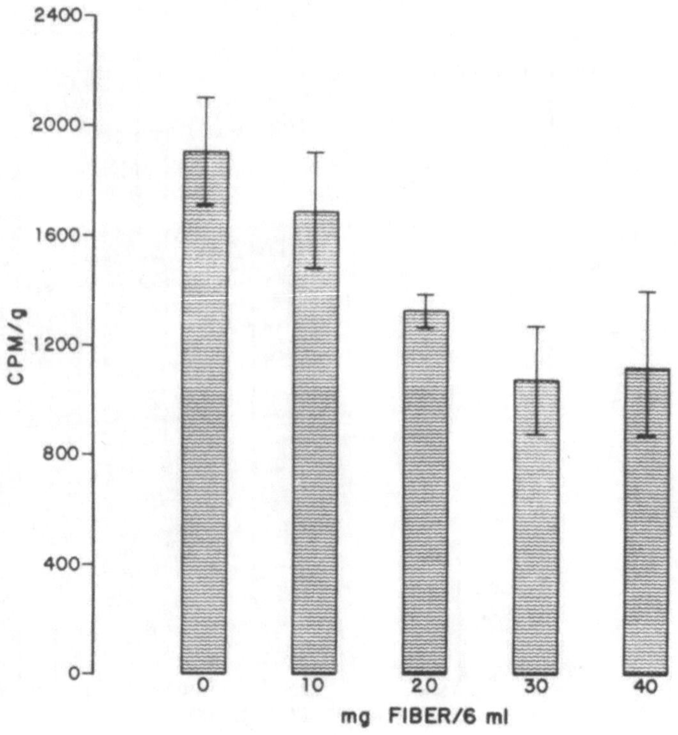

FIGURE 2. Uptake of iron-59 by segments of small intestine when the amount of fiber varied. Inceasing the amount of fiber decreased iron uptake.

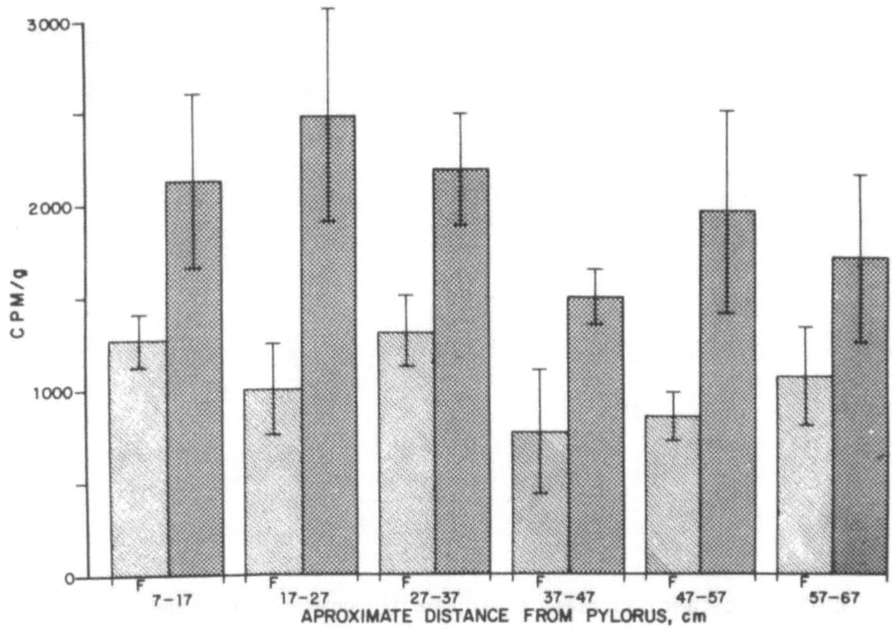

FIGURE 3. The effect of fiber in decreasing uptake of iron-59 along the length of the small intestine. F, fiber present.

be seen that in the presence of fiber, uptake of iron was consistently less than in its absence.

4.2. Fiber Amount and Iron Uptake

In these experiments, uptake of iron was measured in the presence of 10–40 mg of fiber in 10-mg increments (Fig. 2). The presence of 10 mg of fiber caused only a small and statistically insignificant decrease in iron uptake. However, 20 mg produced a statistically significant decrease in uptake by the segments. A further decrease occurred as the amount of fiber rose to 30 and 40 mg.

4.3. Segment Locus and Fiber Effect

Fiber effectively decreases iron uptake along the entire length of the intestine. Differences between fiber-containing and control segments were statistically significant in all except the most distal pair (Fig. 3).

Interference by fiber with iron uptake extends into the cecum and colon. Inhibition is highly significant in the cecal segments, but fails significance for those of the colon (Fig. 4).

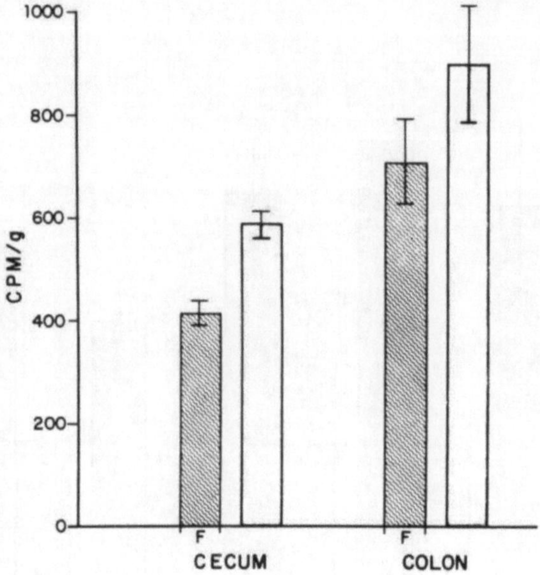

FIGURE 4. Fiber also decreases iron-59 uptake by segments of cecum and colon. F, fiber present.

4.4. Variability of Inhibition of Iron Uptake

The degree of inhibition of iron uptake produced by fiber varies widely when intestinal segments from different rats are examined. The distribution of the degrees of inhibition of some 40 experiments is shown in the lower portion of Fig. 5. Values range from −4 to 70% inhibition, with a median of 38%. One cause of the variation is the source of the dietary fiber, that from maize being less inhibitory than that from wheat. Method of preparation, whether by detergent or enzymatic methods, does not make a difference. The main determinant of inhibition may be the iron status of the segments. The upper portion of Fig. 5 shows that a single preparation of wheat NDF produced inhibitions that varied between 16 and 68% in 14 trials.

4.5. pH and Iron Uptake

The effect of pH upon the action of fiber in inhibiting iron uptake is shown by Fig. 6. Above pH 5.0, the presence of fiber significantly inhibits iron uptake in all experiments. At pH 4.5, the difference in the presence of fiber is no longer consistently significant. The same is true of the experiments done at pH 3.3, where iron uptake has become depressed. At pH 2.5, iron uptake is markedly decreased both in control and fiber-containing flasks, significantly more so, however, in the presence of fiber. Evaluation of the effects of pH was complicated by the speed with which segments removed protons from acidified solutions, a rise of an entire pH unit occurring within 20 min after a segment was added to the bathing solution.

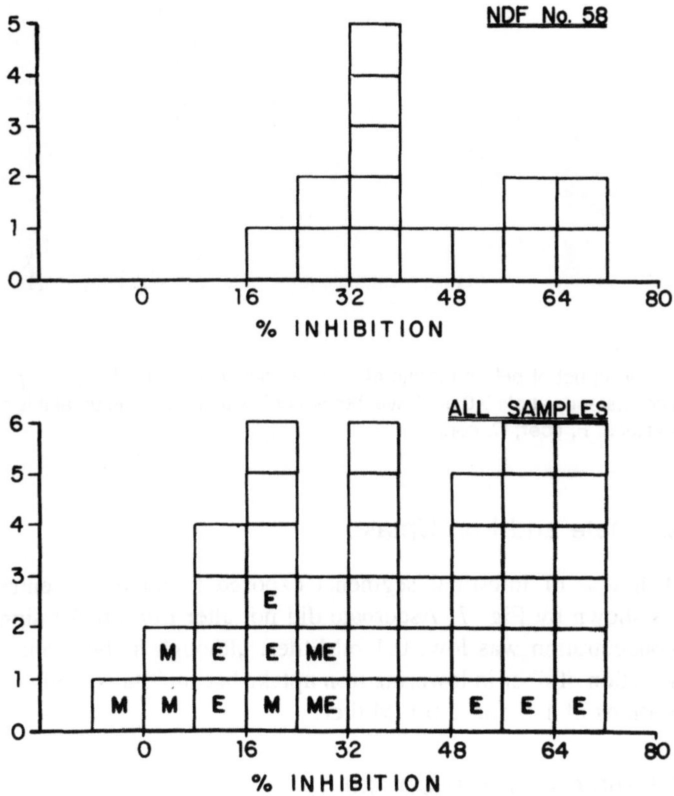

FIGURE 5. Variability of the inhibitory effect of dietary fiber upon iron-59 uptake by segments of small intestine. Maize fiber (M) is less inhibitory than is wheat fiber. Fiber prepared by use of digestive enzymes (E) does not differ from that prepared by the neutral detergent fiber method. The upper set of distributions shows the varied responses of a single preparation of NDF. Probably differences in iron status of the segments explain differences in inhibitory response.

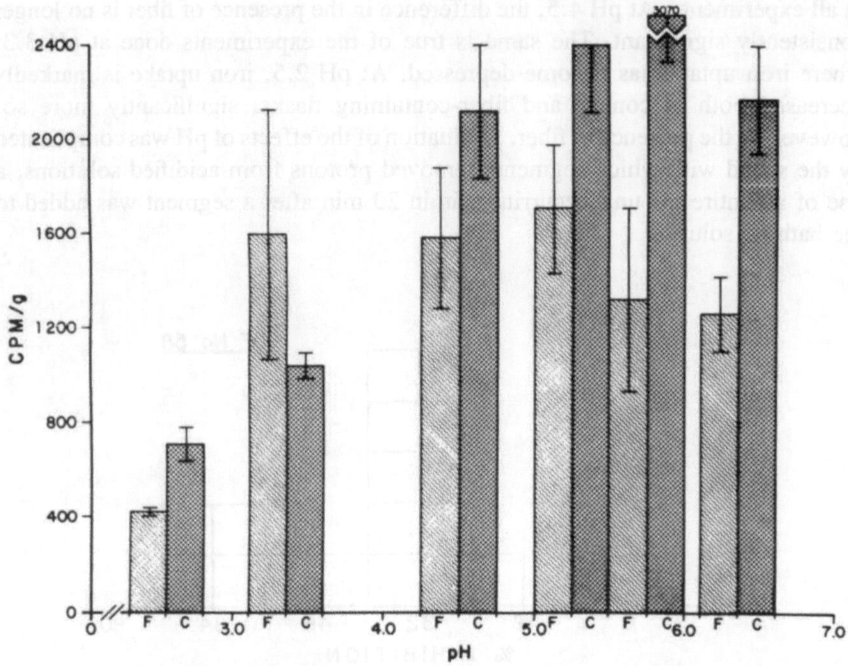

FIGURE 6. The effect of pH on uptake of iron-59. Inhibition of uptake in the presence of fiber occurs consistently at pH 5.0 or above, but not below this limit. Iron uptake is depressed at lower pH values. F, fiber; C, control.

4.6. Ascorbate and Iron Uptake

The behavior of intestinal segments exposed to three concentrations of ascorbate is shown by Fig. 7. Ascorbate did not alter iron uptake significantly when its concentration was low, 0.1 mM/liter, although higher concentrations did so. The action of fiber in lowering iron uptake became statistically significant at concentrations of 1.0 and 2.0 mM/liter.

4.7. Glutamate and Iron Uptake

Glutamate was studied in some detail because it was among the moderate inhibitors of iron binding *in vitro,* and because of its high concentrations in the major proteins of cereals, up to 25% of the total amino acid content. In Fig. 8 it is seen that glutamate has no signficant effect on iron uptake in either the presence or absence of fiber, other than apparent enhancement of the fiber effect when glutamate was present.

4.8. Cysteine and Iron Uptake

Cysteine at a concentration of 1 mM/liter depresses iron uptake markedly (Fig. 9), although not as much as does the presence of dietary fiber. When cysteine and fiber additions were combined, the effect was the same as that produced by fiber alone.

4.9. Citrate and Phytate and Iron Uptake

The behavior of two other strong inhibitors of iron binding *in vitro* is shown by Fig. 10. Four concentrations of each were examined, but the results were combined since concentration had no effect. Both inhibited iron uptake far more

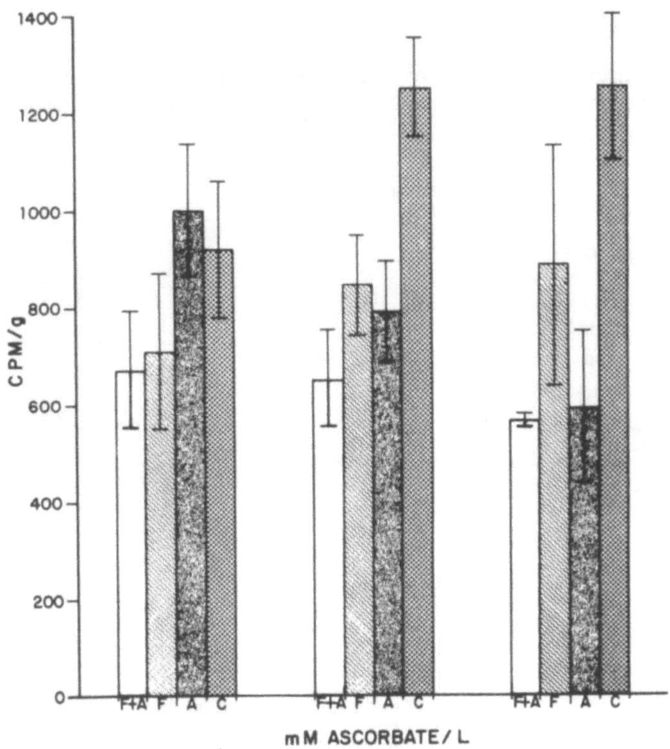

FIGURE 7. Ascorbic acid decreases iron-59 uptake. It fails to release iron from its combination with fiber. C, control; A, ascorbate added; F, fiber, no ascorbate; F + A, fiber plus ascorbate.

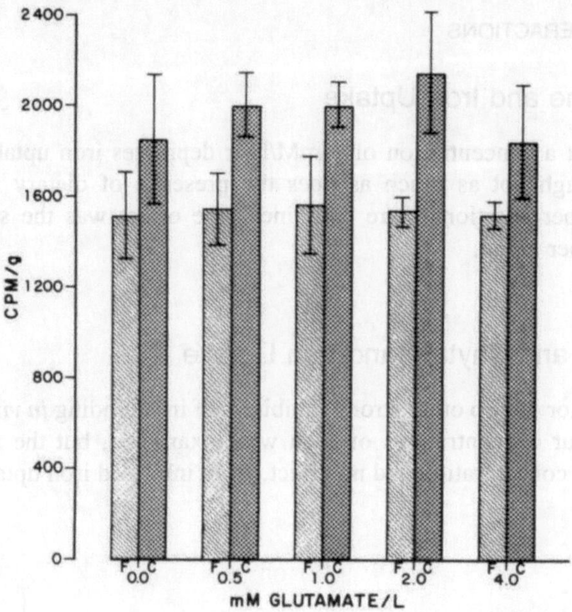

FIGURE 8. Glutamate at four concentrations has no effect on uptake of iron-59 by segments of intestine; neither does it overcome the inhibition produced by fiber. F, fiber; C, control.

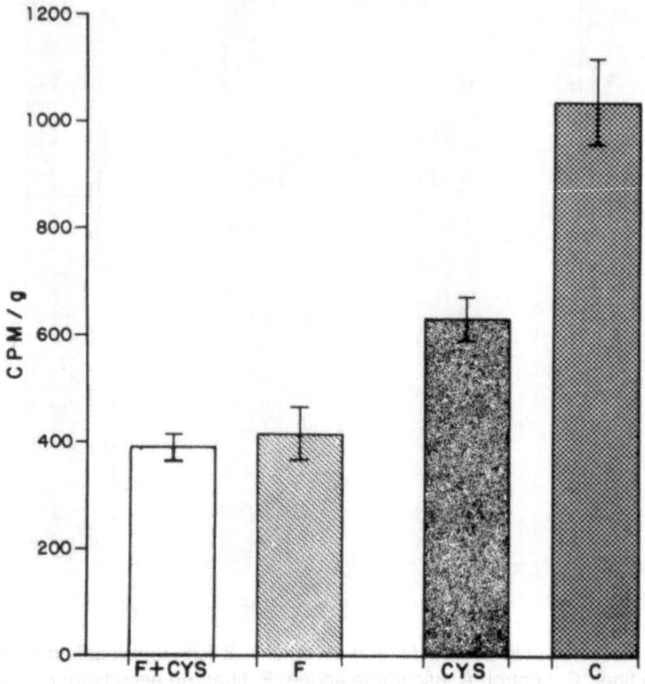

FIGURE 9. Fiber (F) inhibits iron uptake more than does cysteine (CYS); C, control.

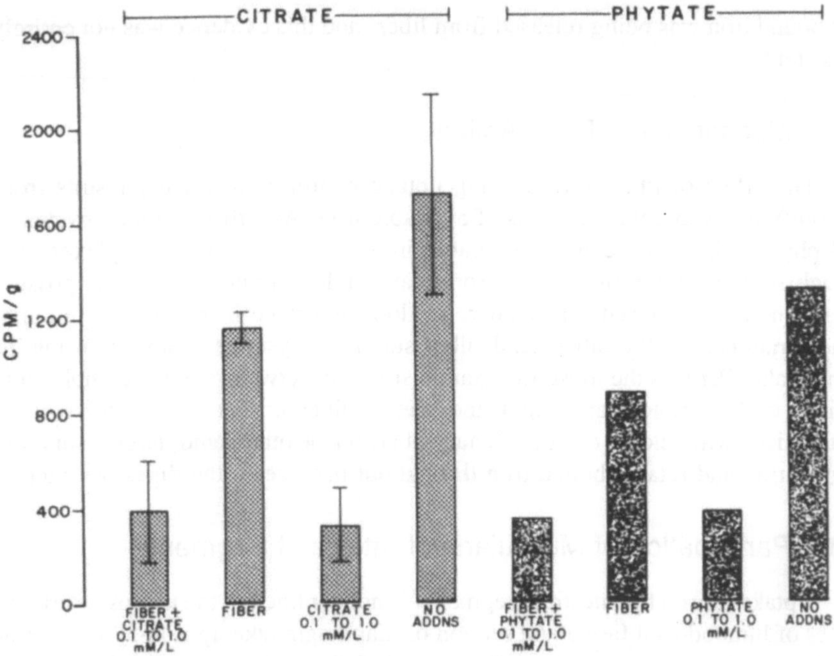

FIGURE 10. Citrate and phytate strongly inhibit iron-59 uptake, more so than fiber.

than did fiber, but neither showed any ability to release iron from combination with fiber.

5. DISCUSSION

5.1. General Comments

Our experiments show that uptake of iron by segments of intestine is inhibited by a substantial decrement in the presence of dietary fiber from wheat. The inhibition by maize fiber is considerably less. Although the extent of the inhibition varied, it occurred throughout jejunum and ileum and extended into the cecum and colon. The inhibition was not lessened by substances that strongly inhibit binding of iron by fiber *in vitro*, namely ascorbate, citrate, cysteine, or phytate. Each markedly decreased uptake of iron by intestinal segments without effecting release of iron from its combination with fiber. Glutamate, a moderate inhibitor of binding of iron by fiber *in vitro*, possibly enhanced the action of fiber. Only when the pH of the bathing fluid was less than 5.0 was there evidence that bound iron was being released from fiber, and this evidence was not entirely consistent.

5.2. Mechanism of Fiber Action

The effect of fiber in decreasing uptake of iron undoubtedly results from its ability to lower concentrations of available iron. Ascorbate, citrate, cysteine, and phytate likewise decrease available iron by forming iron complexes. Although soluble, these also render iron unavailable. Citrate and phytate possess greater inhibitory activity than fiber, as does ascorbate in more than minimal concentrations. On the other hand, fiber surpasses cysteine in ability to inhibit iron uptake. Perhaps the most important distinction between soluble complexants of iron of low molecular weight and dietary fiber is that the former will be metabolized with release of iron. Dietary fiber, on the other hand, largely survives degradation and retains bound iron throughout passage of the digestive tract.

5.3. Participation of Muscularis of Intestinal Segments

Uptake of iron by intestinal segments is not confined to the mucosal surface. Slices of lumbodorsal fascia and section of diaphragm take up iron-59 with great avidity, diaphragm being especially alive. It follows that the muscularis of the segments contributes to the uptake of iron. However, in experiments in which ligated everted intestinal segments were prepared to exclude participation of the muscularis, the response to the presence of fiber was similar to that of segments with both surfaces exposed, except that iron uptakes were lower.

5.4. Dietary Fiber and Iron Availability in Certain Human Diets

Use of the coefficients we have calculated to define the quantity of iron bound per gram of wheat and maize fibers permits estimation of the amount of iron that is made unavailable by complexation with fiber from these sources. In Central Mexico, the NDF intake per person per day averages 25 g, of which half is supplied by maize tortillas (Reinhold and García L., 1979). Using our value of 0.3 mg of iron bound/g maize fiber, this yields only 3.6 mg of iron daily diverted to this form. The remainder of the dietary fiber is derived largely from fruits and vegetables. According to Kelsay *et al.* (1979), fiber from these sources does not impair bioavailability of iron, a finding that is compatible with our finding that ascorbate and citrate, both abundant in such foods, strongly inhibit iron binding *in vitro*. The average daily intake of iron in Central Mexico is about 20 mg/day (Perez Hidalgo *et al.*, 1971); therefore, we are obliged to conclude that the fiber of maize is not a primary cause of the iron deficiency existing in many Mexicans. Nevertheless, poverty may lead to greatly increased consumption of tortillas associated with low intakes of fruits and vegetables. In these circumstances it is possible that fiber intake would be great enough to diminish iron absorption sufficiently to impair iron nutrition.

In the Middle East and Northern India, also, the very large intakes of dietary fiber that result from the consumption of 500 g or more of unleavened wholemeal wheaten flat bread daily per person create conditions in which binding of iron by fiber may be critical. The higher coefficient of iron binding of wheat flour in combination with excessive fiber intakes, 50 g or more daily, would result in 20 mg of iron or more becoming fiber-bound. While iron intakes tend to be high in the Middle East, in many circumstances they may not suffice to provide an adequate margin of available iron to fulfill requirements. In addition, fruit and vegetables may be available only part of the year, with consequent lack of adjuvant action from these sources.

ACKNOWLEDGMENTS. The authors thank Dr. J. Manuel Sotomayor M. del C. and Q.F.B. Laura Elena Domínguez Meza of the Hospital de Especialidades del Centro Médico de Occidente del IMSS for carrying out measurements of radioactivity, Sección de Audiovisual de la Unidad de Investigaciones Biomédicas de Occidente del IMSS for the artwork, and V.M.Z. Jaimes García Johnson for performing nitrogen determinations on fiber preparations. They also thank Q.F.B. Carmen Yolanda Partida O. and M.V.Z. Rodolfo Barba Lopez for facilities and encouragement.

The studies were sponsored by the Subsecretariat of Higher Education and Scientific Investigation and supported in part by Grant 79-04-70.

REFERENCES

Bremner, I., 1970, The nature of trace element binding in herbage and gut contents, in: *Trace Element Metabolism in Animals* (C. F. Mills, J. Bremner, J. Chesters, and J. Quarterman, eds.), Livingston, Edinburgh, pp. 366–369.

Eastwood, M. A., and Kay, R. M. 1979, An hypothesis for the action of fiber along the gastrointestinal tract, *Am. J. Clin. Nutr.* **32:**364–367.

Ismail-Beigi, F., Faraji, B., and Reinhold, J. G., 1977, Binding of zinc and iron to wheat bread, wheat bran and their components, *Am. J. Clin. Nutr.* **30:**1721–1725.

Kelsay, J. L., Behall, K. M., and Prather, E. S., 1979, Effect of fibers from fruits and vegetables on metabolic responses of human subjects. II. Calcium, magnesium, iron and silicon balances, *Am. J. Clin. Nutr.* **32:**1876–1880.

Oberleas, D., Muhrer, M. E., and O'Dell, B. L., 1966, Dietary metal-complexing agents and zinc availability in the rat, *J. Nutr.* **90:**56–62.

Perez Hidalgo, C., Chavez, V., and Madrigal, E. H., 1971, El problem nutricional de hierro en Mexico, *Salud Publ. Mex.* **13:**71–77.

Reilly, C., 1979, Zinc, iron and copper binding by dietary fiber, *Biochem. Soc. Trans.* **7:**202–204.

Reinhold, J. G., and García L., J. S., 1979, Fiber of the maize tortilla, *Am. J. Clin. Nutr.* **32:**1326–1329.

Reinhold, J. G., Parsa, A., Karimian, N., Hammick, J. W., and Ismail-Beigi, F., 1974, Availability of zinc in leavened and unleavened wholemeal wheaten breads as measured by solubility and uptake by rat intestine *in vitro*, *J. Nutr*. **104**:976–982.

Reinhold, J. G., Ismail-Beigi, F., and Faraji, B., 1975, Fiber vs. phytate as determinant of the availability of calcium, zinc and iron in breadstuffs, *Nutr. Rep. Int*. **12**:75–85.

Reinhold, J. G., García L., J. S., and Garzón, P., 1981, Binding of iron by fiber of wheat and maize, *Am. J. Clin. Nutr*. **34**:1384–1391.

Robertson, J. B., 1978, The detergent system of fiber analysis, in: *Topics in Dietary Fiber Research* (G. A. Spiller, ed.), Plenum Press, New York, pp. 11–42.

Robertson, J. B., and Van Soest, P. J., 1977, in: *Proceedings, 69th Meeting, American Society for Animal Science, Madison, Wisconsin*.

Sahagian, B. M., Barlow, H. I., and Perry, Jr., H. M., 1966, Uptakes of zinc, cadmium and mercury by intact strips of rat intestine, *J. Nutr*. **90**:259–267.

Thompson, S. A., and Weber, C. W., 1979, Influence of pH on the binding of copper, zinc and iron in six fiber sources, *J. Food Sci*. **44**:752–754.

12

Fiber, Obesity, and Diabetes

DAVID KRITCHEVSKY

Cleave and Campbell (1966) and Trowell (1974, 1975) have suggested that consumption of fiber-depleted foods plays an important role in the etiology of diabetes. Trowell advanced the argument that the incidence of diabetes in England and Wales could be correlated with the fiber content of flour. He also cited examples of populations (South African Bantu, Yemenite Jews) whose incidence of diabetes increased soon after changing from a high-fiber diet to a low-fiber diet. Jenkins *et al.* (1976, 1977) showed that meals rich in fiber reduced hyperglycemia and improved glucose tolerance in human subjects: Their findings were confirmed in other laboratories (Miranda and Horwitz, 1978; Monmer *et al.*, 1978). Anderson and his colleagues (Anderson, 1980; Anderson and Ward, 1978; Kiehm *et al.*, 1976) have devised a high-carbohydrate, high-fiber diet which contains the same number of calories as the diet recommended by the American Diabetes Association but in which the percentage of calories from carbohydrates is increased from 43 to 70. The increase in carbohydrate is principally in the form of complex carbohydrate. A comparison of the representative diets is shown in Table I. This type of diet reduces serum cholesterol and blood glucose and permits reductions in dosages of insulin or oral hypoglycemic agents needed for control of diabetes. Stone and Connor (1963) fed a similar diet to diabetics that also proved to be hypolipidemic. These diets act by inhibiting glucose absorption and thus reduce serum insulin response. Plasma glucagon levels are also lower in diabetic subjects ingesting a high-fiber diet (Miranda and Horwitz, 1978). Excellent reviews on the effect of fiber on carbohydrate metabolism (Jenkins, 1980) and diabetes (Anderson, 1980) have appeared recently.

The influence of dietary fiber on obesity may also be adduced from epi-

DAVID KRITCHEVSKY • The Wistar Institute of Anatomy and Biology, Philadelphia, Pennsylvania 19104.

TABLE I. Composition of Control and High-Carbohydrate, High-Fiber (HCF) Diet[a]

	Diet,[b] g/day			
	ADA[c]		HCF	
Protein	98	(20)	96	(19)
Carbohydrate	215	(43)	351	(70)
Simple (S)	100		94	
Complex (C)	115		257	
S/C	0.87		0.37	
Fat	83	(37)	23.5	(11)
Saturated (S)	19		4.7	
Monounsaturated (M)	43		9.5	
Polyunsaturated (P)	15		6.9	
P/S	0.79		1.47	
Cholesterol	0.47		0.04	
Fiber	27		82	
Soluble (S)	15		67	
Insoluble (I)	12		15	
S/I	1.25		4.47	

[a] After Anderson (1980).
[b] Figures in parentheses give percent of calories.
[c] American Diabetic Association recommended diet.

demiological observations. Cleave (1956, 1974) has suggested that most of the obesity prevalent in industrialized societies is due to replacement of crude carbohydrate by refined carbohydrate. Heaton (1973, 1978) hypothesizes that dietary fiber inhibits energy intake and thus reduces weight gain. Dietary fiber reduces caloric density of food and it increases the sensation of satiety, which, in turn, reduces food intake. In a comparison of the sensation of satiety in subjects ingesting raw apples or their equivalent in apple juice, the satiety score was found to be almost five times greater in the former group (Haber *et al.*, 1977). A similar difference was seen between subjects ingesting white or wholemeal bread (Grimes and Gordon, 1978). Heaton (1980) has also reported on weight loss in subjects permitted to eat a fiber-rich diet *ad libitum*. These subjects ate to satiety and on the fiber-rich diet they ingested 480 fewer kcal than on a fiber-poor diet (Heaton, 1980). Replacement of sucrose by starches has resulted in weight loss in subjects who were instructed to eat to satiety (Mann *et al.*, 1970; Rifkind *et al.*, 1966). Thus, high-fiber diets appear to exert their effect by increasing satiety (Heaton, 1973; Hodges and Krehl, 1965); decreasing food intake, and enhancing fecal energy loss (Southgate *et al.*, 1976; Kelsay *et al.*, 1978), and possibly by hormonal effects (Albrink, 1978).

Dietary fiber can affect serum and liver lipids and body weight, but the question of its effects on body composition requires investigation. We (Mueller

TABLE II. Influence of Fiber (15%) on Weight Gain, Lipids, and Body Composition in Rats[a]

	Diet[b]				
	Cellulose	Hemicellulose	Pectin	Lignin	Fiber-free
Grams consumed	1129	1107	1126	1068	1146
kcal consumed	292	287	291	277	252
Weight gain, g	15	33	37	10	25
Feed efficiency[c]	1.28	2.96	3.29	1.07	2.16
Cholesterol					
Serum, mg/dl	53	61	56	53	51
Liver, mg/100 g	541	285	335	315	391
Triglycerides					
Serum, mg/dl	97	82	51	53	100
Liver, mg/100 g	297	348	372	264	216
Gut weight, g	14.5	13.4	14.0	17.0	15.0
Carcass					
Weight, g	177	182	189	164	182
Percent lipid	34	28	31	34	25

[a] After Mueller *et al.* (1981).
[b] Casein (25), dextrose (40), fiber (15), corn oil (14), salt mix (5), vitamin mix (1); fiber-free has 55% dextrose.
[c] Weight gain/calories consumed $\times 10^2$.

et al., 1981) have recently compared food intake, lipid metabolism, and carcass composition in rats fed semipurified diets containing 15% cellulose, hemicellulose, pectin, or lignin. A control group ingested a fiber-free diet. The results are summarized in Table II.

The different fibers exerted different effects on weight gain and feed efficiency. Cellulose feeding resulted in significantly higher levels of liver cholesterol. Serum triglycerides were reduced in rats fed pectin or lignin, but liver triglyceride levels ranged from 22 to 72% higher than those observed in the rats fed the fiber-free diet. Carcass lipid was increased significantly in rats fed cellulose and lignin, but was higher than control levels in the other two test groups as well.

Among the areas which must be elucidated for a true understanding of the mechanisms of action of the complex collection of substances which we call fiber are its effects on hormonal actions and on whole-body disposition of lipids and proteins.

ACKNOWLEDGMENTS. Supported in part by a Career Research Award (HL00734) from the National Institutes of Health and a grant-in-aid from the Commonwealth of Pennsylvania.

REFERENCES

Albrink, M. J., 1978, Dietary fiber, plasma insulin and obesity, *Am. J. Clin. Nutr.* **31:**S277–S279.

Anderson, J. W., 1980, Dietary fiber and diabetes, in: *Medical Aspects of Dietary Fiber* (G. A. Spiller and R. M. Kay, eds.), Plenum Medical, New York, pp. 193–221.

Anderson, J. W., and Ward, K., 1978, Long term effects of high carbohydrate, high fiber diets on glucose and lipid metabolism: A preliminary report on patients with diabetes, *Diab. Care* **1:**77–87.

Anderson, J. W., and Ward, K., 1979, High carbohydrate, high fiber diets for insulin-treated men with diabetes mellitus, *Am. J. Clin. Nutr.* **32:**2312–2321.

Cleave, T. L., 1956, The neglect of natural principles in current medical practice, *J. R. Nav. Med. Serv.* **42:**55–83.

Cleave, T. L., 1974, *The Saccharine Disease,* John Wright and Sons, Bristol, England.

Cleave, T. L., and Campbell, G. D., 1966, *Diabetes, Coronary Thrombosis and Saccharine Diseases,* John Wright and Sons, Bristol, England.

Grimes, D. S., and Gordon, C., 1978, Satiety value of wholemeal and white bread, *Lancet* **2:**106.

Haber, G. B., Heaton, K. W., Murphy, D., and Burroughs, L., 1977, Depletion and disruption of dietary fibre. Effects on satiety, plasma-glucose and serum-insulin, *Lancet* **2:**679–682.

Heaton, K. W., 1973, Food fibre as an obstacle to energy intake, *Lancet* **2:**1418–1421.

Heaton, K. W., 1978, Fibre, satiety and insulin—A new approach to overnutrition and obesity, in: *Dietary Fibre, Current Developments of Importance to Health* (K. W. Heaton, ed.), Newman, London, pp. 141–149.

Heaton, K. W., 1980, Food intake regulation and fiber, in: *Medical Aspects of Dietary Fiber* (G. A. Spiller and R. M. Kay, eds.), Plenum Medical, New York, pp. 223–238.

Hodges, R. E., and Krehl, W. A., 1965, The role of carbohydrates in lipid metabolism, *Am. J. Clin. Nutr.* **17:**334–346.

Jenkins, D. J. A., 1980, Dietary fiber and carbohydrate metabolism, in: *Medical Aspects of Dietary Fiber* (G. A. Spiller and R. M. Kay, eds.), Plenum Medical, New York, pp. 175–192.

Jenkins, D. J. A., Leeds, A. R., Gassull, M. A. Wolever, T. M. S., Goff, D. V., Alberti, G. M. M., and Hockaday, T. D. R., 1976, Unabsorbable carbohydrates and diabetes: Decreased postprandial hyperglycemia, *Lancet* **2:**172–174.

Jenkins, D. J. A., Leeds, A. R., Gassull, M. A., Cochet, G., and Alberti, G. M. M., 1977, Decrease in postprandial insulin and glucose concentrations by guar and pectin, *Ann. Intern. Med.* **86:**20–23.

Kelsay, J. L., Behall, K. M., and Prather, E. S., 1978, Effect of fiber from fruits and vegetables on metabolic responses in human subjects. 1. Bowel transit time, number of defecations, fecal weight, urinary excretions of energy and nitrogen and apparent digestibilities of energy, nitrogen and fat, *Am. J. Clin. Nutr.* **31:**1149–1153.

Kiehm, T. G., Anderson, J. W., and Ward, K., 1976, Beneficial effects of a high carbohydrate, high fiber diet on hyperglycemic, diabetic men, *Am. J. Clin. Nutr.* **29:**895–899.

Mann, J. I., Truswell, A. S., Hendricks, D. A., and Manning, E., 1970, Effect on serum-lipids in normal men of reducing dietary sucrose or starch for five months, *Lancet* **1:**870–872.

Miranda, P. M., and Horwitz, D. L., 1978, High-fiber diets in the treatment of diabetes-mellitus, *Ann. Intern. Med.* **88:**482–486.

Monmer, L., Pham, T. C., Aguerre, L., Orsetti, A., and Mirouze, J., 1978, Influence of indigestible fibers on glucose tolerance, *Diab. Care* **1:**83–88.

Mueller, M. A., Cleary, M. P., and Kritchevsky, D., 1981, The effects of various types of dietary fiber on lipid storage in adipose tissue, *Fed. Proc.* **40:**853.

Rifkind, B. M., Lawson, D. H., and Gale, M., 1966, Effect of short term sucrose restriction on serum lipid levels, *Lancet* **2:**1379–1381.

Southgate, D. A. T., Branch, J. W., Hill, M. J., Drasar, B. S., Walters, R. L., Davies, P. S., and McLean Baird, I., 1976, Metabolic responses to dietary supplements of bran, *Metabolism* **25:**1129–1135.

Stone, D. B., and Connor, W. E., 1963, The prolonged effects of a low cholesterol, high carbohydrate diet upon serum lipids in diabetic patients, *Diabetes* **12:**27–132.

Trowell, H., 1974, Diabetes mellitus death rates in England and Wales 1920–1970 and food supplies, *Lancet* **2:**998–1002.

Trowell, H. C., 1975, Dietary-fiber hypotheses of the etiology of diabetes mellitus, *Diabetes* **24:**762–765.

13

Dietary Fiber and Obesity

A Review

R. ALI, H. STAUB, G. A. LEVEILLE, and P. C. BOYLE

1. INTRODUCTION

Burkitt and Trowell (1975) extended their hypothesis concerning the health attributes of dietary fiber to include its role in the development of obesity. Others have discussed the relationship of dietary fiber intake to obesity and suggested a variety of mechanisms (Southgate, 1978; Cleave, 1977; Heaton, 1973, 1979; Van Itallie, 1978). Fiber, with its low caloric availability and capacity to bind water, offers considerable opportunity for caloric dilution of foods, while its gut-filling properties and the chewing required during ingestion may trigger components of the satiety mechanism. Fiber has been proposed as an obstacle to energy intake by Heaton (1973) and as related to insulin balance by Albrink (1978), who maintained that low-fiber diets could lead to insulin resistance and obesity in susceptible individuals. No matter what the mechanism is, recent results support the notion that increased dietary fiber intake can aid in the control and management of dietary obesity. Providing bulk without high caloric density may well prevent excessive energy intake.

Trowell and Burkitt (1978) suggested that the diet of primitive man was high in fiber, like that of today's African Bushman. Consistent with this opinion is their observation that the appearance of obesity in East Africa occurred with

R. ALI • Nutritional Research and Development, Bristol-Myers International Division, New York, New York 10154. H. STAUB, G. A. LEVEILLE, and P. C. BOYLE • Department of Nutrition and Health Sciences, General Foods Technical Center, Tarrytown, New York 10591.

the availability of low-fiber, calorically concentrated foods. Such foods can allow dietary excesses to occur with obesity as the result.

2. EFFECT OF DIETARY FIBER ON DIGESTION AND ABSORPTION OF NUTRIENTS

The digestibility of dietary fiber varies in different subjects. This is probably a reflection of differences in the diets consumed and microbial species in the lower gut capable of fermenting complex fiber ingredients. Because of their limited and varied digestibility in the colon, fibers are generally recognized as poor sources of calories (Ali *et al.*, 1981; Staub and Ali, 1982).

Southgate and Durnin (1970) and Southgate (1973) noted that the subjects with the longest transit times tended to show the greatest digestibility of cellulose. A high-fiber diet had a negative effect on metabolized energy. The digestibility of protein and carbohydrate, but not fat, was significantly reduced. Southgate speculated that increased bulk could be speeding transit time and therefore allowing less time for digestion and absorption; alternatively, through its physicochemical properties such as water binding, fiber might reduce the availability of the products of digestion for absorption. Unfortunately, few data are available to support these speculations. Slavin and Marlett (1980) noted that the daily ingestion of a 16 g solka floc supplement did not reduce the digestibility of fat and nitrogen nor did it interfere with energy utilization. Recent comparative studies in rats have shown that increasing amounts of acid detergent fiber (ADF), i.e., the cellulose and lignin portion of dietary fiber, decreased the digestibility of the diet. The coefficients of digestibility for dry matter, protein, and gross energy all declined with increased fiber intake (Garrison *et al.*, 1978). In a further consideration of the role of dietary fiber in preventing obesity, Southgate (1978) quoted findings that substantial amounts of arabinoxylan did not have much effect on fecal nitrogen or fat. Therefore, not only composition of the diet but also the type and source of dietary fiber may influence activity of colonic microflora and digestibility of nutrients.

3. DIETARY FIBER AND ENERGY UTILIZATION

Dietary fiber in the form of cellulose or agar was shown to decrease nutrient digestibility and energy utilization in animal, but no differences in weight gain or body composition were evident (Schaeffer and Bennick, 1979). The addition of 12 g of wheat bran per day to a low-fiber diet significantly decreased apparent digestibility in man by the end of the 24th day on the regimen. Some 80% and 55% of the ADF (cell wall material) disappeared from the low- and the high-fiber diets, respectively (Farrell *et al.*, 1978).

Different responses were noted in studies in which mice were fed diets containing relatively pure sources of cellulose or hemicellulose and at levels up to 20% of the diets (Keim and Kies, 1979). As the dietary fiber increased, feed consumption and weight gain decreased, whereas feed efficiency increased. Animals fed diets containing 10% hemicellulose fared best in terms of feed intake and weight gain, but their feed efficiency was reduced compared to the 5% and 20% hemicellulose groups. Significant differences in feed efficiency, however, were noted among all the groups. It is possible that the animals responded to the increasing levels of fiber sources as if the diet were more satiating, and therefore they were not able to adjust their food intake in response to caloric dilution of the diet. In weanling rats fed two types of pectin-containing rations at 20% of the diet, food and energy intakes and efficiency of energy utilization were reduced (Rao and Bright-See, 1979). Efficiencies, however, remained constant in diets containing up to 8% pectin, while at the higher levels weight gain was greatly depressed. Guar, locust, tragacanth, or pectin, at only 2% of the diet, resulted in reduced body weights in the chicken. Gum arabic at the same level of intake did not reduce body weight significantly. Results in Japanese quail were similar for pectin and other gums, except that in addition to gum arabic, tragacanth also showed no effect (Vohra *et al.*, 1979). The rate of growth was not influenced by the gel-forming dietary fibers. Available data seem to indicate that various animal species may respond differently to the same source of dietary fiber when added to their diet.

4. EFFECTS OF DIETARY FIBER ON FOOD INTAKE

Dietary fiber can displace available nutrients from the diet. Among the complex of substances that constitute dietary fiber, mucilages, gums, and pectins can bind considerable amounts of water and therefore increase the potential for caloric dilution (James and Theander, 1981). The potential effectiveness of caloric dilution per se, however, in controlling obesity is probably limited. A large amount of literature suggests that daily caloric intake is controlled accurately so as to maintain a constant body weight. When the caloric density of food offered to laboratory animals was diluted with cellulose, the animals increased their food intake such that caloric intake was maintained at premanipulation levels. With very high levels of caloric dilution, however, reductions in total caloric intake and body weight have been reported (Adolph, 1947; Peterson and Baumgardt, 1971a,b; O'Hea *et al.*, 1974). A similar ability to maintain constant food intake despite dilution of the caloric density of the food available has been reported in human volunteers (Campbell *et al.*, 1971; Spiegel, 1973).

Nevertheless, two recent observations suggest that calorically diluted foods may assist at least some overweight individuals in losing body weight. First, it has been reported that covert substitution of the sweetener aspartame for sucrose in the daily diet reduced daily caloric intake in human subjects. Recently, the initial short-term observations have been extended to demonstrate that this reduction in caloric density of the diet reduced caloric intake for at least 12 days, the longest period studied to date (Porikos *et al.*, 1977, 1982). Second, there is some evidence to suggest that certain obese subpopulations may be less sensitive than normal to caloric dilution of the diet. Genetically obese Zucker rats have been reported to show little compensation in intake when their diet was diluted with either fiber (alphacel) or a poorly absorbed lipid material (Sullivan *et al.*, 1978). This situation is different from that observed in another frequently studied animal model of obesity, the rat with ventromedial hypothalamic lesions, which was reported to respond normally to dietary caloric dilution (Smutz *et al.*, 1975).

Low-calorie bulking agents have had some successful use in weight reduction regimens and are used as adjunctive support with some weight control techniques (Beereboom, 1979). Evans and Miller (1975), using both methylcellulose and guar gum at about 10 g per day in human subjects, found a reduction in voluntary food intake which was more pronounced in obese than normal subjects over a two-week test period and some weight loss was also noticed. In a double-blind study with female subjects, 15 g per day of guar resulted in "permanent" weight loss (Tuomilehto *et al.*, 1980). In college-age males, Mickelsen *et al.* (1979) found a high-fiber bread to aid in weight loss with a restricted calorie diet. Overweight subjects following the program lost 8.77 kg in 8 weeks, whereas those eating ordinary enriched bread lost 6.26 kg in the same period. Each group consumed 12 slices of bread per day, but the high-fiber bread supplied 25.5 g daily crude fiber, while the enriched white bread contained only 1.02 g of crude fiber. Those eating the reduced calorie bread also reduced their calorie intake to a slightly greater extent. These findings might be unexpected since, in the rat, dilution of the diet by relatively small amounts of cellulose were readily overcome by increasing food intake (Adolph, 1947). However, when the level of dilution was sufficiently high, growth inhibition was noted and the animals were no longer able to compensate by increasing their intake. Mickelsen (1980) further reported that high-fiber bread included in a rat ration supported a much lower weight gain than a ration containing regular white bread.

5. DIETARY FIBER AND SATIETY—PROPOSED MECHANISM(S) OF ACTION

5.1. Caloric Dilution

A number of workers have reported that high-fiber foods have a greater satiety effect in human subjects than do low-fiber foods (Haber *et al.*, 1977; Grimes and Gordon, 1978; Kay, 1978; Bolton *et al.*, 1981; Wilmshurst and

Crawley, 1980). The effect upon satiety could be mediated either by the effect of dietary fiber on the volume of materials in the gastrointestinal tract or by the effect of fiber upon the postingestive changes in blood glucose and insulin levels. Relatively long-term effects of dietary fiber have been reported upon total food intake (Evans and Miller, 1975) and upon weight loss (Mickelsen *et al.*, 1979). Practically all the human subjects in the bread studies of Mickelsen *et al.* (1979), when questioned, felt hungry at the beginning of the study. A reduction of hunger sensation toward the end of the study was generally reported, but those on the high-fiber bread were least hungry at any point in time. Grimes and Gordon (1978) reported that in ten of twelve subjects, white bread was less effective in reaching a point of comfortable fullness compared to wholemeal bread. Although the fiber content of the bread was not noted, the caloric density of the white bread (calories/gram) was 6% higher than that of the wholemeal bread. In another study, subjects were offered unlimited quantities of bread in a meal setting. The average energy intakes, no matter which bread was offered, were similar over an approximately 24-hr period (Bryson *et al.*, 1979). There was a large variation in individual intakes, but no indication of a reduction in wholemeal bread consumption compared to white bread was noted.

5.2. Modulation of Blood Glucose and Insulin Responses

Haber *et al.* (1977) reported that decreasing or disrupting dietary fiber structure influenced satiety. Using a subjective measure, they found that satiety was more pronounced with an apple test meal compared to an apple puree (disrupted fiber) or an apple juice (fiber-depleted) meal, which scored the lowest among the three meals tested. The satiety scores remained significantly higher than fasting and continued for a longer period of time with both apple and apple puree than with apple juice. No significant differences were found between any of the meals in the rate, height, and duration of blood glucose response curves. However, following the apple meal, plasma glucose was at no time significantly different from fasting level. Following the puree and juice meals, plasma glucose fell below fasting levels during the second hour and remained lower. Serum insulin rose to higher levels following ingestion of juice or puree than whole apple meal. Destruction or depletion of fiber may have influenced satiety by allowing faster and easier consumption of the test meals, which resulted in the observed changes in plasma glucose and insulin levels. On the other hand, Grimes and Gordon (1978) found no significant differences in blood glucose, insulin, or glucagon between groups receiving whole or white bread meals.

When equal quantities of glucose were administered in the form of almost equal weights of peeled orange or orange juice, the mean blood levels of glucose were slightly but insignificantly higher after the juice meal than after the orange meal. However, the level did fall faster following the juice meal (Kay, 1978).

Subjective signs of hunger were reduced in the third hour following fruit ingestion. Thus the presence of dietary fiber apparently prolonged satiety.

In a more extensive test of the hypothesis that the fiber present in the fruit modulates insulin response to the sugar present, prevents rebound hypoglycemia, and is more satiating, healthy adults were fed grape or grape juice meals, as well as orange and orange juice meals. The oranges and grapes both were considerably more satiating than the corresponding juices. With oranges there was a smaller insulin response to the fruit than the juice. The grape juice caused a smaller insulin response than expected (Bolton *et al.*, 1981).

Jenkins *et al.* (1976) had observed that the addition of guar gum and pectin to a 106-g carbohydrate meal presented to a few diabetic, non-insulin-dependent subjects significantly decreased the rise in blood glucose for the first 2 hr following ingestion and resulted in lower serum insulin values over the same time period. In experiments with normal subjects, the results were not so clear-cut. However, the addition of guar and pectin to a test meal of bread and marmalade did decrease the postprandial blood glucose levels and insulin values for a short time (Jenkins *et al.*, 1977). The dietary fiber present may have reduced the rate of availability of the products of digestion for absorption.

To study the effects of dietary fiber on oral glucose tolerance, a variety of substances, including guar, pectin, gum tragacanth, methylcellulose, and wheat bran, were used. The addition of each substance significantly reduced blood glucose at one or more time points and generally reduced serum insulin concentration (Jenkins *et al.*, 1978). Only guar significantly reduced the percentage of maximum blood glucose and serum insulin rise. Based on first appearance of bread hydrogen in a lactose test, all substances except bran and methylcellulose delayed mouth to cecum transit time. In a xylose absorption test, a significant decrease in xylose excretion was observed with guar as the test substance, with slight reductions also observed with pectin and methylcellulose. When the guar was hydrolyzed, its observed effects in the gut were abolished. Increased viscosity of gastric content may delay stomach emptying and extend glucose absorption from the lumen over a longer period of time, thereby sustaining satiety. Wilmshurst and Crawley (1980), studying obese subjects, found that gastric transit time of a low-energy meal could be made to approximate that of a high-energy meal by the addition of 10% guar to the test meal, which also resulted in increased time to maximum hunger. A brief report by Jeffrey (1973) showed bran to improve glucose tolerance, but wood cellulose and bagasse had the reverse effect.

The effects of guar cannot be explained only by its viscosity-modifying properties, since in a study of blood alcohol levels following beverage consumption, guar in the diet resulted in increased mean blood alcohol levels in the face of reduced blood insulin and flattened glucose curves. Therefore, although glucose uptake may have been slowed, appearance of alcohol in the blood was enhanced (Tredger *et al.*, 1979).

The influence of dose level and form of administration of guar were investigated using bread and soup test meals (Wolever *et al.*, 1979). The guar when taken in the form of liquid appeared most active in modulating blood glucose and serum insulin. A dose as little as 5 g was reported as effective following the administration of a 45-g carbohydrate meal. Other experiments with guar, however, showed little difference in postprandial glucose, whether guar was ingested as a dry powder or in the form of a pudding. The carbohydrate meal, however, was a typical cereal and milk breakfast. Under those conditions the effect of guar in the diet on serum glucose was shown to be limited (Williams and James, 1979; Williams *et al.*, 1980).

The effects of pectin and wheat bran were further studied using the drug Paracetamol (Brown *et al.*, 1979). Plasma level and urinary excretion of Paracetamol were significantly higher over the first 8 hr after administration in the pectin-fed groups compared to controls, indicating greater small intestinal absorption. Using ^{14}C-labeled polyethylene glycol, no significant differences were noted in the rate of gastric emptying. Bran showed no effect under the same experimental conditions. The influence of dietary fiber in the form of hydrocolloids on intestinal absorption of sugars has also been studied in the rat (Forster and Haas, 1977). The polysaccharides were introduced directly into the small gut in the form of solutions or gels. Neither glucose absorption nor maltose absorption was influenced by the presence of hydrocolloids in the small intestine.

The effect of dietary fiber on the absorption of nutrients in the gut varies according to the source of the fiber and the nutrient in question. The experimental evidence neither confirms nor denies speculations offered to explain the effect of dietary fiber on serum glucose and insulin levels. However, a definition of the mechanism involved still remains obscure.

5.3. Other Hormonal Responses

Dietary fiber has been reported to influence hormones other than insulin (Jenkins *et al.*, 1977; Miranda and Horowitz, 1978; Wahlquist *et al.*, 1979; Albrink *et al.*, 1979). In recent studies measuring some effects of guar, hormonal measurements were reported by Morgan *et al.* (1979). The decrease in postprandial blood glucose and serum insulin for both a mixed meal and a carbohydrate meal were similar to an earlier report (Jenkins *et al.*, 1976).

Gastric inhibitory polypeptide (GIP) was significantly lowered by the inclusion of guar in the diet, but was more effective with a carbohydrate meal. No differences in GIP levels were observed between normal and diabetic subjects, but significant correlation between serum insulin and GIP responses were reported. Guar resulted in separation of blood glucose and serum insulin curves in time. While total plasma glucagonlike immunoreactivity (GLI) rose in all groups, no significant differences in "pancreatic-specific" GLI were evident. The GLI response to guar-supplemented meals, however, was smaller than to control meals. No correlation was observed between peak insulin and total plasma GLI among the groups.

In the clinical studies of Munoz *et al.* (1979), observations with a variety of fiber sources led the investigators to suggest that dietary fiber may improve insulin release or peripheral insulin activity. The insulin–glucose ratio improved, and after chronic feeding of some fibers, the fasting plasma level of glucagon and glucagon responses to an oral glucose load were lower. Anderson and Chen (1979) also speculated that alteration in gut hormone or glucagon secretion may have a role in the response to dietary fiber.

Both the GIP and GLI are capable of stimulating insulin release. They are also subject to feedback control by insulin. It is possible, therefore, that the diminished insulin release in response to guar meals results from change in GIP activity. We speculate that the effect of dietary fiber on insulin response and gastrointestinal motility to influence satiety may be mediated by the gastrointestinal hormones GIP or GLI. More research, however, is required to verify this proposed mechanism of action under a variety of feeding conditions.

6. SUMMARY AND CONCLUSIONS

1. Fiber in the diet can interfere with digestion and absorption in the gut and may result in reduced efficiency of energy utilization.
2. Not all dietary fiber sources are equally effective in this regard and the mechanism of action may depend on the fiber's physicochemical properties and resistance to microbial action in the gut.
3. Dietary fiber may influence satiety, although the exact mechanisms remain unclear.
4. Reduced caloric density and modified blood glucose tolerance and hormonal response associated with high-fiber foods may make fiber a useful adjunct measure in controlling food intake in obese subjects.
5. A more precise definition and description of dietary fiber needs to be developed for use in obesity treatment/management programs.

ACKNOWLEDGMENTS. The authors wish to acknowledge Ms. C. Riley and Mrs. C. Mondello for their valuable assistance in preparation of this manuscript.

REFERENCES

Adolph, E. F., 1947, Urges to eat and drink in rats, *Am. J. Physiol.* **151:**110–125.
Albrink, M. J., 1978, Dietary fiber, plasma insulin, and obesity, *Am. J. Clin. Nutr.* **31:**S277–279.
Albrink, M. J., Newman, T., and Davidson, P. C., 1979, Effect of high- and low-fiber diets on plasma lipids and insulin, *Am. J. Clin. Nutr.* **32:**1486–1491.

Ali, R., Staub, H., Coccodrilli, G., Jr., and Schanbacher, L., 1981, Nutritional significance of dietary fiber: Effect on nutrient bioavailability and selected gastrointestinal functions, *J. Agric. Food Chem.* **29**:465–472.

Anderson, J. W., and Chen, W. L., 1979, Plant fiber. Carbohydrate and lipid metabolism, *Am. J. Clin. Nutr.* **32**:346–363.

Beereboom, J. J., 1979, Low calorie bulking agents, *Crit. Rev. Food Sci. Nutr.* **1979**(May):401–413.

Bolton, R. P., Heaton, K. W., and Burroughs, L. F., 1981, The role of dietary fiber in satiety, glucose and insulin: Studies with fruit and fruit juices, *Am. J. Clin. Nutr.* **34**:211–217.

Brown, R. C., Kelleher, J., Walker, B. E., and Losowsky, M. S., 1979, The effect of wheat bran and pectin on paracetamol absorption in the rat, *Br. J. Nutr.* **41**:455–464.

Bryson, E., Dore, C., and Garrow, J. S., 1979, Wholemeal bread and satiety, *Lancet* **2**:260–261.

Burkitt, D. F., and Trowell, H. C. (eds.), 1975, *Refined Carbohydrate Foods and Disease*, Academic Press, New York.

Campbell, R. G., Hashim, S. A., and Van Itallie, T. B., 1971, Studies of food intake regulation in man. Responses to variation in nutritive density in lean and obese subjects, *N. Engl. J. Med.* **285**:1402–1407.

Cleave, T. L., 1977, Over-consumption, now the most dangerous cause of disease in Westernized countries, *Public Health (London)* **91**:127–131.

Evans, E., and Miller, D. S., 1975, Bulking agents in the treatment of obesity, *Nutr. Metab.* **18**:199–203.

Farrell, D. J., Girle, L., and Arthur, J., 1978, Effects of dietary fiber on the apparent digestibility of major food components and on blood lipids in men, *Aust. J. Exp. Biol. Med. Sci.* **56**:469–479.

Forster, H., and Haas I., 1977, Influence of gums on intestinal absorption, *Nutr. Metab.* **21**(Suppl. I):262–264.

Garrison, M. V., Reel, R. L., Fawley, P., and Breidenstem, C. P., 1978, Comparative digestibility of acid detergent fiber by laboratory albino and wild polynesian rats, *J. Nutr.* **108**:191–195.

Grimes, D. S., and Gordon, C., 1978, Satiety value of wholemeal and white bread, *Lancet* **2**:106.

Haber, G. R., Heaton, K. W., Murphy, D., and Burroughs, L. F., 1977, Depletion and disruption of dietary fiber effects on satiety, plasma-glucose and serum insulin, *Lancet* **2**:679–682.

Heaton, K. W., 1973, Food fiber as an obstacle to energy intake, *Lancet* **2**:1418–1421.

Heaton, K. W., 1979, Fibre, satiety and insulin, a new approach to overnutrition and obesity, in: *Dietary Fibre: Current Developments of Importance to Health* (K. W. Heaton, ed.), Food and Nutrition Press, Westport, Connecticut, pp. 141–149.

James, W. P. T., and Theander, O., 1981 (eds.), *The Analysis of Dietary Fiber in Food*, Marcel Dekker, New York.

Jeffrey, D. B., 1973, The effect of dietary fiber on the response to orally administered glucose, *Proc. Nutr. Soc.* **33**:11A–12A.

Jenkins, D. J. A., Goff, D. V., Leeds, A. R., Alberti, K. G. M. M., Wolever, T. M. S., Gassull, M. A., and Hockaday, T. D. R., 1976, Unabsorbable carbohydrate and diabetic: Decreased post-prandial hyperglycemia, *Lancet* **2**:172–174.

Jenkins, D. J. A., Leeds, A. R., Gassull, M. A. Cochet, B., and Alberti, K. G. M. M., 1977, Decrease in postprandial insulin and glucose concentrations by guar and pectin, *Ann. Intern. Med.* **86**:20–23.

Jenkins, D. J. A., Wolever, T. M. S., Leeds, A. R., Gassull, M. A., Haisman, P., Dilawari, J., Goff, D. V., Metz, G. L., and Alberti, K. G. M. M., 1978, Dietary fibers, fiber analogues, and glucose tolerance importance of viscosity, *Br. Med. J.* **1**:1392–1394.

Kay, R. M., 1978, Food form, postprandial glycemia and satiety, *Am. J. Clin. Nutr.* **31**:738–741.

Keim, K., and Kies, C., 1979, Effects of dietary fiber on nutritional status of weanling mice, *Cereal Chem.* **56**:73–78.

Mickelsen, O., 1980, Prevention and treatment of obesity, in: *Nutrition, Physiology and Obesity*, (R. Schemmel, ed.), CRC Press, Boca Raton, Florida, pp. 167–184.

Mickelsen, O., Makdani, D. P., Cotton, R. H., Titcomb, S. T., Colmey, J. C., and Gatty, R., 1979, Effects of a high fiber bread on weight loss in college-age males, *Am. J. Clin. Nutr.* **32:**1703–1709.

Miranda, P. M., and Horowitz, D. L., 1978, High-fiber diets in the treatment of diabetic mellitus, *Ann. Intern. Med.* **88:**482–486.

Morgan, L. M., Goulder, T. J., Tsiolakis, D., Marks, V., and Alberti, K. G. M. M., 1979, The effect of unabsorbable carbohydrate on gut hormones modification of post-prandial GIP secretion by guar, *Diabetologia* **17:**85–89.

Munoz, J. M., Sandstead, H. H., Jacob, R. A., Johnson, L., and Mako, M. E., 1979, Effects of dietary fiber on glucose tolerance of normal men, *Diabetes* **28:**495–502.

O'Hea, E. K., Aldis, A. E., Skidmore, G. B., Smith, B. H., Stevenson, S., Strongitharm, D., and Stuart, G. C. E., 1974, Effect of dietary dilution on food intake, weight gain, plasma, glucose and insulin and adipose tissue insulin sensitivity in rats, *Comp. Biochem. Physiol.* **48A:**21–26.

Peterson, A. D., and Baumgardt, B. R., 1971a, Food and energy intake of rats fed diets varying in energy concentration and density, *J. Nutr.* **101:**1057–1068.

Peterson, A. D., and Baumgardt, B. R., 1971b, Influence of level of energy demand on the ability of rats to compensate for diet dilution, *J. Nutr.* **101:**1069–1074.

Porikos, K. P., Booth, G., and Van Itallie, T. B., 1977, Effect of covert nutritive dilution on the spontaneous food intake of obese individuals: A pilot study, *Am. J. Clin. Nutr.* **30:**1638–1644.

Porikos, K. P., Hesser-Saulle, M., and Van Itallie, T. B., 1982, Caloric regulation in normal-weight humans on a palatable diet, *Physiol. Behav.* (in press).

Rao, A. V., and Bright-See, E., 1979, Effect of graded amounts of dietary pectin on growth parameters of rats, *Nutr. Rep. Int.* **19:**411–417.

Schaeffer, A. M. C., and Bennick, M., 1979, Nutrition and energy utilization as influenced by dietary fiber, *Fed. Proc.* **38:**767.

Slavin, J. L., and Marlett, J. A., 1980, Effect of refined cellulose on apparent energy, fat and nitrogen digestibilities, *J. Nutr.* **110:**2020–2026.

Smutz, E. R., Hirsch, E., and Jacobs, H. L., 1975, Caloric compensation in hypothatamie obese rats, *Physiol. Behav.* **14:**305–309.

Southgate, D. A. T., 1973, Fiber and other unavailable carbohydrates and their effects on the energy value of the diet, *Proc. Nutr. Soc.* **1973:**131–136.

Southgate, D. A. T., 1978, Has dietary fiber a role in the presentation and treatment of obesity? *Bibl. Nutr. Dieta* **26:**70–76.

Southgate, D. A. T., and Durnin, J. V. G. A., 1970, Calorie conversion factors. An experimental reassessment of the factors used in the calculation of the energy values of human diets, *Br. J. Nutr.* **24:**517–535.

Spiegel, T. S., 1973, Caloric regulation of food intake in man, *J. Comp. Physiol. Psychol.* **84:**24–37.

Staub, H., and Ali, R., 1982, Nutritional and physiological values of gums, in: *Food Hydrocolloids,* (M. Glicksman, ed.), CRC Press, Boca Raton, Florida, (in press).

Sullivan, A. C., Triscari, J., and Comai, K., 1978, Caloric compensatory responses to diets containing either nonabsorbable carbohydrate or lipid by obese and lean Zucker rats, *Am. J. Clin. Nutr.* **31:**S261–S266.

Tredger, J., Wright, J., and Marks, V., 1979, The effect of guar gum on blood alcohol levels following gin and tonic consumption, *Proc. Nutr. Soc.* **38:**70A.

Trowell, H. C., and Burkitt, D. P., 1978, Nutrient deficiencies in man: Dietary fibers, in: *Handbook Series of Nutrition and Food,* Section E, *Nutritional Disorders* (M. Rechegl, ed.), CRC Press, Boca Raton, Florida, pp. 313–330.

Tuomilehto, J., Voutilainen, E., Huttumen, J., Vinni, S., and Homan, K., 1980, Effect of guar gum on body weight and serum lipids in hypercholesterolemic females, *Acta Med. Scand.* **208:**45–48.

Van Itallie, T. B., 1978, Dietary fiber and obesity, *Am. J. Clin. Nutr.* **31:**S43–S52.

Vohra, P., Shariff, G., and Fratzer, F. H., 1979, Growth inhibitory effect of some gums and pectin for *Tribolium castaneum* larvae, chickens and Japanese quail, *Nutr. Rep. Int.* **19:**463–469.

Wahlquist, M. L., Morris, M. J., Littlejohn, G. O., Bond, A., and Jackson, R. V. J., 1979, The effects of dietary fiber on glucose tolerance in healthy males, *Aust. N. Z. J. Med.* **9:**154–158.

Williams, D. R. R., and James, W. P. T., 1979, Fiber and diabetes, *Lancet* **1:**271–272.

Williams, D. R. R., James, W. P. T., and Evans, I. E., 1980, Dietary fibre supplementation of a normal breakfast, *Diabetologia* **18:**379–383.

Wilmshurst, P., and Crawley, J. C. W., 1980, The measurement of gastric transit time in obese subjects using 24Na and the effects of energy content and guar gum on gastric emptying and satiety, *Br. J. Nutr.* **44:**1–6.

Wolever, T. M. S., Jenkins, D. J. A., Nineham, R., and Alberti, K. G. M. M., 1979, Guar gum and reduction of post-prandial glycemia: Effect of incorporation into solid food, liquid food and both, *Br. J. Nutr.* **41:**505–510.

14

Dietary Fiber and Diabetes

JAMES W. ANDERSON

1. INTRODUCTION

The proposed interrelationships between dietary fiber and certain diseases continue to enkindle debate and spark scientific research (Trowell and Burkitt, 1981). Among the fiber–disease connections, diabetes probably is more strongly related to the intake of dietary fiber than any other disease from epidemiological and experimental standpoints. The concept that a generous intake of dietary fiber may protect humans from diabetes has emerged over the last two decades from clinical, experimental, and epidemiological observations (Trowell, 1975). More recently the link between fiber and diabetes has been strengthened by well-documented studies (Jenkins *et al.*, 1976; Kiehm *et al.*, 1976) demonstrating that fiber intake has therapeutic benefits for diabetic patients. I will try to cite all of the full-length reports on the effects of dietary fiber on glucose metabolism of diabetic patients.

Two approaches are used to study the effects of dietary fiber on glucose metabolism of diabetic patients. Jenkins and colleagues (1976) use *fiber-supplemented diets;* they add highly purified products such as guar to the usual diets of their patients. We (Kiehm *et al.*, 1976; Anderson and Ward, 1979) use *high-fiber diets* to treat diabetic patients; we change the diet by substituting commonly available foods rich in dietary fiber for low-fiber items. I will compare the effects of these two types of interventions on the glucose control of lean diabetic patients. Subsequently I will briefly review our experience using high-fiber diets for obese or hyperlipidemic patients and summarize our long-term studies with high-fiber

JAMES W. ANDERSON • Endocrine–Metabolic Section, Veterans Administration Medical Center, and Departments of Medicine and Clinical Nutrition, University of Kentucky College of Medicine, Lexington, Kentucky 40511.

diets. The mechanisms potentially responsible for these changes in glucose metabolism will be reviewed.

2. FIBER-SUPPLEMENTED DIETS

Jenkins and colleagues (1976) clearly documented the favorable effects of fiber supplements on postprandial blood glucose concentrations for lean diabetic patients. When guar and/or pectin was incorporated into a large breakfast, post-meal glycemia was significantly less than when the same diabetic patient ate a fiber-free meal containing similar nutrients on another day. When patients with mild diabetes were given a water-soluble fiber along with an oral glucose test solution, postprandial glucose and insulin concentrations were significantly lower than when these patients were given a similar glucose test solution without fiber on another day (Jenkins *et al.*, 1977a, 1978a). To examine the practical implications of these observations, Jenkins and colleagues (1977b) supplemented the diet with guar for 5–7 days and documented a 50% reduction in glycosuria. For long-term home use, Jenkins and colleagues (1978b) developed a crispbread preparation containing guar gum. When diabetic patients ate between 14 and 26 g/day of guar in this form, they required less insulin and had less glycosuria. This guar crispbread preparation has been well tolerated by their patients and not accompanied by demonstrable adverse effects on mineral metabolism (Jenkins *et al.*, 1980).

Many groups (Gold *et al.*, 1980; Goulder *et al.*, 1978; Hall *et al.*, 1980; Levitt *et al.*, 1980; Monnier *et al.*, 1978; Morgan *et al.*, 1979; Williams *et al.*, 1980) have confirmed the short-term beneficial effects of fiber supplements on glucose or meal tolerance tests. The long-term beneficial effects of guar supplements, specifically, have not been confirmed (Cohen *et al.*, 1980), although others have demonstrated long-term improvements in glucose tolerance or metabolism with fiber-supplemented diets (Bosello *et al.*, 1980; Brodribb and Humphreys, 1976; Kanter *et al.*, 1980; Munoz *et al.*, 1979; Scuro *et al.*, 1978). Miranda and Horwitz (1978) clearly documented the effects of fiber supplements on postprandial glucose concentration and reported that insulin requirements were lowered when patients were receiving fiber-supplemented diets.

3. HIGH-FIBER DIETS FOR LEAN DIABETICS

We have evaluated the effects of high-carbohydrate, high-fiber (HCF) diets for lean diabetic patients (Kiehm *et al.*, 1976; Anderson and Ward, 1979; Anderson, 1980a; Anderson *et al.*, 1980a). Our diets differ from the fiber-supplemented diets described above in that we use commonly available high-fiber foods and have not relied on highly purified products such as guar or utilized items such as cellulose-containing bread. HCF diets are remarkably effective in decreasing insulin requirements of lean diabetic patients. With these diets average insulin doses can be decreased to less than one-third of values on control diets, and we have discontinued insulin therapy in approximately 60% of the lean patients we have treated. These studies indicate that diets generous in complex carbohydrate and fiber while being restricted in fat are very effective in managing insulin-treated diabetic patients.

3.1. HCF Diet Protocol

The HCF diets we have used for initial treatment of hospitalized patients provide 70% of energy as carbohydrate, with three-fourths in the complex form, 19% protein, 11% fat, 50 mg of cholesterol per day, and 35–40 g of dietary fiber/1000 kcal (Anderson and Ward, 1979). Available carbohydrate comes from

TABLE I. Representative 1800-kcal HCF Diet for Hospital Use[a]

Breakfast	Orange juice	1/3 cup
	Oats, whole, dry	1/4 cup
	Milk, skim	1/2 cup
	Bran muffin	1
	Margarine	1 tsp
Snack	Corn bread muffin	1
	Tomato, sliced	1/2
Noon meal	Coleslaw	1 1/3 cup
	Kidney beans	1/2 cup
	Brown rice	1 cup
	Cauliflower	1/2 cup
	Whole wheat bread	2 slices
	Margarine	1 tsp
Snack	Graham crackers	4 squares
	Peaches	1/2 cup
Evening meal	Tossed salad	2 cups
	Chicken	1 oz
	Potatoes, boiled	1 cup
	Margarine	1 tsp
	Green beans	3/4 cup
	Peas	3/4 cup
	Rye bread	2 slices
Snack	Grapenuts	3 tbsp
	Applesauce	1/2 cup

[a] Beverages such as tea, coffee, and diet cola can be used with meals and snacks.

grain products (50% of total), vegetables and fruits (48%), and skim milk (2%). Protein is provided by vegetables and fruits (50%), grain products (36%), and milk and lean meat (14%). The fat was derived from grain products (60%), vegetables and fruits (28%), and milk and meat (12%). The dietary fiber measured by our modification (Chen and Anderson, 1981) of the Southgate technique (1976) originated from vegetables (51%), grain products (40%), and fruits (9%). A representative diet is provided in Table I. Our control diets provide 43% of energy as carbohydrate, with half in the complex form, 19% protein, 38% fat, 450 mg cholesterol/day, and 8–12 g of dietary fiber/1000 kcal.

We admit patients to the hospital and usually feed them control diets for 7–10 days to stabilize their diabetic control. We then switch them to weight-maintaining HCF diets and attempt to maintain the fasting and 3-hr postprandial plasma glucose concentration at approximately 150 mg/dl and urine glucose excretion less than 10 g/day. Utilizing a standardized protocol, we decrease insulin doses by two units every 1–2 days if plasma glucose values are less than 160 mg/dl. The response of a representative patient to this protocol is shown in Fig. 1.

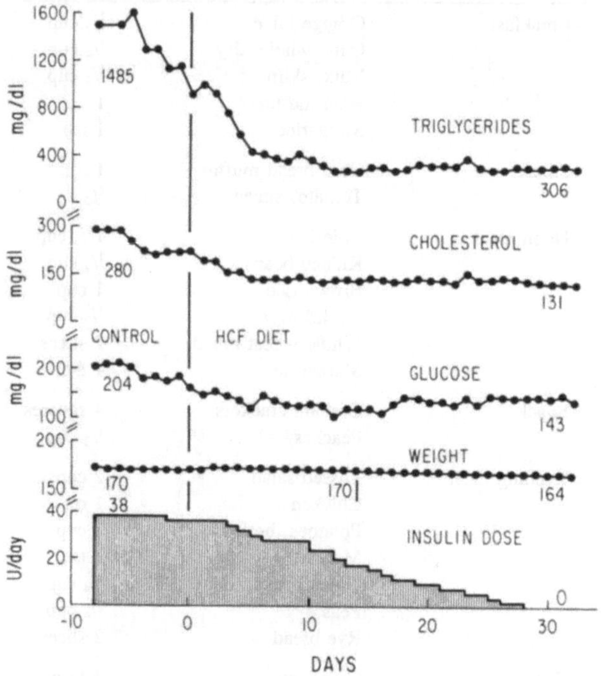

FIGURE 1. Response of lean diabetic man to HCF diet. Fasting plasma concentrations of glucose, cholesterol, and triglycerides are given. After day 14 on the HCF diet he increased his level of exercise. (Reprinted from Anderson, 1981, with permission.)

TABLE II. Insulin Doses of Lean Patients with Non-Insulin-Dependent (NIDDM) or Insulin-Dependent (IDDM) Types of Diabetes Mellitus on Control and HCF Diets

Group	Insulin dose range, U/day	No.	Insulin dose,[a] U/day: Control diet	HCF diet
NIDDM	15–20	12	17 ± 2	0
	25–38	10	31 ± 2	0
IDDM	22–32	10	28 ± 2	15 ± 3
	40–55	6	44 ± 2	31 ± 4

[a] Mean ± SEM.

3.2. Short-Term Effects of HCF Diets

The changes in insulin requirements of 38 lean patients treated with control and HCF diets are presented in Table II. These patients can be classified as insulin-dependent diabetes mellitus (IDDM) or non-insulin-dependent (NIDDM) types. In the NIDDM group, we have discontinued insulin therapy in all 12 patients receiving insulin doses of 20 units or less per day. These patients have done quite well without subsequent insulin therapy for up to 5 years. We also discontinued insulin therapy in ten additional patients of the NIDDM group who were receiving between 25 and 38 units/day. This group subsequently has done equally well and has not required insulin therapy if they closely followed the diet after discharge from the hospital.

Of the 16 patients in the IDDM category, ten have had insulin requirements of 22–32 units and could not be distinguished historically from the ten patients who were receiving 25–38 units of insulin/day on the control diet. Nevertheless, on the HCF diet the ten patients of the IDDM group had a reduction of insulin requirements to 46% of values on the control diet. We have treated only six patients who have the classical features of the juvenile-onset diabetes with proneness to ketoacidosis (typical for IDDM). With HCF diets, average insulin doses were 20–35% lower than requirements on control diets; half of these patients were distinctly easier to manage on HCF diets than control diets and had fewer hypoglycemic episodes. Our studies consistently document that diets generous in both complex carbohydrate and dietary fiber significantly lower insulin requirements of adult diabetic patients.

3.3. Long-Term Effects of High-Fiber Diets

For home use we instruct patients in high-fiber maintenance diets. These diets are more practical for long-term use and some patients have used these diets for over 6 years. These high-fiber diets provide 55–60% of energy from carbohydrate, 20% protein, 20–25% fat, 75–200 mg cholesterol/day, and 50 g dietary fiber/day (approximately 25 g/1000 kcal). The long-term effects of these diets on glucose and lipid metabolism are provided elsewhere (Anderson and Ward, 1978; Anderson *et al.*, 1980b). Dietary compliance has been fair to good in almost 90% of patients (Anderson and Sieling, unpublished observations). Basically, when patients closely follow high-fiber maintenance diets after discharge from the hospital, their glucose metabolism and insulin requirements remain stable. If we discontinue insulin in the hospital, none of our patients who closely follow the high-fiber diet subsequently require insulin therapy. When patients resume conventional diabetic diets their insulin requirements return to doses used before hospitalization. Average serum cholesterol concentrations on high-fiber maintenance diets are approximately 180 mg/dl and average high-density lipoprotein cholesterol concentrations are approximately 10 mg/dl higher than values before initial hospitalization. Fasting serum triglyceride concentrations appear to be the most sensitive indicator of dietary compliance. When hypertriglyceridemic patients (see Fig. 1, for example) closely adhere to high-fiber diets their serum triglyceride concentrations remain at satisfactory levels; deviations from the diet are heralded by recurrence of hypertriglyceridemia. Our overall experience indicates that these diets are practical, palatable, and safe for long-term therapy of adults with diabetes.

3.4. Confirmatory Studies

Simpson and colleagues (1981) also have noted that high-carbohydrate, high-fiber diets improve the control of insulin-treated diabetic patients and lower insulin requirements slightly. They documented improvement in glycemic control in 18 non-insulin-dependent and 9 insulin-dependent diabetics who were fed a high-carbohydrate, high-fiber diet for 6 weeks in comparison to a low-carbohydrate diet for 6 weeks. The high-carbohydrate, high-fiber diet provided 61% of energy as carbohydrate, 21% protein, and 18% fat with 97 g of dietary fiber/day. The control diet provided 40% of energy as carbohydrate, 21% protein, 39% fat, and 18 g of dietary fiber/day. The effects of these diets on blood glucose, urine glucose, and glycohemoglobin concentrations are shown in Fig. 2. They observed significant reductions in blood glucose concentrations in both groups of patients. Glycohemoglobin concentrations were significantly lower in the non-insulin-dependent group and average insulin doses of the insulin-treated group declined from 50 to 47 units/day on the high-carbohydrate, high-fiber diet.

These studies of Simpson and colleagues (1981) and their previous reports (Simpson *et al.*, 1979a,b) differ from ours in that their diets were lower in carbohydrate than the HCF diets (61% versus 70%, respectively) and, except for one day of intensive metabolic study on a hospital ward, their diets were fed to outpatients. These differences in protocol may account for their failure to

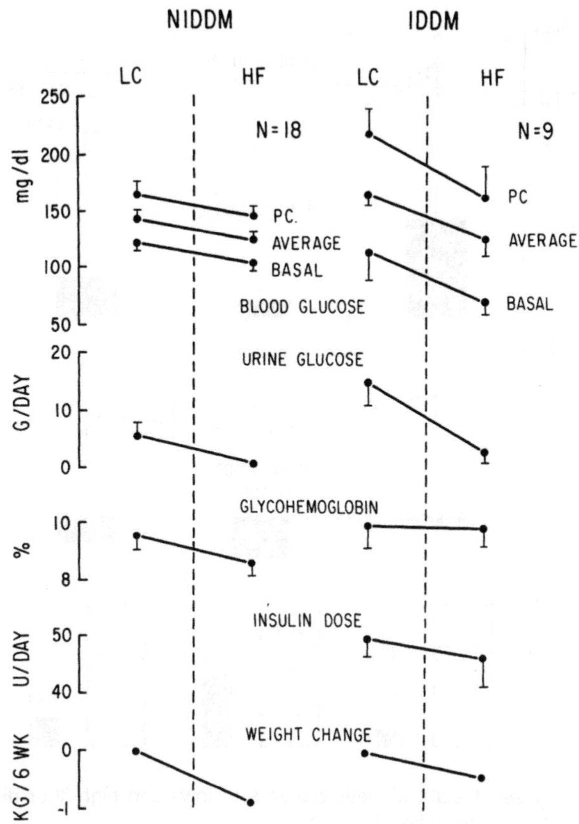

FIGURE 2. Response of patients with NIDDM or IDDM types of diabetes to control (LC) and high-carbohydrate, high-fiber (HF) diets. (Data from Simpson *et al.*, 1981, as described in text.)

substantially lower insulin requirements with the high-carbohydrate, high-fiber diets they employed, whereas we see significant reductions in insulin requirements utilizing HCF diets in the hospital (Table II).

Rivellese and colleagues (1980) clearly documented that high-fiber diets improve blood glucose control of diabetic patients. On a metabolic ward they fed eight patients three consecutive diets of 10 days duration each. The first diet was low in dietary fiber (16 g/day), while the second diet was high in fiber (54 g/day); both diets provided 53% of energy as carbohydrate, 17% protein, and 30% fat. The third diet also was low in fiber (20 g/day) and provided slightly less carbohydrate (see Fig. 3). With the high-fiber diet, fasting and average blood glucose concentrations were significantly lower and urine glucose excretion also

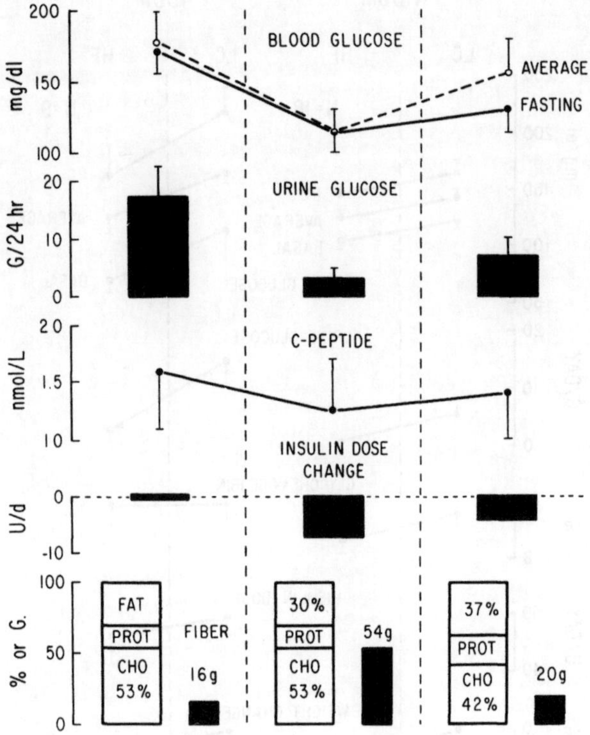

FIGURE 3. Response of eight diabetic patients to low- and high-fiber diets. (Data from Rivellese *et al.*, 1980, as described in text.)

was significantly lower than values on the first diet. Glucose metabolism was improved with the high-fiber diet despite lower exogenous insulin doses and decreased secretion of endogenous insulin as measured by C-peptide concentrations. These observations indicate that the high-fiber diets were accompanied by substantially improved sensitivity to the action of exogenous or endogenous insulin. When these patients were subsequently placed on a low-fiber diet, the improvements observed with the high-fiber diet were reversed and average glucose concentrations were similar to values observed on the initial low-fiber diet.

Kay and colleagues (1981) also documented that high-fiber diets consisting of commonly available foods are accompanied by improvement in carbohydrate metabolism of elderly diabetic patients. After 14 days on a high-fiber diet providing 30 g of dietary fiber/day, the response to a high-fiber test meal was substantially better than the response to a low-fiber test meal after two weeks on a low-fiber diet providing 10 g of dietary fiber/day. Plasma insulin and gastric inhibitory polypeptide responses also were significantly lower after the high-fiber test meal than after the low-fiber test meal.

4. HIGH-FIBER DIETS FOR OBESE PATIENTS

Since 60–90% of maturity-onset diabetics (type II or NIDDM) are obese, we examined the effectiveness and acceptability of high-fiber diets for patients of this type. We used two approaches. First we evaluated the response of obese patients receiving large doses of insulin to high-fiber diets that were either moderately restricted or severely restricted in energy. Second, we assessed the satiety and hunger of obese diabetic patients receiving 800-kcal diets that were either low or high in dietary fiber. We have reported some of these results previously (Anderson and Sieling, 1980).

4.1. Moderate versus Severe Caloric Restriction

Twelve obese patients who exceeded their desirable body weight by more than 20% and were receiving 40 units or more of insulin per day were studied. After a control period of 5–7 days on a weight-maintaining diet, patients were allocated to high-fiber diets (40 g of dietary fiber/1000 kcal) that provided an average of either 7.3 or 3.3 kcal/lb for a 2-week period. After moderate caloric restriction (7.3 kcal/lb), patients lost an average of 2 lb per week and average insulin doses dropped from 49 to 29 units/day (Table III). With severe caloric

TABLE III. Response of Obese Diabetic Patients to High-Fiber, Calorically Restricted Diets[a]

Measure	Restricted diets			
	Moderate		Severe	
	Mean	Range	Mean	Range
Body weight, % desirable	137	123–163	154	123–197
Energy provided, kcal/lb	7.3	6.2–8.3	3.3	2.2–4.3
Weight loss, lb/day	0.32	0.21–0.38	0.84	0.37–1.6
Insulin dose, U/day				
Initial	49	40–70	58	40–82
Final	29	0–45	4	0–8
Plasma glucose, mg/dl				
Initial	199	139–239	229	86–290
Final	179	142–249	150	90–196

[a] Patients were fed control diet (initial values) followed by moderately (N = 4) or severely (N = 8) restricted diets (final values) for 14 days.

restriction patients lost an average of 6 lb/week and average insulin doses dropped from 58 to 4 units/day. Whereas insulin could be discontinued in only one patient receiving 7.3 kcal/lb, insulin was discontinued in five of eight patients receiving 3.3 kcal/lb. This study suggests that high-fiber diets that are severely restricted in available energy are much more effective in reducing insulin requirements than are conventional diets providing approximately 6–8 kcal/lb or 1000–1600 kcal/day.

4.2. High-Fiber versus Low-Fiber Diets

During the preceding study we were surprised that our obese diabetic patients did not complain of hunger on high-fiber diets providing only 800 kcal/day. To examine the effects of fiber content on hunger and satiety we are feeding obese patients with both high- and low-fiber diets for 10 days in an alternating sequence. We are assessing satiety and hunger before and after each of their main meals

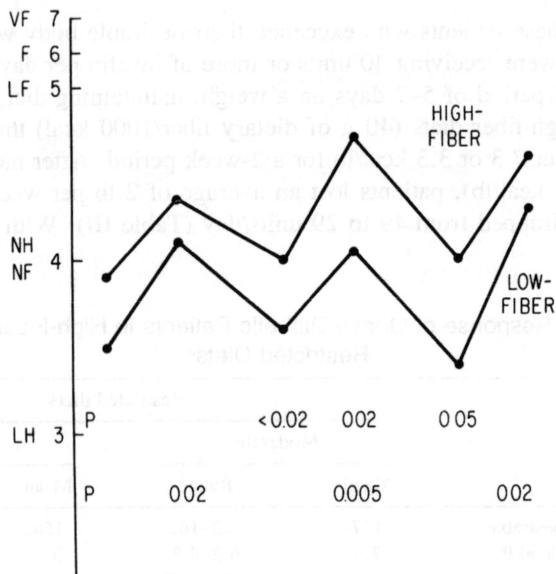

FIGURE 4. Hunger and satiety rating of obese patient on 800-kcal diets either low in fiber or high in fiber. Values represent mean of measurements from 10 days on each diet. *p* Values represent significant differences reported using verbal scale (above) or linear scale (below), as described in text. VF, very full; F, full; LF, little full; NH–NF, not hungry, not full; LH, little hungry; H, hungry; VH, very hungry.

(6 times per day) using a verbal and linear scale. The verbal scale is printed vertically with terms ranging from very full (VF) at the top, full, a little full, not hungry–not full, a little hungry, hungry, to very hungry (VH) at the bottom. The patient circles the term that most closely describes how he feels. Also we give patients a linear scale which has a vertical line with the terms very full at the top and very hungry at the bottom. The patient draws a slash across the line in the region that indicates how he feels. Preliminary results with four patients who have been treated for 10 days with the low-fiber and the high-fiber, 800-kcal diets in an alternating sequence indicate that the high-fiber diet is consistently accompanied by a significantly higher satiety rating than is the low-fiber, 800-kcal diet. The response of one representative patient is indicated in Fig. 4. These studies document that high-fiber, calorically restricted diets are accompanied by significantly greater satiety and less hunger than are associated with low-fiber, calorically restricted diets.

5. SERUM LIPID RESPONSES

Although we developed HCF diets specifically to lower insulin requirements and plasma glucose concentrations, an important by-product is the reduction in serum lipid concentrations. With HCF diets, average serum cholesterol concentrations of lean diabetic patients are 32% lower than values on control diets (Anderson *et al.*, 1980a). Three features of these diets contribute to the cholesterol-lowering effects: first, the diets are restricted in both total and saturated fats; second, the diets provide only 50 mg of cholesterol per day; and third, the dietary fiber content is almost threefold higher than that of the control diets.

To examine the specific hypocholesterolemic effects of dietary fiber we studied the effects of diets supplemented with oat bran, a rich source of water-soluble fiber (Chen and Anderson, 1979). When we fed oat-bran-supplemented diets which were identical in carbohydrate, protein, fat, and cholesterol content to control diets (Kirby *et al.*, 1981) we observed that serum cholesterol concentrations were 15% lower than values on the control diet. Thus we feel that the dietary-fiber content of the HCF diets has an independent role in lowering serum cholesterol concentrations. Since our diabetic patients now have average serum cholesterol concentrations below 200 mg/dl, we feel the risk for ischemic heart disease is lower than when their average values were above 200 mg/dl.

Although these HCF diets provide 70% of energy as carbohydrate, average fasting serum triglyceride concentrations are 14% lower than values on control diets (Anderson *et al.*, 1980). We seldom observe "carbohydrate-induced lipemia" with these diets, which are generous in both carbohydrate and fiber. To determine if dietary-fiber intake protected our patients from developing hypertriglyceridemia we evaluated 70% carbohydrate diets which were either high in fiber (HCF diets with 64 g fiber/day) or low in fiber (HCLF diets with 23 g fiber/day). Eleven patients were treated with both HCF and HCLF diets for 10 days in an alternating sequence. On HCF diets average serum triglyceride concentrations were slightly lower than values on control diets, while serum triglyceride values increased in every patient on the HCLF diet and average values were 28% higher ($p < 0.01$) than control values. Because of the triglyceride-lowering effects of HCF diets, we have used them effectively to treat diabetic patients with moderate to severe hypertriglyceridemia (Anderson, 1980b).

6. PROPOSED MECHANISM

Leeds (Chapter 6) has reviewed the effects of dietary fiber on glucose metabolism and I will only briefly discuss the possible mechanisms by which dietary fiber improves diabetic control and lowers insulin requirements. Delayed absorption of glucose undoubtedly contributes importantly to the reduction in postprandial glycemia seen after high-fiber meals. The prolonged gastric emptying time (Holt *et al.*, 1979; Wilmshurst and Crawley, 1980) appears to make a major contribution to the slow absorption of carbohydrate. Fiber also may slow the absorption of carbohydrate by entrapment of carbohydrate or digestive enzymes, by altered digestive enzyme activity (Schneemann, Chapter 7), or by influencing the transport across the unstirred water layer (Elsenhans *et al.*, 1980). Fiber intake does not appear to cause malabsorption of simple carbohydrate (Jenkins *et al.*, 1978a), but may accentuate the delivery of certain forms of starch to the colon (Anderson *et al.*, 1981).

Hormonal changes after fiber intake are of considerable interest. Decreases in apparent insulin secretion despite improvement in apparent glucose disposal suggests that fiber enhances sensitivity to insulin (Jenkins *et al.*, 1977a). Reductions in apparent secretion of glucagon (Miranda and Horwitz, 1978), a counterregulatory hormone which stimulates hepatic glucose output, would act to enhance glucose disposal. Reductions in serum gastric inhibitory polypeptide concentrations (Morgan *et al.*, 1979; Kay *et al.*, 1981) also may enhance hepatic disposal of glucose since gut hormones in this family may share some of the anti-insulin actions of glucagon on hepatic glucose metabolism.

The increase in tissue sensitivity to insulin associated with high-fiber diets is further substantiated by the observed increase in insulin binding to circulating monocytes and the increased number of insulin receptors on these cells. With HCF diets we see an 80% increase in the number of receptor sites on circulating monocytes as well as increased affinity of these receptors for insulin for adult diabetic patients (Anderson, 1979). Pedersen *et al.* (1980) observed a 25% increase in the number of insulin receptors on circulating monocytes of juvenile diabetic patients. The insulin binding and number of insulin receptors of insulin-sensitive tissue such as fat cells should be measured to further evaluate these observations. Other procedures designed to assess insulin sensitivity of the entire organism could be used to quantitate changes in insulin sensitivity on low-fiber and on high-fiber diets.

We recently have observed that fermentation products of fiber alter hepatic production of glucose (Anderson and Bridges, 1980). In the colon most fibers are fermented to short-chain fatty acids such as acetate, butyrate, and propionate. Whereas long-chain fatty acids antagonize glucose metabolism in the liver, propionate stimulates glucose utilization. We have measured the effects of fatty acids on rates of glycolysis and gluconeogenesis in hepatocytes isolated from the rat. For glycolysis we used glucose as a substrate and measured lactate production, while for gluconeogenesis we used lactate as well as fructose for substrates and measured glucose production. Long-chain fatty acids such as oleate inhibited glycolysis by approximately 16%, but propionate produced a 61% increase in glycolytic rates. Glycolysis was stimulated by propionate concentrations as low as 0.25 mM and maximal effects were seen with 1–2 mM. While oleate stimulated gluconeogenesis by 74%, propionate inhibited gluconeogenesis from both lactate and fructose by approximately 28%. Propionate significantly inhibited gluconeogenesis at concentrations of 0.25 mM and had maximal effects at concentrations of 1–2 mM. Acetate, the major short-chain fatty acid metabolite of carbohydrate fermentation, however, had actions similar to oleate. Thus, while propionate clearly has the potential to enhance glucose metabolism in the liver, we cannot, at present, assess its physiologic significance.

HCF diets are higher in carbohydrate and fiber than control diets and are much lower in fat content. The fiber content of these diets has a major impact on cholesterol and triglyceride metabolism but does not appear to be the major dietary alteration that reduces insulin requirements (Anderson, 1979). High-carbohydrate, low-fat diets that are low in fiber are accompanied by substantial reductions in apparent insulin requirement and, at present, we feel that fiber plays only a modest role in the improvement in diabetic control and the reduction in insulin doses which we observe (Anderson *et al.*, 1980a). The long-term studies of Jenkins *et al.* (1980) as well as the shorter-term studies of Rivellese *et al.* (1980) indicate that fiber-supplemented diets or high-fiber diets are accompanied by significant improvements in glucose metabolism and reductions in insulin requirements. However, the magnitude of the response seen by these workers is not as great as that which we observed with the HCF diets. Furthermore, the studies of Simpson and colleagues (1979a,b, 1981) indicate that a very-high-fiber diet that is quite generous in carbohydrate produces only minimal reduction in insulin requirements in an outpatient setting. In my assessment, the combination of all three factors—high-carbohydrate, high-fiber, and low-fat contents of the diet—is required to achieve the major changes in insulin requirements which we have observed. Additional studies obviously are required to delineate further the specific effects of alterations in dietary fiber, carbohydrate, and fat on glucose metabolism of diabetic patients. Nonetheless, our studies with HCF diets clearly indicate that vigorous dietary intervention has the potential of producing major improvements in diabetic control and major reductions in insulin requirements. Further work is required to translate these metabolic-ward studies into practical therapeutic approaches for diabetic patients in the free-living state.

7. CONCLUSIONS

Epidemiological observations have linked fiber-depleted diets to development of diabetes and high-fiber diets to the prevention of diabetes (Trowell, 1975). Recently, clinical studies have documented that fiber-supplemented diets lower postprandial glucose concentrations and decrease glycosuria in diabetic patients. High-fiber diets lower insulin requirements of lean diabetic patients and allow insulin to be discontinued in certain patients. We have examined the effects of high-carbohydrate, high-fiber (HCF) diets on patients with insulin-dependent diabetes, non-insulin-dependent diabetes, obesity and diabetes, hypercholesterolemia, and hypertriglyceridemia.

HCF diets lowered average insulin requirements of 38 lean diabetic patients by 67% and allowed insulin to be discontinued in 58%. With HCF diets we were able to discontinue insulin therapy in 22 lean patients taking between 15 and 38 units of insulin per day. When 16 insulin-dependent diabetic patients were treated with HCF diets, average insulin requirements dropped by 38% (from 34 to 21 units/day). Fasting plasma glucose values, postprandial glucose values, and urine glucose excretion were lower on HCF diets despite much lower insulin doses than were values on control diets.

High-fiber, weight-reducing diets are very effective in treating obese diabetic patients. The high satiety value of these high-fiber diets decreases the hunger associated with these weight-reducing diets. We have treated eight obese diabetic patients with 800-kcal high-fiber diets. On control diets these patients were taking an average of 58 units of insulin/day. With the high-fiber, weight-reducing diet insulin doses could be reduced to an average of 4 units/day and, in fact, insulin could be discontinued in five of these patients. Preliminary studies indicate that high-fiber, weight-reducing diets are acceptable for long-term use because of the high degree of satiety and the resultant high degree of patient acceptance of these diets.

With HCF diets fasting serum cholesterol concentrations are 32% lower and triglyceride values 14% lower than values on the control diet. Other studies indicate that dietary-fiber supplements or high-fiber diets will lower fasting serum cholesterol and triglyceride concentrations. Our preliminary studies indicate that high-fiber diets are useful in the long-term management of patients with either hypercholesterolemia or hypertriglyceridemia. Because of their beneficial effects on glucose metabolism and serum lipid concentrations, high-fiber diets have considerable promise for the management of many adults with diabetes mellitus.

ACKNOWLEDGMENTS. The assistance of Linda Story, R.D., and Sue Satterly is appreciated. This work was supported by a grant (AM20889) from the National Institute of Metabolic and Digestive Diseases and by the Veterans Administration.

REFERENCES

Anderson, I. J., Levine, A. S., and Levitt, M. D., 1981, Incomplete absorption of the carbohydrate in all-purpose wheat flour, *N. Engl. J. Med.* **304:**892–992.

Anderson, J. W., 1979, High carbohydrate, high fiber diets for patients with diabetes, in: *Treatment of Early Diabetes* (R. A. Camerini-Davalos and B. Hanover, eds.), Plenum Press, New York, pp. 263–273.

Anderson, J. W., 1980a, High-fiber diets in diabetes and hypertriglyceridemia, *Can. Med. Assoc. J.* **123:**975–979.

Anderson, J. W., 1980b, Dietary fiber in diabetes, in: *Medical Aspects of Dietary Fiber* (G. A. Spiller and R. Kay, eds.), Plenum Medical, New York, pp. 193–221.

Anderson, J. W., 1981, Plant fiber treatment for metabolic diseases, in: *Special Topics in Endocrinology and Metabolism* (M. P. Cohen and P. P. Foa, eds.), Alan R. Liss, New York.

Anderson, J. W., and Bridges, S., 1980, Short chain fatty acids alter rates of glycolysis and gluconeogenesis in isolated rat hepatocytes, *Clin. Res.* **28:**776A.

Anderson, J. W., and Sieling, B., 1980, High-fiber diets for obese diabetic patients, *J. Obes. Bar. Med.* **9:**109–113.

Anderson, J. W., and Ward, K., 1978, Long term effects of high carbohydrate, high fiber diets on glucose and lipid metabolism: A preliminary report on patients with diabetes, *Diabetes Care* **1:**77–82.

Anderson, J. W., and Ward, K., 1979, High carbohydrate, high fiber diets for insulin-treated men with diabetes mellitus, *Am. J. Clin. Nutr.* **32:**2312–2321.

Anderson, J. W., Chen, W. L., and Sieling, B., 1980a, Hypolipidemic effects of high carbohydrate, high fiber diets, *Metabolism* **29:**551–558.

Anderson, J. W., Ferguson, S. K., Karonous, D., O'Malley, L., Sieling, B., and Chen, W. L., 1980b, Mineral and vitamin status on high-fiber diets: Long term studies of diabetic patients, *Diabetes Care* **3:**38–40.

Bosello, O., Ostuzzi, R., Armellini, F., Micciolo, R., and Scuro, L. A., 1980, Glucose tolerance and blood lipids in bran-fed patients with impaired glucose tolerance, *Diabetes Care* **3:**46–49.

Brodribb, A. J. M., and Humphreys, D. M., 1976, Diverticular disease: Three studies, *Br. Med. J.* **2:**424–430.

Chen, W. L., and Anderson, J. W., 1979, Effects of plant fiber in decreasing plasma total cholesterol and increasing high density lipoprotein cholesterol, *Proc. Soc. Exp. Biol. Med.* **162:** 310–313.

Chen, W.-J. L., and Anderson, J. W., 1981, Soluble and insoluble plant fiber in selected cereals and vegetables, *Am. J. Clin. Nutr.* **34:**1077–1082.

Cohen, M., Leong, V. W., Salmon, E., and Martin, F. I. R., 1980, Role of guar and dietary fibre in the management of diabetes mellitus, *Med. J. Aust.* **1:**59–61.

Elsenhans, B., Sufke, U., Blume, R., and Caspary, W. F., 1980, The influence of carbohydrate gelling agents on rat intestinal transport of monosaccharides and neutral amino acids *in vitro*, *Clin. Sci.* **59**:373–380.

Gold, L. A., McCourt, J. P., and Merimee, T. J., 1980, Pectin: An examination in normal subjects, *Diabetes Care* **3**:50–52.

Goulder, T. J., Alberti, K. G. M. M., and Jenkins, D. A., 1978, Effects of added fiber on the glucose and metabolic response to a meal in normal and diabetic subjects, *Diabetes Care* **1**:351–355.

Hall, S. E. H., Bolton, T. M., and Hetenyi, G., Jr., 1980, The effect of bran on glucose kinetics and plasma insulin in non-insulin-dependent diabetes mellitus, *Diabetes Care* **3**:520–525.

Holt, S., Heading, R. C., Carter, D. C., Prescott, L. F., and Tothill, P., 1979, Effect of gel fibre on gastric emptying and absorption of glucose and paracetamol, *Lancet* **1**:636–639.

Jenkins, D. J. A., Leeds, A. R., Gassull, M. A., Wolever, T. M. S., Goff, D. V., Alberti, K. G. M. M., and Hockaday, T. D. R. 1976, Unabsorbable carbohydrates and diabetes: Decreased post-prandial hyperglycaemia, *Lancet* **2**:172–174.

Jenkins, D. J. A., Leeds, A. R., Gassull, M. A., Cochet, B., and Alberti, K. G. M. M., 1977a, Decrease in postprandial insulin and glucose concentrations by guar and pectin, *Ann. Intern. Med.* **86**:20–23.

Jenkins, D. J. A., Wolever, T. M. S., Hockaday, T. D. R., Leeds, A. R., Howarth, R., Bacon, S., Apling, E. C., and Dilawari, J., 1977b, Treatment of diabetes with guar gum, *Lancet* **2**:779–780.

Jenkins, D. J. A., Wolever, T. M. S., Leeds, A. R., Gassull, M. A., Haisman, P., Dilawari, J., Goff, D. V., Metz, G. K., and Alberti, K. G. M. M., 1978a, Dietary fibres, fibre analogues and glucose tolerance: Importance of viscosity, *Br. Med. J.* **1**:1392–1394.

Jenkins, D. J. A., Wolever, R. M. S., Nineham, R., Taylor, R., Metz, G. L., Bacon, S., and Hockaday, T. D. R., 1978b, Guar crispbread in the diabetic diet, *Br. Med. J.* **2**:1744–1746.

Jenkins, D. J. A., Wolever, T. M. S., Taylor, R. H., Reynolds, D., Nineham, R., and Hockaday, T. D. R., 1980, Diabetic glucose control, lipids, and trace elements on long-term guar, *Br. Med. J.* **1**:1353–1354.

Kanter, Y., Eitan, N., Brook, G., and Barzilai, D., 1980, Improved glucose tolerance and insulin response in obese and diabetic patients on a fiber-enriched diet, *Isr. J. Med. Sci.* **16**:1–6.

Kay, R. M., Grobin, W., and Tract, N. S., 1981, Diets rich in natural fibre improve carbohydrate tolerance in maturity-onset, non-insulin dependent diabetics, *Diabetologia* **20**:18–21.

Kiehm, T. G., Anderson, J. W., and Ward, K., 1976, Beneficial effects of a high carbohydrate, high fiber diet on hyperglycemic diabetic men, *Am. J. Clin. Nutr.* **29**:895–899.

Kirby, R. W., Anderson, J. W., Sieling, B., Rees, E. D., Chen, W. L., Miller, R. E., and Kay, R. M., 1981, Oat-bran intake selectively lowers serum low-density lipoprotein cholesterol concentrations, *Am. J. Clin. Nutr.* **34**:824–829.

Levitt, N. A., Vinik, A. I., Sive, A. A., Child, P. T., and Jackson, W. P. U., 1980, The effect of dietary fiber on glucose and hormone responses to a mixed meal in normal subjects and in diabetic subjects with and without autonomic neuropathy, *Diabetes Care* **3**:515–519.

Miranda, P. M., and Horwitz, D. L., 1978, High-fiber diets in the treatment of diabetes mellitus, *Ann. Intern. Med.* **88**:482–486.

Monnier, L., Pham, T. C., Aguirre, L., Orsetti, A., and Mirouze, J., 1978, Influence of indigestible fibers on glucose tolerance, *Diabetes Care* **1**:83–88.

Monnier, L. H., Blotman, M. J., Colette, C., Monnier, M. P., and Mirouze, J., 1981, Effects of dietary fibre supplementation in stable and labile insulin-dependent diabetics, *Diabetologia* **20**:12–17.

Morgan, L. M., Goulder, T. J., Tsiolakis, D., Marks, V., and Alberti, K. G. M. M., 1979, The effect of unabsorbable carbohydrate on gut hormones, *Diabetologia* **17**:85–89.

Munoz, J. M., Sandstead, H. H., and Jacob, R. A., 1979, Effects of dietary fiber on glucose tolerance of normal men, *Diabetes* **28**:496–502.

Pedersen, O., Hgollund, E., Lindskov, H. O., and Sorensen, N. S., 1980, Increased insulin receptors on monocytes from insulin-dependent diabetes after a high-starch, high-fiber diet, *Diabetologia* **19**:306.

Rivellese, A., Giacco, A., Genovese, S., Riccardi, G., Pacioni, D., Mattioli, P. L., and Mancini, M., 1980, Effect of dietary fibre on glucose control and serum lipoproteins in diabetic patients, *Lancet* **1**:447–450.

Scuro, L. A., Bosello, O., and Ostuzzi, R., 1978, Effect of a high fibre diet upon serum lipids and glucose tolerance, *Ital. J. Gastroenterol.* **10**:56–57.

Simpson, R. W., Mann, J. I., Eaton, J., Carter, R. D., and Hockaday, T. D. R., 1979a, High-carbohydrate diets and insulin-dependent diabetics, *Br. Med. J.* **2**:523–525.

Simpson, R. W., Mann, J. I., Eaton, J., Moore, R. A., Carter, R., and Hockaday, T. D. R., 1979b, Improved glucose control in maturity-onset diabetes treated with high-carbohydrate-modified fat diet, *Br. Med. J.* **1**:1753–1756.

Simpson, H. C. R., Lousley, S., Geekie, M., Simpson, R. W., Carter, R. D., Hockaday, T. D. R., and Mann, J. I., 1981, A high carbohydrate leguminous fibre diet improves all aspects of diabetic control, *Lancet* **1**:1–5.

Southgate, D. A. T., 1976, *Determination of Food Carbohydrate*, Applied Science Publishers, London.

Trowell, H. C., 1975, Dietary-fiber hypothesis of the etiology of diabetes mellitus, *Diabetes* **24**:762–765.

Trowell, H. C., and Burkitt, D. P., 1981, *Western Diseases: Their Emergence and Prevention*, Edward Arnold, London.

Williams, D. R. R., James, W. P. T., and Evans, I. E., 1980, Dietary fibre supplementation of a "normal" breakfast administered to diabetics, *Diabetologia* **18**:379–383.

Wilmshurst, P., and Crawley, J. C. W., 1980, The measurement of gastric transit time in obese subjects using ^{24}Na and the effects of energy content and guar gum on gastric emptying and satiety. *Br. J. Nutr.* **44**:1–6.

15

Effect of Dietary Fiber on Lipids and Glucose Tolerance of Healthy Young Men

MARGARET J. ALBRINK and IRMA H. ULLRICH

1. INTRODUCTION

Epidemiologic evidence supports a protective effect of dietary fiber against diabetes (Trowell 1974), obesity (Trowell 1975), and coronary artery disease (Trowell 1972). However, the diet of peoples consuming large amounts of fiber is likely to differ in other important respects, i.e., to be high in carbohydrate, low in fat and cholesterol, and low in sugars. The effect of dietary fiber must be examined in the context of these other variables. Since other chapters in this book deal with the effect of fiber on diabetes (Anderson, Chapter 14) and lipids (Story and Kelley, Chapter 20), these will be only briefly discussed here. The chief emphasis will be on the effect of diet on glucose tolerance of normal young men. Young women appear to be immune to some of the effects of very high carbohydrate diets (Reiser *et al.*, 1979a). Our own studies have therefore been confined to young men.

2. EFFECT OF DIET ON GLUCOSE TOLERANCE

A beneficial effect of dietary carbohydrate on glucose tolerance has been debated in the past few decades. Himsworth (1935) demonstrated that of dietary

MARGARET J. ALBRINK and IRMA H. ULLRICH • Department of Medicine, West Virginia University Medical Center, Morgantown, West Virginia 26505.

factors, carbohydrate was the sole determinant of glucose tolerance; the higher the carbohydrate intake the better the glucose tolerance. Epidemiologic evidence has been cited both for and against this hypothesis. Cleave *et al.* (1969) and Cohen *et al.* (1961) have suggested the converse if sucrose is the source of carbohydrate, that is, a diabetogenic effect of sucrose. West and Kalbfleisch (1971) could not confirm sugar as diabetogenic in epidemiologic studies; rather, two closely interrelated factors, dietary fat and obesity, appeared to be the culprits. Evidence is hardly universal that high-carbohydrate diet is always beneficial. Grey and Kipnis (1971) showed that a low-carbohydrate diet led to improved glucose tolerance in obese young women.

The older literature on the effect of high-carbohydrate diets on glucose tolerance has been reviewed (Albrink, 1973, 1975). In general, a slight improvement was noted in most cases, but the type of carbohydrate was often not stated, and insulin response was often not recorded.

3. IMPORTANCE OF INSULIN RESPONSE

While diets high in carbohydrate may lead to improved glucose tolerance, the improvement may be at the expense of hyperinsulinism, i.e., increased insulin response to a glucose challenge (Farquhar *et al.*, 1966; Grey and Kipnis, 1971), and later to compensatory insulin resistance. Hyperinsulinemia may be particulary relevant to obesity. Obese persons are characterized by hyperinsulinism (Karam *et al.*, 1963) and hypertriglyceridemia and decreased HDL (Albrink *et al.*, 1980), all of which may presage arteriosclerotic cardiovascular disease (Stout, 1979; Goldstein *et al.*, 1973; Miller and Miller, 1975). The insulin resistance and hyperinsulinemia which are common in obesity and in the early stages of non-insulin-dependent diabetes mellitus may play a role in the genesis of these conditions (Yalow and Berson, 1960). Whether hyperinsulinism is a primary metabolic defect or is secondary to insulin resistance is uncertain; however, a cycle of hyperinsulinism, compensatory insulin resistance, and further hyperinsulinism is common in the progression of obesity, hypertriglyceridemia, and diabetes in experimental animals (Assimacopoulos-Jeannet and Jeanrenaud, 1976). The cause of the initial hyperinsulinism may be of importance in the etiology of these conditions.

Because of the importance of hyperinsulinemia and insulin resistance, the effect of dietary carbohydrate on insulin response as well as on glucose tolerance becomes important. Moreover, insulin response to nutrient challenge is greater than the glucose response in persons with normal fasting blood sugars (Ullrich and Albrink, 1982a) and is more sensitive to change in diet (Potter *et al.*, 1981).

The insulin-stimulating effect of a high-carbohydrate meal could be one factor initiating hyperinsulinemia and consequent insulin resistance. The type and form of carbohydrate may well be of critical importance in determining the insulin response to a given high-carbohydrate diet. Reiser *et al.* (1979b) have reported that substitution of sucrose in the amount of 30% of calories for an equal number of starch calories (total carbohydrate calories being 43%) caused increase in both glucose and insulin response to glucose challenge in normal adults. On the other hand, fructose, up to 17% of calories, replacing dextro-maltose in diets of which 85% was composed of carbohydrate, had no effect on glucose tolerance or insulin response of hypertriglyceridemic men (Turner *et al.*, 1979), perhaps because the amount of fructose was small. Brunzell *et al.* (1971) found that 85% of calories as carbohydrate led to improved glucose tolerance in normal persons, but the insulin response was unclear. In a study of normal persons by Anderson *et al.* (1973), as glucose in the diet was increased from 20 to 80% of calories (replacing corn oil), glucose tolerance improved and insulin levels were lower, but not different from those on a solid diet of 40% carbohydrate calories. Plasma triglyceride concentration increased on the 80% glucose diet. It thus appeared that a liquid formula glucose diet was more insulinogenic than a solid diet of equal carbohydrate content. When sucrose was given as 80% of calories there was a slight improvement of glucose tolerance compared to the control solid diets (Anderson *et al.*, 1973). However, sucrose or glucose replaced corn oil in these experiments, so that total carbohydrate was also high on the high-glucose and sucrose diets, and low on the 20% glucose diet (Anderson *et al.*, 1973). Change of daily intake from a balanced weight-maintaining diet to one of 500 cal of which 85 g was sucrose did not lead to impaired glucose tolerance or change in insulin response in healthy young men (Anderson and Herman, 1972), while reduction of carbohydrate calories to 600 cal, full caloric intake being maintained by fat, led to impaired glucose tolerance (Anderson *et al.*, 1973). The implication is that dietary fat might have a diabetogenic effect. Substitution of 22% sucrose calories for 22% starch calories led to a slight increase in fasting blood sugar in normal men (Irwin *et al.*, 1964). In five obese males consuming diets in which total carbohydrate constituted 40–50% of calories, when 30% of calories was given as complex carbohydrate the insulin but not the glucose response was lower than when 30% of calories was fed as simple carbohydrate (Rodger *et al.*, 1971). Dunnigan *et al.* (1970) found that 30% sucrose calories compared to 30% starch calories in middle-aged men led to increase in fasting blood sugar but had no effect on insulin levels.

Evidence, while conflicting, suggests that sucrose may lead to impairment of glucose tolerance and to hyperinsulinism, indications of insulin resistance, while starch may improve glucose tolerance, but that these effects are contingent upon feeding rather large amounts of sucrose or starch.

4. EFFECT OF DIETARY FIBER

Only recently has the possible effect of dietary fiber on glucose tolerance been studied. Jenkins (1979) and Jenkins *et al*. (1978) reported that addition of guar and pectin to glucose meal resulted in lower glucose and insulin response, indication of increased insulin sensitivity. The role of high-fiber diet in the treatment of diabetes (Kiehm *et al*., 1976) is considered elsewhere in this volume. Many studies comparing the effects of complex versus simple carbohydrate and of starch versus sucrose may have differed in the amount of fiber, since fiber is a natural part of many starchy foods.

Most studies of the effect of diet on glucose tolerance have measured the response to a standard glucose challenge. However, if the day-long levels of insulin are important in the genesis of insulin resistance and insulin-resistant states, the response to a meal representative of that diet becomes of importance.

5. MEAL TOLERANCE TESTS

To assess the insulin and glucose response to various types of high-carbohydrate foods, we have performed meal tolerance tests (MTTs) using various sources of carbohydrate, each designed to contain equal equivalents of hexose as such or as the product of digestion. Insulin response varied greatly, while glucose response was only slightly affected by type of carbohydrate. Glucose in solution caused the greatest rise in insulin. White bread (Swan *et al*., 1966) and brown rice (Potter *et al*., 1981) were only slightly less insulinogenic than glucose (Potter *et al*., 1981), while bran, pinto beans (Potter *et al*., 1981), and mung beans (Ullrich and Albrink, 1982b) were the least insulinogenic. The responses of plasma glucose were in the same direction but smaller. Fructose caused only a slight increase in insulin and glucose (Swan *et al*., 1966), while sucrose caused a response midway between those of fructose and glucose. The small effect of fructose reflects the fact that fructose does not stimulate insulin secretion unless it is first converted to glucose, although it does enter metabolic pathways by insulin-independent steps (Zakim and Herman, 1968). Crapo *et al*. (1977, 1981) have reported large differences in insulin response to various meals. Behall *et al*. (1976) found a greater increase in plasma glucose and insulin when either glucose or sucrose was added to a basic breakfast. Bolton *et al*. (1981) and Haber *et al*. (1977) found a lesser insulin response to meals of whole fruit than pureed fruit. Meyer *et al*. (1971) reported a greater response of blood sugar of normal adults to sucrose and starch than to glucose, and lower blood sugar response in women than in men. Addition of protein to a carbohydrate meal caused greater insulin release in eight women (Rabinowitz *et al*., 1966).

6. EFFECT OF DIETARY CARBOHYDRATE ON SERUM LIPIDS

6.1. Triglycerides

A possible disadvantage of a very high-carbohydrate diet is its triglyceride-raising effect (Ahrens *et al.*, 1961). Most studies of carbohydrate-induced hypertriglyceridemia have used liquid formula diets in which the carbohydrate source was often not stated. While there is some evidence that sucrose is more hypertriglyceridemic than glucose and glucose more than starch, others have found no difference attributable to the source of carbohydrate [reviewed by Bierman (1979)]. Since hyperinsulinemia and insulin resistance may be critical elements in the genesis of hypertriglyceridemia (Farquhar *et al.*, 1966; Tobey *et al.*, 1981; Davidson and Albrink, 1965) the degree to which a particular high-carbohydrate food raises plasma insulin may determine its hypertriglyceridemic effect. Up to 17% fructose calories in 85% carbohydrate diets changed neither the triglycerides nor the insulin response of healthy young men (Turner *et al.*, 1979). On the other hand, 30% of calories as sucrose, when replacing an equal amount of starch in otherwise identical diets, raised triglycerides (and cholesterol) in normal persons (Reiser *et al.*, 1979a). Palumbo *et al.* (1977) postulated that persons with atherosclerosis were more sensitive to the triglyceride-raising effect of sucrose (about 30 g daily in their experiments) than were normal persons. The measurement of the insulin response not only to glucose challenge but to a representative meal is critical to testing the hypothesis that triglyceride levels are related to degree of day-long hyperinsulinemia.

6.2. Cholesterol

The cholesterol-lowering effect of low-fat, low-cholesterol diets is well known, but here again the type of carbohydrate may be important. A cholesterol-raising effect of sucrose and cholesterol-lowering effect of legumes and pectin was reviewed by McGandy *et al.* (1967). They concluded that while sucrose might have a small cholesterol-raising effect and starch, particularly legumes, and certain high-fiber foods a cholesterol-lowering effect, the effect was small compared to the large effects of changes in dietary fat, type of fat, and dietary cholesterol. A review of the older literature on the effect of carbohydrate on triglycerides concluded that moderately high-carbohydrate diets (60% of calories) had little effect on triglycerides, though type of carbohydrate was usually not stated (Albrink, 1974). More recent work has suggested that dietary fiber might have a cholesterol-lowering effect (Kelsay, 1978).

6.3. High-Density Lipoproteins

Yet another effect of high-carbohydrate diet on serum lipids is a decrease in high-density lipoproteins (HDL), the purported protective factor against arteriosclerotic heart disease (Miller and Miller, 1975; Blum *et al.*, 1977; Schonfeld *et al.*, 1976; Wilson and Lees, 1972; Macdonald 1978). Whether sucrose potentiates or fiber protects against this effect is not known. Connor *et al.* (1978) reported lower HDL in Tarahumara Indians, a group consuming high-fiber diets, their low levels of low-density lipoproteins presumably protecting them against arteriosclerotic heart disease.

HDL and triglycerides are inversely related (Miller and Miller, 1975). It is not surprising, therefore, that HDL concentration has a correlation with insulin that is the opposite to its correlation with triglycerides. We have recently reported that HDL concentration is negatively correlated with insulin response to glucose challenge and conclude, therefore, that low HDL, like high triglyceride concentration, is an indication of insulin resistance (Riales and Albrink, 1981; Ullrich and Albrink, 1982b). The effect of dietary carbohydrate on cholesterol as well as triglycerides may be in part related to its effect on glucose tolerance (Bennion and Grundy, 1977).

7. EFFECT OF DIET ON CORONARY ARTERY DISEASE

Recent epidemiologic evidence from three carefully studied populations, while finding no correlation between sucrose intake and future coronary artery disease, also found little correlation between intake of fat, cholesterol, or P/S ratio and future coronary disease. Starch intake, however, proved to be protective (Gordon *et al.*, 1981). Dietary fiber associated with the starch or some effect of starch itself could have mediated this benefit. The interaction of dietary fat, starch, sucrose, and fiber must be considered in efforts to define the effect of diet on glucose tolerance and serum lipids.

8. INTERACTION OF DIETARY FIBER AND CARBOHYDRATE WITH PLASMA GLUCOSE AND INSULIN

In recent studies we have examined the effect of a variety of high-carbohydrate diets, either high or low in fiber, on glucose, insulin, and lipids of normal young men. In all the experiments calories were adjusted to maintain weight, and protein, fat, and carbohydrate constituted 15, 15, and 70% of calories, respectively. In the first diet seven young men were randomly assigned to a diet either high (>60 g neutral detergent fiber) or low (<1 g neutral detergent fiber) in fiber and then to the opposite diet (Albrink *et al.*, 1979). The low-fiber diet was fed as a liquid formula diet consisting of glucose, casein, and corn oil in solution in a milk base, while the high-fiber diet consisted largely of All Bran, pinto beans, brown rice, and whole wheat bread. The insulin response was markedly lower following meal tolerance tests representative of the high-fiber than the low-fiber diet. There was, however, no difference between the glucose responses to the two meals. While fasting plasma triglyceride concentration increased during the seven days on the low-fiber diet, it decreased slightly following the high-fiber diet. Cholesterol decreased after both diets, slightly more so following the high-fiber diet. It was concluded that a higher ambient insulin concentration during the low-fiber than the high-fiber diet led to the increase in fasting triglyceride concentration following the low-fiber diet. The low ambient insulin concentration might explain the lack of carbohydrate-induced hypertriglyceridemia following the high-fiber diet, even though the carbohydrate content was equally high in both diets. The insulin-stimulating effect of a particular high-carbohydrate diet might thus determine its potential for raising fasting triglyceride concentration.

9. EFFECT OF HIGH- AND LOW-FIBER SOLID DIETS

Since the difference between the high- and low-fiber diets of the experiment just cited could have been due to the difference between liquid and solid states, another experiment was carried out to test the difference between high- and low-fiber solid foods in a four-day study of seven young men. Carbohydrate was present chiefly in starchy foods. The foods were matched as closely as possible for types of starch, the low-fiber diet consisting mainly of white rice, spaghetti made from mung bean flour (bean threads), white bread, and low-fiber wheat cereal, while the high-fiber diet consisting largely of brown rice, mung beans, and whole wheat bread. To our surprise the response of plasma insulin and glucose to the meals was not different between the high- and low-fiber diets. It thus appeared that some of the difference between the high- and low-fiber diets of the earlier study might have reflected the different response to cooked starch compared to glucose in solution (Ullrich and Albrink, 1982b). The lack of difference in the mung bean study, on the other hand, could have been due to inadequate proportion of the high-fiber foods (mung beans, All Bran) compared to lower fiber foods (rice, bread, fruit) in the high-fiber diet. An alternative explanation is that even the low-fiber diet may have contained enough fiber to cause maximal beneficial effect on glucose tolerance (16 versus 59 g neutral detergent fiber).

10. INTERACTION OF DIETARY FIBER AND SUCROSE

A series of experiments was next carried out to examine the interaction of dietary fiber and sucrose (Albrink and Ullrich, 1982). To test the hypothesis that dietary fiber might protect against the increase in insulin and triglycerides and decrease in HDL expected with high-sucrose diets, three groups of seven or eight young men were assigned in random order to 10 days of either a high-fiber (>68 g neutral detergent fiber) diet or a low-fiber (<15 g) diet, then to 10 days of the opposite diet. Each group of men was assigned to one of three levels of sucrose: 36%, 18%, or 0% of calories. As before, protein, fat, and carbohydrate constituted 15, 15, 70% of calories. The dietary fiber content, as neutral detergent fiber, was measured by the method of Van Soest and Robertson (1976). In order to maintain the usual cholesterol intake of about 500 mg daily, two eggs daily were included in all diets. The high-fiber foods consisted chiefly of pinto beans, All Bran, and whole wheat bread, while the equivalent low-fiber foods were spaghetti, low-fiber wheat, and white bread. The results showed a difference between high- and low-fiber diets only at the 36% level of sucrose intake (Ullrich and Albrink, 1980). In that diet triglycerides increased 73% above baseline following the low-fiber diet, compared to 26% following the high-fiber diet ($p < 0.02$). Insulin response to MTTs measured as the sum of the 0, $^1/_2$, 1, and 2 hr levels following test meals was significantly higher in response to the low-fiber meal than the high-fiber meal ($p < 0.05$). There was no difference in oral glucose tolerance tests (OGTTs). There was no difference in glucose or insulin response to MTTs or OGTTs between the high- or low-fiber diets of either 18 or 0% sucrose diets. Triglycerides did not change following the 18% sucrose diet, whether high or low in fiber. Following the 0% sucrose diet, triglycerides decreased 22% below baseline in both the high-fiber and low-fiber diets.

Thus the triglyceride and insulin-raising effect of sucrose was observed only at the highest sucrose level tested, 36% of calories, and only at this level of sucrose was there a protective effect of fiber against the rise in triglyceride and insulin.

The different levels of sucrose were compared using nonpaired *t* tests. Glucose sums and insulin sums (sum of the 0, ½, 1, and 2 hr values) after 7–10 days on each diet were expressed as percent of glucose or insulin sums observed at a baseline OGTT. The only significant change in insulin sums from baseline was a decrease in insulin response to both the OGTT and MTT during the 0% sucrose diet to 65 and 71% of baseline for the low- and high-fiber diets, respectively. The responses to the 18 and 36% sucrose diets were quite variable. While the MTT insulin sums for the 0% sucrose diet were significantly lower than for the 18 or 36% sucrose diets, the latter were not significantly different from each other. Similarly, MTT glucose sums were lowest following the 0% sucrose diet at both levels of fiber. An adaptive improvement in glucose tolerance was indicated by the decrease in OGTT glucose sums and insulin sums following the 0% sucrose diet but not the 18 or 32% sucrose diets.

Total cholesterol decreased following the 0 and 18% sucrose diets regardless of fiber content. Cholesterol did not decrease following the 36% sucrose diet. HDL decreased following all diets, significantly so for the 18 and 36% sucrose diets. There was no clear effect of fiber on cholesterol. The only effect of fiber on HDL was a blunting of the decrease in HDL which followed the 36% sucrose diet.

The effects of fiber must be interpreted in the setting of these particular experiments, where three other major changes in the diet were also taking place: fat was drastically decreased, carbohydrate was greatly increased, and sucrose was being substituted for varying amount of complex carbohydrate. The cholesterol-lowering effect of all diets was probably due to the decrease in dietary fat, no further beneficial effect of dietary fiber being evident that could not be explained by the low fat. Sucrose in substantial amounts appeared to prevent the decrease in cholesterol, and fiber had no influence on this effect.

In the same vein, while high-carbohydrate diets may lead to improved glucose tolerance (Himsworth, 1935), high-sucrose diets may lead to impaired glucose tolerance. Here again the two opposite effects seem to have canceled each other out, the improvement being seen only in the high-carbohydrate, sucrose-free diets. No impairment from baseline glucose tolerance was seen with sucrose substitution—rather, failure of improvement. In our studies the beneficial effect of fat deletion on glucose tolerance (Anderson *et al.*, 1973) may have been counteracted by a deleterious effect of sucrose on glucose tolerance. Perhaps it is glucose, or glucose-yielding carbohydrates, which lead to adaptive improvement in glucose tolerance, fructose (from sucrose digestion) merely displacing glucose-yielding foods rather than having a specific effect per se.

11. CONCLUSIONS

The beneficial effect of fiber may be mediated in part by its effect of lowering plasma insulin and improving glucose tolerance when these are abnormal because of insulin-resistant states. Those diseases or abnormalities related to insulin resistance and hyperinsulinemia might be expected to respond the most to dietary fiber. The beneficial effects of fiber, as demonstrated in our study, include the prevention of the triglyceride-raising effect and the HDL-lowering effect of sucrose. Persons with normal lipids and glucose tolerance and normal insulin sensitivity would be expected to be less likely to be benefited than persons with insulin resistance.

The benefits of fiber can probably be extended to other insulin-resistant states, including non-insulin-dependent diabetes mellitus, obesity, and hyperlipidemia of the hypertriglyceridemic type. All of these conditions are strongly correlated with hyperinsulinemia.

Cholesterol, which is little if at all related to insulin resistance, is powerfully influenced by levels of dietary fat, with little additional effect of fiber not explained by the low-fat diet characteristic of high-fiber diets of natural foods. A weak cholesterol-lowering effect through its effect on bile acid metabolism (Ullrich *et al.*, 1981) could, however, theoretically lead to a further reduction of cholesterol in some persons.

Research is needed to elucidate the type of fiber most effective in the interaction of dietary fiber with glucose tolerance and insulin resistance in the range of fiber between the extremes used in the studies reported.

12. SUMMARY

1. A high-carbohydrate diet composed chiefly of glucose in liquid form evoked a much greater meal-stimulated rise in insulin and an increase in fasting triglycerides after several days, compared to a high-fiber starchy diet.

2. In diets composed of starchy foods with no sugar in which carbohydrate constituted 70% of calories, dietary fiber had little or no effect on glucose or insulin response or on triglycerides.

3. A high-carbohydrate diet rich in starchy foods led to improvement in glucose tolerance, evidenced by lower glucose and particularly lower insulin levels in response to a standard glucose challenge and to representative test meals. Substantial (36% of calories) substitution of sucrose reversed this beneficial effect, causing increased insulin and glucose levels, and dietary fiber (over 60 g daily) protected against this effect.

4. A very low-fat, high-complex-carbohydrate isocholesterol diet had a powerful cholesterol-lowering effect, and dietary fiber had little or no further effect. Sucrose (18–36% of calories) prevented this cholesterol-lowering effect regardless of fiber content.

5. A very high-carbohydrate, high-starch diet in which simple sugar (sucrose) constituted a substantial (36%) part of calories caused an increase in triglycerides and a decrease in HDL, and fiber prevented these unfavorable effects.

REFERENCES

Ahrens, E. H., Hirsch, J., Oette, K., Farquhar, J. W., and Stein, Y., 1961, Carbohydrate-induced and fat-induced lipemia, *Trans. Assoc. Phys.* **74**:134–46.

Albrink, M. J., 1973, Dietary carbohydrates in lipid disorders in man, *Progr. Biochem. Pharmacol.* **8**:242–70.

Albrink, M. J., 1974, Dietary and drug treatment of hyperlipidemia in diabetes, *Diabetes*
Albrink, M. J., 1974, Dietary and drug treatment of hyperlipidemia in diabetes, *Diabetes* **23**:913–918.

Albrink, M. J., 1975, The effects of low-fat, high-carbohydrate diet, *Comprehensive Therapy* **1**:30–33.

Albrink, M. J., and Ullrich, I. H., 1982, Interaction of dietary fiber and sucrose on glucose tolerance of healthy young men (in preparation).

Albrink, M. J., Newman, T., and Davidson, P. C., 1979, Effect of high- and low-fiber diets on plasma lipids and insulin, *Am. J. Clin. Nutr.* **32**:1486–91.

Albrink, M. J., Krauss, R. M., Lindgren, F. T., Von Der Groeben, J., Pan, S., and Wood, P. D., 1980, Intercorrelations among plasma high density lipoprotein, obesity and triglycerides in a normal population, *Lipids* **15**:668–76.

Anderson, J. W., and Herman, R. H., 1972, Effect of fasting, caloric restriction, and refeeding on glucose tolerance of normal men, *Am. J. Clin. Nutr.* **25**:41–52.

Anderson, J., Herman, R. H., and Zakim, D., 1973, Effect of high glucose and high sucrose diets on glucose tolerance of normal men, *Am. J. Clin. Nutr.* **26**:600–607.

Assimacopoulos-Jeannet, F., and Jeanrenaud, B., 1976, The hormonal and metabolic basis of experimental obesity, *Clin. Endocrinol. Metab.* **5**:337–365.

Behall, K. M., Kelsay, J. L., and Clark, W. M., 1976, Response of serum lipids and insulin in men to sugars ingested with or without other foods, *Nutr. Rep. Int.* **14**:485–94.

Bennion, L. J., and Grundy, S., 1977, Effects of diabetes on cholesterol metabolism, *N. Engl. J. Med.* **296**:1365–71.

Bierman, E. L., 1979, Carbohydrates, sucrose, and human diseases, *Am. J. Clin. Nutr.* **32**:2712–22.

Blum, B., Levy, R. I., Eisenberg, S., Hall, M., III, Goebel, R. H., and Berman, M., 1977, High density lipoprotein metabolism in man, *J. Clin. Invest.* **60**:795–807.

Bolton, R. P., Heaton, K. W., and Burroughs, L. F., 1981, The role of dietary fiber in satiety, glucose, and insulin: Studies with fruit and fruit juice, *Am. J. Clin. Nutr.* **34**:211–17.

Brunzell, J. D., Lerner, L., Hazzard, R., Porte, Jr., and Bierman, E. L., 1971, Improved glucose tolerance with high carbohydrate feeding in mild diabetes, *N. Engl. J. Med.* **284**:521–4.

Cleave, T. L., Campbell, G. D., and Painter, N. S., 1969, *Diabetes, Coronary Thrombosis and the Saccharine Disease,* John Wright and Sons, Bristol, England.

Cohen, A. M., Bavly, S., and Poznanski, R., 1961, Change of diet of Yemenite Jews in relation to diabetes and ischemic heart disease, *Lancet* **2**:1399–1401.

Connor, W. E., Cerqueira, M. T., Connor, R. W., Wallace, R. B., Malinow, R., and Casdorph, H. R., 1978, The plasma lipids, lipoproteins, and diet of the Tarahumara Indians of Mexico, *Am. J. Clin. Nutr.* **31**:1131–1142.

Crapo, P. A., Reaven, G., and Olefsky, J., 1977, Postprandial plasma glucose and insulin responses to different complex carbohydrates, *Diabetes* **26**:1178–84.

Crapo, P. A., Reaven, G. M., and Olefsky, J. M., 1981, Hormonal and substrate responses to a standard meal in normal and hypertriglyceridemic subjects, *Metabolism* **30**:331–34.

Davidson, P. C., and Albrink, M. J., 1965, Insulin resistance in hypertriglyceridemia, *Metabolism* **14**:1059–70.

Dunnigan, M. G., Fyfe, T., McKiddie, M. T., and Crosbie, S. M., 1970, The effects of isocaloric exchange of dietary starch and sucrose on glucose tolerance, plasma insulin and serum lipids in man, *Clin. Sci.* **38**:1–9.

Farquhar, J. W., Frank, A., Gross, R. C., and Reaven, G. M., 1966, Glucose, insulin and triglyceride response to high and low carbohydrate diets in man, *J. Clin. Invest.* **45**:1648–56.

Goldstein, J. L., Hazzard, W. R., Schrott, H. G., Bierman, E. L., and Motulsky, A. G., 1973, Hyperlipidemia in coronary heart disease. Lipid levels in 500 survivors of myocardial infarction, *J. Clin. Invest.* **52**:1533–43.

Gordon, T., Kagan, W. B., Zukel, W. J., Tilloston, J., Sorlie, P., and Hjortland, M., 1981, Diet and its relation to coronary heart disease and death in three populations, *Circulation* **63:**500–15.

Grey, N., and Kipnis, D., 1971, Effect of diet composition on the hyperinsulinemia of obesity, *N. Engl. J. Med.* **285:**827–31.

Haber, G. B., Heaton, K. W., Murphy, D., and Burroughs, L., 1977, Depletion and disruption of dietary fibre. Effects on satiety, plasma-glucose and serum-insulin, *Lancet* **2:**679–82.

Himsworth, H. P., 1935, The dietetic factor determining the glucose tolerance and sensitivity to insulin of healthy men, *Clin. Sci.* **2:**67–94.

Irwin, M. I., Taylor, D. D., and Feeley, R. M., 1964, Serum lipid levels, fat, nitrogen, and mineral metabolism of young men associated with kind of dietary carbohydrate, *J. Nutr.* **82:**338–42.

Jenkins, D. J. A., 1979, Dietary fibre, diabetes, and hyperlipidemia, Progress and prospects, *Lancet* **2:**1287–90.

Jenkins, D. J. A., Leeds, A. R., Gassull, M. A., Cochet, B., and Alberti, K. G., 1978, Dietary fibers, fiber analogues and glucose tolerance: Importance of viscosity, *Br. Med. J.* **1:**1392–1394.

Karam, J. H., Grodsky, G. M., and Forsham, P. H. 1963, Excessive insulin response to glucose in obese subjects as measured by immunochemical assay, *Diabetes* **12:**196–204.

Kelsay, J. L., 1978, A review of research on effects of fiber intake on man, *Am. J. Clin. Nutr.* **31:**142–59.

Kiehm, T. G., Anderson, J. W., and Ward, K., 1976, Beneficial effects of a high carbohydrate, high fiber diet on hyperglycemic diabetic men, *Am. J. Clin. Nutr.* **29:**985–99.

Macdonald, I., 1978, The effects of dietary carbohydrates on high density lipoprotein levels in serum, *Nutr. Rep. Int.* **17:**663–68.

McGandy, R. B., Hegsted, D. M., and Stare, F. J., 1967, Dietary fats, carbohydrates and atherosclerotic vascular disease, *N. Engl. J. Med.* **277:**186–92, 242–47.

Meyer, F. L., Mattox, H., Bolick, M., and MacDonald, C., 1971, Metabolic changes after test meals with different carbohydrates, *Am. J. Clin. Nutr.* **24:**615–21.

Miller, G. J., and Miller, N. E., 1975, Plasma-high-density-lipoprotein concentration and development of ischaemic heart disease, *Lancet* **1:**16–19.

Palumbo, P. J., Briones, E. R., Nelson, R. A., and Kottke, B. A., 1977, Sucrose sensitivity of patients with coronary-artery disease, *Am. J. Clin. Nutr.* **30:**394–410.

Potter, J. G., Coffman, K. P., Reid, R. L., Krall, J. M., and Albrink, M. J., 1981, Effect of dietary fiber content of foods on insulin and glucose response to test meals, *Am. J. Clin. Nutr.* **34:**328–34.

Rabinowitz, D., Merimee, T. J., Maffezzoli, R., and Burgess, J. A., 1966, Patterns of hormonal release after glucose, protein, and glucose plus protein, *Lancet* **2:**454–56.

Reiser, S., Hallfrisch, J., Michaelis, O. E., IV, Lazar, F. L., Martin, R. E., and Prather, E. S., 1979a, Isocaloric exchange of dietary starch and sucrose in humans. I. Effects on levels of fasting blood lipids, *Am. J. Clin. Nutr.* **32:**1659–1668.

Reiser, S., Handler, H. B., Gardner, L. B., Hallfrisch, J. G., Michaelis, O. E., IV, and Prather, E. S., 1979b, Isocaloric exchange of dietary starch and sucrose in humans. II. Effect on fasting blood insulin, glucose, and glucagon on insulin and glucose response to a sucrose load, *Am. J. Clin. Nutr.* **32:**2206–16.

Riales, R., and Albrink, M. J., 1981, Effect of chromium chloride supplementation on serum lipids including HDL and on glucose tolerance of adult men, *Am. J. Clin. Nutr.*, **34:**2670–2678.

Rodger, N. W., Squires, B. P., and Du, E. L., 1971, Changes in plasma insulin related to the type of dietary carbohydrate in overweight hyperlipidemic male patients, *Can. Med. Assoc. J.* **105:**923–34.

Schonfeld, G., Weidman, S. W., Witztum, J. L., and Bowen, R. M., 1976, Alterations in levels and interrelations of plasma apolipoproteins induced by diet, *Metabolism* **25:**261–275.

Stout, R. W., 1979, Diabetes and atherosclerosis—The role of insulin, *Diabetologia* **16:**141–150.

Swan, D. C., Davidson, P. C., and Albrink, M. J., 1966, The effects of simple and complex carbohydrates on plasma non-esterified fatty acids, plasma sugar and insulin during oral carbohydrate tolerance tests, *Lancet* **1**:60–63.

Tobey, T. A., Greenfield, M., Kraimer, F., and Reaven, A. G., 1981, Relationship between insulin resistance, insulin secretion, very low density lipoprotein kinetics, and plasma triglyceride levels in normotriglyceridemic man, *Metabolism* **30**:165–71.

Trowell, H., 1972, Ischemic heart disease and dietary fiber, *Am. J. Clin. Nutr.* **25**:926–32.

Trowell, H. C., 1974, Diabetes mellitus death rates in England and Wales, 1920–70 and food supplies, *Lancet* **2**:998–1002.

Trowell, H. C., 1975, Obesity in the Western World, *Plant Foods for Man* **1**:157–68.

Turner, J. L., Bierman, E. L., Brunzell, J. D., and Chait, A., 1979, Effect of dietary fructose on triglyceride transport and glucoregulatory hormones in hypertriglyceridemic men, *Am. J. Clin. Nutr.* **32**:1043–50.

Ullrich, I. H., and Albrink, M. J., 1980, Dietary fiber protects against sucrose-induced hypertriglyceridemia, *Clin. Res.* **28**:816A.

Ullrich, I. H., and Albrink, M. J., 1982a, Effect of dietary fiber and other factors on insulin response: Role in obesity, in: *Advances in Modern Human Nutrition* (R. D. Tobin and M. A. Mehlman, eds.), Pathotox Publishers, Park Forest South, Illinois (in press).

Ullrich, I. H., and Albrink, M. J., 1982b, Lack of effect of dietary fiber on serum lipids, glucose and insulin in healthy young men fed high sterol diets, *Am. J. Clin. Nutr.* (in press).

Ullrich, I. H., Lai, H.-Y., Vona, L., Reid, R. L., and Albrink, M. J., 1981, Alterations of fecal steroid composition induced by changes in dietary fiber consumption, *Am. J. Clin. Nutr.* **34**:2054–2060.

VanSoest, P. J., and Robertson, J. B., 1976, Chemical and physical properties of dietary fiber, in: *Miles Symposium on Dietary Fiber, Nutrition Society of Canada,* Miles Laboratories, Ltd., Rexdale, Ontario.

West, K. M., and Kalbfleisch J. M., 1971, Influence of nutritional factors on prevalence of diabetes, *Diabetes* **20**:99–108.

Wilson, D. E., and Lees, R. S., 1972, Metabolic relationships among the plasma lipoproteins. Reciprocal changes in the concentrations of very low and low density lipoproteins in man, *J. Clin. Invest.* **51**:1051–1057.

Yalow, R. S., and Berson, S. A., 1960, Immunoassay of endogenous plasma insulin in man, *J. Clin. Invest.* **39**:1157–75.

Zakim, D., and Herman, R. H., 1968, Fructose metabolism. II. Regulatory control to the triose level, *Am. J. Clin. Nutr.* **21**:315–19.

16

Does Simple Substitution of Fiber-Rich Foods for Refined Foods Aid in the Treatment of Diabetes Mellitus?

K. W. HEATON, A. MANHIRE, C. L. HENRY, and M. HARTOG

1. INTRODUCTION

There is now considerable evidence that high-carbohydrate, high-fiber diets lead to improved diabetic control compared with conventional 40% carbohydrate diets (Kiehm *et al.*, 1976; Anderson and Ward, 1978, 1979; Simpson *et al.*, 1979a,b). However, it is not clear whether these diets achieve their beneficial effect through their high carbohydrate intake, their high fiber intake, or the fiber-rich nature of their carbohydrate, or, perhaps, through a combination of these properties.

It is generally accepted that refined sugars are harmful to the diabetic, whereas sweet-tasting fruits and vegetables are harmless, at least in their raw state.

We were attracted by the idea (Cleave and Campbell, 1966; Cleave, 1974) that carbohydrates of all kinds become diabetogenic only when they are stripped of their natural complement of fiber or cell wall material. This led to the hypothesis that diabetic control can be improved beyond that achieved with the traditional restriction of refined sugar and of total carbohydrate simply by eating

K. W. HEATON, A. MANHIRE, C. L. HENRY, and M. HARTOG • University Department of Medicine, Bristol Royal Infirmary, Bristol BS2 8HW, England.

all carbohydrate foods in unrefined or fiber-rich form. To test this hypothesis we carried out a randomized crossover trial.

2. PATIENTS, METHODS, AND DESIGN OF STUDY

Sixteen nonobese diabetics aged 27–70 took part in the study—seven men and three women requiring insulin, four men and two women receiving oral hypoglycemic drugs. Their diabetic control had been relatively stationary and their treatment unchanged for three months. After assessment by an experienced dietician (CLH) to establish that they understood their present dietary prescription and were trying to comply with it, they were allocated at random into two groups of eight subjects. One group was reinstructed in the conventional diet of modest carbohydrate restriction and was maintained on this diet for 6 weeks. These subjects were also prescribed one placebo capsule of calcium lactate daily to give an impression of novel treatment. The other group was instructed to avoid all refined (fiber-depleted) carbohydrate foods, namely, all foods which contain manufactured sugar (white or brown), also flour and other cereal products of any sort other than 100% wholemeal. They were allowed unlimited amounts of unrefined (fiber-intact) carbohydrate foods, such as wholemeal bread and pasta, brown rice, wholegrain breakfast cereals, vegetables, and fruit. This diet was also followed for 6 weeks. Both groups then entered a 6-week, observation-free, "washout period" on their original (usual) diet before being instructed in the alternative diet and following this for a further 6 weeks.

Every sixth day during the test period the patient recorded on a special form everything he or she ate or drank. Amounts were estimated and recorded as accurately as possible using standard household measures, including slices of bread, numbers of biscuits, spoonfulls, and cupfulls. Though not asked to do so, two patients weighed everything and several others weighed some items. All patients were interviewed with their records to clarify any ambiguities. Completed record forms were subsequently assessed for energy, carbohydrate, refined sugar, protein, fat, and dietary fiber content using McCance and Widdowson's tables of food composition (Paul and Southgate, 1978). Refined sugar was defined as any fiber-depleted sugar, including brown sugar and syrup.

At the beginning of each test period the dose of insulin was reduced by 20% and that of oral hypoglycemics by 25% to avoid hypoglycemia. Insulin and drug treatment were subsequently altered according to home urine tests, hypoglycemic episodes, and clinic assessments by a physician (AM) who did not know which diet the patient was taking.

Body weight, plasma glucose, and cholesterol $1^1/_2$–$2^1/_2$ hr after breakfast, blood glycosylated hemoglobin (Flückiger *et al.*, 1977), and 24-hr urinary glucose were measured after 3 and 6 weeks of each test period. Statistical analysis was by the paired *t* test.

TABLE I. Daily Intake of Food Components on Two Diets

	Traditional diet, largely refined	Diet wholly unrefined	
Energy, kcal	2163 ± 189	2139 ± 165	ns
Protein, g	94.0 ± 8.1	94.1 ± 6.4	ns
Fat, g	115.0 ± 11.0	113.0 ± 10.0	ns
Total available carbohydrate, g	190 ± 18	190 ± 16	ns
Refined sugar, g	20.3 ± 4.6	7.5 ± 0.9	$p < 0.01$
Dietary fiber, g	18.3 ± 1.6	32.6 ± 2.7	$p < 0.001$
Cereal fiber, g	7.8 ± 1.0	17.2 ± 1.5	$p < 0.001$

3. RESULTS AND DISCUSSION

Patients complied well with the unrefined diet in that they reduced their already low intake of refined sugar to almost zero and increased their intake of dietary fiber by 78% (Table I). However, despite the liberal instructions, they did not increase their intake of total available carbohydrate. This remained at the low level of 190 g/day, which provided only 36% of energy intake. This was unexpected but can be explained by the fact that, as long-standing diabetics, these conscientious subjects had been thoroughly indoctrinated in the traditional regime of carbohydrate restriction. In any event, the constancy of total carbohydrate intake provided a model for testing the hypothesis that it is the refinement of carbohydrate, as opposed to the amount, which is what matters in improving diabetic control.

As shown in Table II, there was little difference in indices of diabetic control between the two dietary periods. On the unrefined diet, postbreakfast plasma glucose was modestly reduced after 6 weeks, but this was not accom-

TABLE II. Indices of Diabetic Control after 3 and 6 Weeks on Two Diets

	After 3 weeks		After 6 weeks	
	Refined	Unrefined	Refined	Unrefined
Urinary glucose, g/24 hr	60 ± 14	59 ± 16	68 ± 14	55 ± 11
Postbreakfast plasma glucose, mg/dl	234 ± 23	239 ± 16	261 ± 18	209 ± 25[a]
Glycosylated hemoglobin, %	10.8 ± 0.7	10.3 ± 0.8	11.4 ± 0.8	11.3 ± 0.7

[a] $p < 0.05$.

panied by any significant fall in urinary glucose nor in glycosylated hemoglobin levels.

The main change in diet was the use of wholegrain rather than cereal products, as indicated by the 9.4-g/day increase in cereal fiber (as opposed to a 4.9-g/day increase in fruit and vegetable fiber). Hence the present findings suggest that simply switching from refined to unrefined cereal starch is of little benefit in the control of diabetes.

We conclude that the beneficial effect obtained by high-fiber, high-carbohydrate diets is not due simply to the fiber-rich nature of their carbohydrate. It is true that such diets have contained two to three times as much fiber as our unrefined diet, but very high intakes of dietary fiber can be achieved only by increasing the intake of total carbohydrate (at the expense of fat), unless artificial supplements of fiber or fiber components are used. For the practicing diabetologist dietary fiber may be helpful, but it is not a simple answer to all his problems.

REFERENCES

Anderson, J. W., and Ward, K., 1978, Long-term effects of high-carbohydrate, high-fiber diets on glucose and lipid metabolism: A preliminary report on patients with diabetes, *Diabetes Care* **1**:77–82.

Anderson, J. W., and Ward, K., 1979, High-carbohydrate, high-fiber diets for insulin-treated men with diabetes mellitus, *Am. J. Clin. Nutr.* **32**:2312–2321.

Cleave, T. L., 1974, *The Saccharine Disease,* Wright, Bristol, England.

Cleave, T. L., and Campbell, G. D., 1966, *Diabetes, Coronary Thrombosis and the Saccharine Disease,* Wright, Bristol, England.

Flückiger, R., Berger, W., and Winterhalter, K. H., 1977, Haemoglobin A_{IC}, a reliable index of diabetic control, *Diabetologia* **13**:393.

Kiehm, T. G., Anderson, J. W., and Ward, K., 1976, Beneficial effects of a high carbohydrate, high fiber diet on hyperglycemic diabetic men, *Am. J. Clin. Nutr.* **29**:895–899.

Paul, A. A., and Southgate, D. A. T., 1978, *McCance and Widdowson's The Composition of Foods,* 4th rev. ed., Medical Research Council Special Report No. 297, HMSO, London.

Simpson, R. W., Mann, J. I., Eaton, J., Moore, R. A., Carter, R., and Hockaday, T. D. R., 1979a, Improved glucose control in maturity-onset diabetes treated with high-carbohydrate-modified fat diet, *Br. Med. J.* **1**:1753–1756.

Simpson, R. W., Mann, J. I., Eaton, J., Carter, R. D., and Hockaday, T. D. R., 1979b, High-carbohydrate diets and insulin-dependent diabetics, *Br. Med. J.* **2**:523–525.

17

Fiber and Lipids

DAVID KRITCHEVSKY

In recent years there have been several comprehensive reviews on the effects of dietary fiber on lipid metabolism in animals and man (Kay and Truswell, 1980; Story, 1980; Story and Kritchevsky, 1980). The purpose of this discussion is not to present yet another review, but rather to give a brief overview which will set the stage for the chapters which follow.

Ershoff and Wells (1962; Wells and Ershoff, 1961, 1962) were among the first investigators to study the effects of fiber on lipid metabolism in experimental animals. They found that addition of (1%) cholesterol to the diet of rats increased plasma cholesterol by 15% and liver cholesterol by a factor of ten. When 5% pectin was added to the cholesterol-containing diet, plasma cholesterol levels were unaffected and liver cholesterol only rose threefold. Cellulose, agar, and sodium alginate led to significant increases in serum cholesterol, and cellulose and agar increased liver cholesterol levels by 26 and 48% over those seen in rats fed cholesterol alone. The hypercholesterolemic effect of cellulose and agar has been confirmed by Kiriyama *et al.* (1969), Tsai *et al.* (1976), and Story *et al.* (1977). Wells and Ershoff (1961) found that both citrus and apple pectin were effective in lowering cholesterol levels and observed a limited dose response. Later experiments (Ershoff and Wells, 1962) showed that guar gum, locust bean gum, and carrageenan were also effective in inhibiting the rise in liver cholesterol seen in rats fed 1% cholesterol. Pectin did not protect against cholesterol accumulation in the rabbit, hamster, or guinea pig (Wells and Ershoff, 1962) (Table I). Other workers, however, have found pectin to be hypocholesterolemic in rabbits (Berenson *et al.*, 1975; Hamilton and Carroll, 1976). Lignin

DAVID KRITCHEVSKY • The Wistar Institute of Anatomy and Biology, Philadelphia, Pennsylvania 19104.

TABLE I. Influence of Citrus Pectin (5%) on Plasma and Liver Cholesterol Levels in Cholesterol-Fed Rabbits, Hamsters, Guinea Pigs, and Rats[a]

	Diet[b]		
Species	Basal (B)	BC	BCP
Rabbit			
Plasma, mg/dl	159 ± 30 ab	1363 ± 244 a	1563 ± 192 b
Liver, mg/g	3.4 ± 0.4 cd	21.1 ± 6.2 c	31.8 ± 5.2 d
Hamster			
Plasma, mg/dl	132 ± 7 ef	495 ± 32 e	435 ± 39 f
Liver, mg/g	2.3 ± 0.1 gh	66.3 ± 1.8 g	71.6 ± 3.3 h
Guinea pig			
Plasma, mg/dl	53 ± 4 ij	398 ± 24 j	326 ± 27 j
Liver, mg/g	2.3 ± 0.1 kl	27.6 ± 2.6 k	25.9 ± 3.3 l
Rat			
Plasma, mg/dl	95 ± 4 m	145 ± 8 mn	104 ± 7 n
Liver, mg/g	2.3 ± 0.1 pq	31.1 ± 3.4 pr	10.8 ± 1.8 qr

[a] After Wells and Ershoff (1962).
[b] BC, basal + 1% cholesterol; BCP, BC + 5% pectin. Ten animals per group. Guinea pigs fed 4 weeks, other species 6 weeks. Values bearing same letter are significantly different ($p \leq 0.05$).

(Judd *et al.*, 1976; Story *et al.*, 1977) and psyllium seed (Beher and Casazza, 1971) have been shown to prevent cholesterol accumulation in rats.

Kay and Truswell (1980) have summarized the effects of fiber on blood and biliary lipids in man. Of the identifiable components of fiber, cellulose has no apparent effect on plasma cholesterol when fed at moderate levels. A summary of the effects of pectin encompasses nine studies in which an average of 12 subjects were fed about 18 g of pectin for an average period of 4 weeks. Two of the studies showed no effect; the average reduction in plasma cholesterol in the other seven studies was 13%. Guar gum has also been shown to lower cholesterol levels in man (Fahrenbach *et al.*, 1965; Jenkins *et al.*, 1975).

Wheat bran, which is more a fiber-rich food supplement than an actual type of fiber, has been studied more extensively than any other fiber-rich substance. A review of 20 studies (Kay and Truswell, 1980) reveals that plasma cholesterol levels were reduced in only two. These studies used soft white wheat bran. Bran prepared from hard red spring wheat (Munoz *et al.*, 1979) is hypocholesterolemic.

Fiber-rich foods may also affect plasma cholesterol levels. Grande (1974) summarized the results of 16 studies in which dietary sucrose was replaced isocalorically by various sources of starch. The replacement foods used included fruits, vegetables, legumes, rice, potatoes, cereals, and bread. Average starting serum cholesterol levels were 195 ± 8 mg/dl and final levels were 186

± 7 mg/dl. The reduction in serum cholesterol was significant in seven of the studies.

Dietary fiber affects the course of experimental atherosclerosis. Collation of the literature in 1964 (Kritchevsky, 1964) showed that the addition of up to 35% saturated fat to a stock diet did not affect cholesterolemia or atherosclerosis in rabbits. When saturated fat was added to a semipurified diet, however, as little as 8% fat was sufficient to induce atherosclerosis (Table II). Experiments designed to test the effect of the nonlipid portion of stock diet showed that it could inhibit atherogenesis (Kritchevsky and Tepper, 1965, 1968). Moore (1967) fed rabbits a semipurified diet containing 20% butterfat and 19% fiber. When the fiber used was cellulose the rabbits exhibited hypercholesteremia and atherosclerosis; wheat straw was significantly less cholesterolemic and atherogenic. Alfalfa (Cookson *et al.*, 1967; Kritchevsky *et al.*, 1977a) also inhibits atherogenesis in rabbits. Pectin will reduce the severity of atherosclerosis in cholesterol-fed chickens (Fisher *et al.*, 1966). Baboons or vervet monkeys fed a cholesterol-free, semipurified diet containing 15% cellulose develop more severe aortic

TABLE II. Effect of Stock and Semipurified Diets on Development of Atherosclerosis in Rabbits[a]

Fat added	Diet %			Duration, months	Atheroma score
	Fat	Protein	Carbohydrate		
Stock diet					
Cream	35	10	32	6	0
Corn oil	11	14	44	2	0.10
Shortening	11	14	44	2	0.10
Coconut oil	13	13	43	3	0
Safflower oil	13	13	43	3	0
Semipurified diet					
Coconut oil	20	25	40	3	1.90
Safflower oil	20	25	40	3	0.03
Coconut oil	8	25	38	4	1.30
Corn oil	8	25	38	4	0
Trilaurin	8	25	38	4	1.30
Coconut oil	10–15	20	45	12	1.14
Corn oil	10–15	20	45	12	0.13
Butterfat	20	25	22	9	2.0
Corn oil	20	25	22	9	0

[a] After Kritchevsky (1964).

TABLE III. Binding *in Vitro* of Bile Acids and Salts to Fiber[a]

Bile acid or salt	Fiber			
	Alfalfa[b]	Bran	Cellulose	Lignin
Cholic	1.00	0.51	0.15	2.20
Chenodeoxycholic	1.25	0.91	0.10	1.17
Deoxycholic	0.52	0.27	0.01	0.87
Taurocholic	1.00	0.20	0.15	3.20
Taurochenodeoxycholic	2.18	1.42	0	3.68
Taurodeoxycholic	1.65	0.49	0.10	4.48
Glycocholic	1.00	0.33	0.10	1.96
Glycochenodeoxycholic	1.30	1.86	0.02	2.19
Glycodeoxycholic	2.42	0.68	0.41	4.57

[a] After Story and Kritchevsky (1976).
[b] For each group alfalfa–cholic acid or salt is set as 1.00. Actual percent binding to alfalfa: cholic, 19.9; taurocholic, 6.9; glycocholic, 11.5.

sudanophilia than when fed stock diet or a control diet of bread, fruit, and vegetables (Kritchevsky *et al.*, 1974; 1977b).

What is the mechanism of fiber action? One possibility is an effect on bile acid metabolism. Portman (1960) found that rats fed a semipurified diet excreted less cholic acid and β-hydroxysterol than when fed a stock diet. Their bile acid turnover time was doubled and pool size was decreased. Leveille and Sauberlich (1966) found that addition of pectin to the diet of rats fed 1% cholesterol increased bile acid excretion and decreased serum cholesterol levels. A review of the effect of various types of fiber on bile acid excretion in man (Kay and Truswell, 1980) finds an average increase in fecal bile acid excretion of 60%.

Animal data (Kyd and Bouchier, 1972; Kritchevsky *et al.*, 1974) suggested that the fiber effect might be mediated via bile acid-binding. *In vitro* studies have shown that various types of dietary fiber do indeed bind bile acids and bile salts (Kritchevsky, 1964; Birkner and Kern, 1974; Kritchevsky and Story, 1974; Story and Kritchevsky, 1976) as well as other lipids (Vahouny *et al.*, 1980). The extent of binding is a function of both the type of fiber and the particular bile acid or salt being used (Story and Kritchevsky, 1976) (Table III).

There are other mechanisms by which fiber may exert its hypocholesterolemic properties, including intestinal transit time, lipoprotein formation, transport and metabolism, and sites of absorption. The discussions in the following chapters will present an exposition of the current state of the art.

ACKNOWLEDGMENTS. Supported in part by a Career Research Award (HL00734) from the National Institutes of Health and a grant-in-aid from the Commonwealth of Pennsylvania.

REFERENCES

Beher, W. T., and Casazza, K. K., 1971, Effects of psyllium hydrocolloid on bile acid metabolism in normal and hypophysectomized rats, *Proc. Soc. Exp. Biol. Med.* **136**:253–256.

Berenson, L. M., Bhandaru, R. R., Radakrishnamurthy, B., Srinivasan, S. B., and Berenson, G. S., 1975, The effect of dietary pectin on serum lipoprotein cholesterol in rabbits, *Life Sci.* **16**:1533–1544.

Birkner, H. J., and Kern, F., Jr., 1974, *In vitro* adsorption of bile salts to food residues, salicylazosulfapyridine and hemicellulose, *Gastroenterology* **67**:237–244.

Cookson, F. B., Altschul, R., and Fedoroff, S., 1967, The effects of alfalfa on serum cholesterol in modifying or preventing cholesterol-induced atherosclerosis in rabbits, *J. Atheroscler. Res.* **7**:69–81.

Eastwood, M. A., and Hamilton, D., 1968, Studies on the absorption of bile salts to non-absorbed components of diet, *Biochim. Biophys. Acta* **152**:165–173.

Ershoff, B. H., and Wells, A. F., 1962, Effects of gum guar, locust bean gum and carrageenan on liver cholesterol of cholesterol-fed rats, *Proc. Soc. Exp. Biol. Med.* **110**:580–582.

Fahrenbach, M. J., Riccardi, B. A., Saunders, J. C., Lourie, I. N., and Heider, J. G., 1965, Comparative effects of guar gum and pectin on human serum cholesterol levels, *Circulation* **32**(Suppl. II):11.

Fisher, H., Soller, W. G., and Griminger, P., 1966, The retardation by pectin of cholesterol-induced atherosclerosis in the fowl, *J. Atheroscler. Res.* **6**:292–298.

Grande, F., 1974, Sugars in cardiovascular disease, in: *Sugars in Nutrition* (H. L. Sipple and K. W. McNutt, eds.), Academic Press, New York, pp. 401–437.

Hamilton, R. M. G., and Carroll, K. K., 1976, Plasma cholesterol levels in rabbits fed low fat, low cholesterol diets: Effect of dietary proteins, carbohydrates and fibre from different sources, *Atherosclerosis* **24**:47–62.

Jenkins, D. J. A., Leeds, A. R., Newton, C., and Cummings, J. H., 1975, Effect of pectin, guar gum and wheat fibre on serum cholesterol, *Lancet* **1**:1116–1117.

Judd, P. A., Kay, R. M., and Truswell, A. S., 1976, Cholesterol lowering effect of lignin in rats, *Proc. Nutr. Soc.* **35**:71A.

Kay, R. M., and Truswell, A. S., 1980, Dietary fiber: Effects on plasma and biliary lipids in man, in: *Medical Aspects of Dietary Fiber* (G. A. Spiller and R. M. Kay, eds.), Plenum Medical, New York, pp. 153–173.

Kiriyama, S., Okazaki, Y., and Yoshida, A., 1969, Hypocholesterolemic effect of polysaccharides and polysaccharide-rich foodstuffs in cholesterol-fed rats, *J. Nutr.* **97**:382–388.

Kritchevsky, D., 1964, Experimental atherosclerosis in rabbits fed cholesterol-free diets, *J. Atheroscler. Res.* **4**:103–105.

Kritchevsky, D., and Story, J. A., 1974, Binding of bile salts *in vitro* by non-nutritive fiber, *J. Nutr.* **104**:458–462.

Kritchevsky, D., and Tepper, S. A., 1965, Factors affecting atherosclerosis in rabbits fed cholesterol-free diets, *Life Sci.* **4**:1467–1471.

Kritchevsky, D., and Tepper, S. A., 1968, Experimental atherosclerosis in rabbits fed cholesterol-free diets: Influence of chow components, *J. Atheroscler. Res.* **8**:357–369.

Kritchevsky, D., Davidson, L. M., Shapiro, I. L., Kim, H. K., Kitagawa, M., Malhotra, S., Nair, P. P., Clarkson, T. B., Bersohn, I., and Winter, P. A. D., 1974, Lipid metabolism and experimental atherosclerosis in baboons: Influence of cholesterol-free, semisynthetic diets, *Am. J. Clin. Nutr.* **27**:29–50.

Kritchevsky, D., Tepper, S. A., Williams, D. E., and Story, J. A., 1977a, Experimental atherosclerosis in rabbits fed cholesterol-free diets. 7. Interaction of animal or vegetable protein with fiber, *Atherosclerosis* **26**:397–403.

Kritchevsky, D., Davidson, L. M., Kim, H. K., Krendel, D. A., Malhotra, S., VanderWatt, J. J., DuPlessis, J. P., Winter, P. A. D., Ipp, T., Mendelsohn, D., and Bersohn, I., 1977b, Influence of semipurified diets on atherosclerosis in African green monkeys, *Exp. Molec. Pathol.* **26**:28–51.

Kyd, P. A., and Bouchier, I. A. D., 1972, Cholesterol metabolism in rabbits with oleic acid-induced cholelithiasis, *Proc. Soc. Exp. Biol. Med.* **141**:846–849.

Leveille, G. A., and Sauberlich, H. E., 1966, Mechanism of the cholesterol-depressing effect of pectin in the cholesterol-fed rat, *J. Nutr.* **88**:209–214.

Moore, J. H., 1967, The effect of the type of roughage in the diet on plasma cholesterol levels and aortic atheroses in rabbits, *Br. J. Nutr.* **21**:207–215.

Munoz, J. M., Sandstead, H. H., Jacob, R. A., Logan, G. M., Reck, S. J., Klevay, L. M., Dintzis, F. R., Inglett, G. F., and Shuey, W. C., 1979, Effect of some cereal brans and TVP on plasma lipids, *Am. J. Clin. Nutr.* **32**:580–592.

Portman, O. W., 1960, Nutritional influences on the metabolism of bile acids, *Am. J. Clin. Nutr.* **8**:462–470.

Story, J. A., 1980, Dietary fiber and lipid metabolism: An update, in: *Medical Aspects of Dietary Fiber* (G. A. Spiller and R. M. Kay, eds.), Plenum Medical, New York, pp. 137–152.

Story, J. A., and Kritchevsky, D., 1976, Comparison of the binding of various bile acids and bile salts *in vitro* by several types of fiber, *J. Nutr.* **106**:1292–1294.

Story, J. A., and Kritchevsky, D., 1980, Nutrients with special functions: Dietary fiber, in: *Human Nutrition—A Comprehensive Treatise,* Volume 3A (R. B. Alfin-Slater and D. Kritchevsky, eds.), Plenum Press, New York, pp. 259–279.

Story, J. A., Czarnecki, S. K., Baldino, A., and Kritchevsky, D., 1977, Effect of components of fiber on dietary cholesterol in the rat, *Fed. Proc.* **36**:1134.

Tsai, A. C., Elias, J., Kelly, J. J., Lin, R. S. C., and Robson, J. R. K., 1976, Influence of certain dietary fibers on serum and tissue cholesterol levels in rats, *J. Nutr.* **106**:118–123.

Vahouny, G. V., Tombes, R., Cassidy, M. M., Kritchevsky, D., and Gallo, L. L., 1980, Dietary fibers. V. Binding of bile salts, phospholipids and cholesterol from mixed micelles by bile acid sequestrants and dietary fibers, *Lipids* **15**:1012–1018.

Wells, A. F., and Ershoff, B. H., 1961, Beneficial effects of pectin in prevention of hypercholesterolemia and increase in liver cholesterol in cholesterol-fed rats, *J. Nutr.* **74**:87–92.

Wells, A. F., and Ershoff, B. H., 1962, Comparative effects of pectin NF administration on the cholesterol-fed rabbit, guinea pig, hamster and rat, *Proc. Soc. Exp. Biol. Med.* **111**:147–149.

18

Modification of Bile Acid Spectrum by Dietary Fiber

JON A. STORY and JAMES N. THOMAS

1. INTRODUCTION

Among the disorders originally implicated as being related to intake of dietary fiber was the incidence of gallstones (Burkitt *et al.*, 1974). Modification of the bile acid spectrum of bile modifies its saturation and is currently being used as a treatment in gallstone patients (Hofmann, 1980).

Interest in the use of dietary fiber to modify the spectrum of biliary bile acids was stimulated by the observation that some sources of dietary fiber adsorbed bile acid *in vitro*, and that this adsorption was not equal for all bile acids (Eastwood and Hamilton, 1968; Kritchevsky and Story, 1974; Story and Kritchevsky, 1976). Adsorption by dietary fiber is greater with dihydroxy bile acids than trihydroxy, which could result in modification of individual bile acid pool sizes by causing differential excretion rates.

Changes in bile acid pool sizes can influence several phases of lipid metabolism in addition to bile saturation. Chenodeoxycholic acid inhibits cholesterol synthesis by altering the activity of 3-hydroxy-3-methylglutaryl coenzyme A reductase, the rate-limiting step in cholesterol synthesis (Coyne *et al.*, 1976; Cooper, 1976). Cholesterol absorption also requires bile salts in the small intestine, and variation in the bile acids available can alter absorption. Chenodeoxycholic acid has been shown to inhibit cholesterol absorption both in experimental animals (Wilson, 1972) and man (Ponz de Leon *et al.*, 1979).

JON A. STORY and JAMES N. THOMAS • Department of Foods and Nutrition, Purdue University, West Lafayette, Indiana 47907.

Phospholipid synthesis and secretion into bile is also controlled, to some extent, by bile acids (Hofmann, 1980). Thus modification of bile acid spectrum could conceivably alter (1) cholesterol absorption, either directly or through phospholipid availability; (2) cholesterol synthesis, directly or by altering availability in the intestine as a result of biliary secretion; (3) synthesis of chylomicrons by altered availability of phospholipids or a change in lipid absorption site; and (4) other lipid metabolism variables. These changes would have effects not only on bile saturation but also on cholesterol and lipoprotein metabolism.

2. CHANGES IN BILE ACID SPECTRUM

2.1. Human Studies

Modification of biliary bile acids in humans by some sources of dietary fiber has been examined by several groups (Table I). Pomare and Heaton (1973) fed 33 g wheat bran per day for 6–10 weeks to women with functioning gallbladders and others after cholecystectomy. Patients with normal gallbladders had a 49% decrease in deoxycholate (DC) and a 44% increase in chenodeoxycholate (CDC) in their bile salt pools, while patients without gallbladders displayed little change in bile salt pool composition. In a subsequent experiment, Pomare *et al.* (1976) observed similar changes in DC and CDC in response to 57 g wheat bran per day for 4–6 weeks in gallstone patients. Reduced dehydroxylation of cholate was suggested as a mechanism for the decrease in DC pool, and, as a result of

TABLE I. Dietary Fiber and Biliary Bile Acids in Man

		Bile salt pools,[a] percent change with fiber			
	Treatment	C	CDC	DC	Others
Pomare and Heaton (1973)	Wheat bran, 33 ± 10 g/day, 6–10 weeks	NC	+44	−49	—
Pomare *et al.* (1976)	Wheat bran, 57 g/day (av), 4–6 weeks	NC	+27	−33	—
Watts *et al.* (1978)	Wheat bran, 30 g/day, 2 months	+6	+6	−24	UDC, −9; LC, −27
Miettinen and Tarpila (1977)	Pectin, 40–50 g/day, 2 weeks	NC	−16	+22	—

[a] C, cholic acid; CDC, chenodeoxycholic acid; DC, deoxycholic acid; UDC, ursodeoxycholic acid; LC, lithocholic acid.

this reduction in DC, the size of the CDC pool increased. The increase in CDC pool size would have beneficial effects on bile saturation.

Watts *et al.* (1978) examined the effects of wheat bran in normal persons without gallstones. Twenty grams of wheat bran was given to the subjects for 2 months and samples of duodenal contents were taken after gallbladder contraction with cholecystokinin. A decrease in DC was again observed along with a modest increase in CDC and cholic. Decreases in ursodeoxycholate and lithocholate, minor constituents of biliary bile, were also observed. Coupled with these changes were decreases in the saturation index of those patients with supersaturated bile prior to bran treatment. These changes help to verify the importance of bile acid spectrum in bile saturation and potentially in formation of gallstones.

Supplementation of human diets with pectin does not appear to cause similar changes in bile acid spectrum (Miettinen and Tarpila, 1977). Average concentrations of DC were increased and of CDC decreased by supplementation with 20 g of pectin per day for 2 weeks, but these changes were "inconsistent" and not statistically significant. As with bran, supersaturated bile observed in one subject was brought into normal ranges by treatment with pectin.

Changes in fecal bile acid excretion and the spectrum of bile acids excreted have been observed in response to dietary fiber feeding in humans. Aries *et al.* (1971) reported a much higher rate of excretion of bile acids and higher levels of tri- and dihydroxy bile acids than monohydroxy and nonsubstituted bile acids in vegetarians as compared to persons consuming a mixed diet.

Wheat bran has been shown to have variable effects on bile acid excretion (Kay and Truswell, 1980), ranging from a decrease of 50% to a 240% increase in daily excretion. Eastwood *et al.* (1973) observed a small decrease (4%) in total bile acid excretion in response to 8 g of wheat bran twice daily for 3 weeks. This decrease occurred in both DC and lithocholic (LC) acids. In a second experiment, a somewhat larger decrease was observed during bran treatment (11%), but during the control period following bran treatment, daily excretion increased 50% over original levels. The increase in DC excretion was slightly larger (35%) than LC (23%).

Pectin causes consistent increases in bile acid excretion in humans (Table

TABLE II. Effects of Pectin on Human Fecal Steroid Excretion

	Kay and Truswell (1977)	Miettinen and Tarpila (1977)
Treatment	15 g/day; 3 weeks	40–50 g/day; 2 weeks
Total bile acids, mg/day	+106	+196
Neutral steroids, mg/day	+55	+58
Percent chenodeoxycholic derivatives	−2%	−3%

II). Miettinen and Tarpila (1977) found a 57% increase in bile acid excretion in normal and hypercholesterolemic subjects fed 40–50 g/day for 2 weeks. Normal individuals appeared to respond more dramatically than hypercholesterolemics. Kay and Truswell (1977) similarly found a 40% increase in bile acid excretion in response to a smaller dose of pectin (15 mg/day), given for 3 weeks. In both cases, there was little change in the spectrum of bile acids excreted. The two studies reported a 3 and 2% decrease in the fraction of bile acids found to be LC, the major excretory metabolite of CDC in humans.

2.2. Experimental Animal Studies

Experiments with hamsters have offered the only direct evidence concerning the effects of some isolated components of dietary fiber on gallstone formation. Using a lithogenic diet, Bergman and Van der Linden (1975) examined gallstone formation when lignin, pectin, or psyllium hydrocolloid was added (Table III). Lignin and psyllium completely prevented gallstone formation, but pectin only modestly reduced the number of animals with gallstones. Changes in biliary bile acids included a small decrease in C and an increase in CDC in pectin-fed animals, an increase in C and decrease in DC in lignin-fed animals, and an increase in DC and a decrease in CDC in psyllium-fed animals. No consistent change was observed in the groups in which gallstone formation was inhibited. Lignin has also been shown to desaturate bile in hamsters by decreasing cholesterol concentration (Kay *et al.*, 1979).

Modification of fecal bile acid excretion in rats has also been examined and provides some interesting data concerning the role of dietary fiber in determining the spectrum of bile acids excreted. Brydon *et al.* (1980) found that wheat bran (20%) reduced the concentration and daily excretion of bile acids in rats in comparison to a low-fiber diet (Table IV). The basal diet also had higher rates of excretion than a commercial ration. Changes in the spectrum of bile acids

TABLE III. Effects of Pectin, Lignin, and Psyllium Hydrocolloid on Diet-Induced Gallstones in Hamsters[a]

	Percent with stones	Bile composition,[b] percent			Bile acid pool, mg
		C	DC	CDC	
Control[c]	58	39	17	44	2.4
5% Pectin	42	31	19	50	—
5% Lignin	0	50	10	40	—
5% Psyllium	0	37	51	12	2.9

[a] Bergman and Van der Linden (1975).
[b] C, cholic acid; DC, deoxycholic acid; CDC, chenodeoxycholic acid.
[c] Lithogenic diet: 74.3% glucose, 20% casein, 5% salt mix, 0.5% vitamin mix, and 0.2% choline chloride.

TABLE IV. Effects of Bran on Colonic Bile Acids in the Rat[a]

	Low-fiber (control)	Low fiber + bran (20%)	Commercial diet
Total bile acids			
μmol/g dry contents	31.1	8.6	8.9
μmol/day	39.8	13.1	14.2
Bile acids, percent			
Cholic	16.3	10.7	13.8
Chenodeoxycholic	2.4	2.7	2.7
Deoxycholic	3.5	9.5	18.7
Lithocholic	1.1	3.6	10.5
Hyodeoxycholic	2.5	13.7	25.9
α-Muricholic	11.4	5.2	1.9
β-Muricholic	27.3	18.4	9.2
ω-Muricholic	35.5	36.2	17.1

[a] Brydon *et al.* (1980).

excreted in response to wheat bran included a shift from C to DC, indicating increased microbial degradation and changes in the metabolites of CDC, and an increase in LC and hyodeoxycholic at the expense of α- and β-muricholic acids. The distribution of bile acids between metabolites of C and CDC was not changed by bran feeding, but the percentage present as CDC derivatives was reduced in the rats fed a commercial ration.

We have recently examined the effects of alfalfa and whole oats on bile acid excretion in rats fed a semipurified diet with added cholesterol (Kelley *et al.*, 1981). As can be seen in Table V, the concentration of bile acids excreted was increased by both alfalfa and oats, while the total excretion per day was not increased, except when cholesterol was included in the oat diet. Cholesterol

TABLE V. Alteration of Bile Acid Excretion in Rats by Alfalfa and Oats[a]

	Control[b]		Alfalfa		Oats	
	−	+	−	+	−	+
Fecal bile acids						
mg/g	2.52	2.88	3.75	3.63	6.16	8.67
mg/day	22.53	25.92	15.77	17.55	23.12	31.64
Distribution						
Cholic	51	31	39	28	40	37
Chenodeoxycholic	49	69	61	72	60	63

[a] Kelley *et al.* (1981). Minus and plus signs indicate, respectively, the absence and addition (0.25%) of cholesterol at the expense of sucrose.

[b] Basal diet: 45% sucrose, 25% casein, 10% corn oil, 5% salt and vitamin mixes, and 15% cellulose. Alfalfa or whole ground oats added in place of cellulose.

feeding increased bile acid excretion in all groups, a primary mechanism by which rats prevent increases in serum cholesterol. This change in total excretion in response to cholesterol results from an increase in metabolites of CDC, resulting in decreased reabsorption and possibly changes in cholesterol synthesis and absorption as described earlier. Addition of alfalfa or oats to these diets also causes a change in the distribution of bile acids from 51 : 49 (C : CDC) in basal-fed animals to 39 : 61 and 40 : 60 in alfalfa- and oat-fed animals, respectively. This change in the nature of bile acids excreted may be a second mechanism (in addition to adsorption) by which bile acid excretion is increased.

We have also been examining the importance of this change in bile acid spectrum in excreting large amounts of accumulated cholesterol. Rats were fed the diet described above, but with 1% cholesterol added for 4 weeks, resulting in accumulation of 5–6 times control levels of liver cholesterol (Thomas *et al.*, 1980). Fecal excretion of bile acids after cholesterol feeding remained high until liver cholesterol levels returned to near normal (Table VI). The distribution of bile acids likewise remained high in CDC derivatives (75%) until liver cholesterol levels reached near normal, at which time they had returned to normal (65%). Addition of pectin to the diet during this regression phase resulted in higher, more prolonged bile acid excretion. However, this change did not result in a lower final liver cholesterol level. The percent of bile acids excreted as derivatives of CDC was also higher in pectin-fed animals. These changes also suggest an ability of some isolated polysaccharides to increase bile acid excretion in rats by altering the spectrum of bile acids produced, as normally occurs in cholesterol-fed rats. These changes could also result in modification of cholesterol synthesis and absorption.

TABLE VI. Regression of Liver Cholesterol Accumulation in Rats[a,b]

	Time,[c] weeks	B	BC	BC → B	BC → B + pectin
Liver cholesterol, mg/g	4	3.72	17.17	—	—
	6	—	—	5.31	4.47
	8	—	—	3.45	4.39
Fecal bile acids, mg/g	4	3.80	9.81	—	—
	6	5.37	—	10.37	14.05
	8	7.01	—	5.43	6.85
Percent CDC	4	70	85	—	—
	6	69	—	74	76
	8	64	—	64	65

[a] Thomas *et al.* (1980).
[b] B, basal; BC, basal + cholesterol. Basal diet: 60% sucrose, 25% casein, 10% corn oil, 5% salt and vitamin mix. Cholesterol added at 1% and pectin at 5% in place of sucrose.
[c] Basal or basal + cholesterol diets fed 4 weeks; animals on BC diet switched to basal or basal + pectin for weeks 5–8.

3. CONCLUSIONS

Alteration of the sizes of the bile acid pools can alter several phases of lipid metabolism and several disease states. Chenodeoxycholic acid appears to inhibit cholesterol synthesis and absorption and, possibly through these changes, desaturate bile. This change in bile saturation would decrease susceptibility of gallstone formation. Changes in cholesterol synthesis and absorption could also result in changes in serum cholesterol level, an important risk factor for coronary heart disease. The reduced reabsorption of the less soluble derivatives of chenodeoxycholic could increase colonic bile acid concentration, a positive risk factor in susceptibility to colon cancer (Hill and Aries, 1971). Thus changes in the spectrum of bile acids could potentially have wide-ranging effects on lipid metabolism and the diseases commonly associated with modifications in lipids.

Changes in individual bile acid pool sizes in response to dietary fiber could be caused by one or more of several effects of dietary fiber on bile acid metabolism. Adsorption of bile acids by dietary fiber is not uniform for all bile acids and may lead to a disproportionate excretion of some bile acids while conserving others. Adsorption of the dihydroxy bile acids is usually reported as greater than that of the trihydroxy bile acids. The decreased size of the DC pool in bran-fed humans could be explained in terms of these observations, but the concomitant increase in CDC could not. Our knowledge of the adsorption process in both the large and small intestine and of adsorption of minor bile acid constituents limits our ability to interpret all the data available. Other factors, such as solubility, will greatly influence reabsorption of some bile acids independent of adsorption by dietary fiber.

Several other controls of bile acid metabolism are possible as a result of consumption of dietary fiber. Changes in the regulation of the enzymes involved in bile acid synthesis, especially the microsomal 12-α hydroxylation of 7α-hydroxy-cholest-4-en-3-one, which results in synthesis of cholic acid rather than chenodeoxycholic acid, are of obvious interest. Little is known concerning the regulation of this enzyme except that it is responsive, in some species, to changes in thyroid hormone status (Beher, 1976). Modification of hormonal status by dietary fiber, as has been observed with insulin, could alter the amount of each bile acid synthesized. Dietary fiber also may alter distribution of cholesterol on lipoproteins, which could, in turn, affect the amount and type of bile acids synthesized.

We have observed changes in the spectrum of bile acids in bile and in feces in response to some sources or isolated components of dietary fiber. Direct interaction of dietary fiber with bile acids in the small and large intestines is partially responsible for these changes. The net effect of dietary fiber in regulating bile acid spectra is, however, a result of its actions in several phases of lipid metabolism. Our understanding of these effects is dependent on elucidation of the role of dietary fiber in lipoprotein metabolism, hormonal homeostasis, and many other regulatory processes. These changes in bile acid spectra help in our understanding of the effects of dietary fiber on human disease, but their part in the mechanism of these effects will require further evaluation.

ACKNOWLEDGMENTS. Supported, in part, by the Indiana Agricultural Experiment Station (paper No. 8601), the Northeast Chapter of the Indiana Affiliate of the American Heart Association, and the Showalter Trust.

REFERENCES

Aries, V. C., Crowther, J. S., Drasar, B. S., Hill, M. J., and Ellis, F. R., 1971, The effect of a strict vegetarian diet on the faecal flora and faecal steroid concentration, *J. Pathol.* **103:**54–56.

Beher, W. T., 1976, *Bile Acids: Chemistry and Physiology of Bile Acids and Their Influence on Atherosclerosis,* S. Karger, Basel, Switzerland.

Bergman, F., and Van der Linden, W., 1975, Effect of dietary fibre on gallstone formation in hamsters, *Z. Ernährungswiss.* **14:**218–224.

Brydon, W. G., Tadesse, K., Eastwood, M. A., and Lawson, M. E., 1980, The effect of dietary fibre on bile acid metabolism in rats, *Br. J. Nutr.* **43:**101–106.

Burkitt, D. P., Walker, A. R. P., and Painter, N. S., 1974, Dietary fiber and disease, *J. Am. Med. Assoc.* **229:**1068–1074.

Cooper, A. D., 1976, The regulation of 3-hydroxy-3-methylglutaryl coenzyme A reductase in the isolated purfused rat liver, *J. Clin. Invest.* **57:**1461–1470.

Coyne, M. J., Bonorris, G. G., Goldstein, L. I., and Schoenfield, L. J., 1976, Effect of chenodeoxycholic acid and phenobarbital on the rate limiting enzymes of hepatic cholesterol and bile acid synthesis in patients with gallstones, *J. Lab. Clin. Med.* **87:**281–291.

Eastwood, M. A., and Hamilton, D., 1968, Studies on the adsorption of bile salts to non-absorbed components of diet, *Biochim. Biophys. Acta* **152:**165–173.

Eastwood, M. A., Kirkpatrick, J. R., Mitchell, W. D., Bone, A., and Hamilton, T., 1973, Effects of dietary supplements of wheat bran and cellulose on faeces and bowel function, *Br. Med. J.* **4:**392–394.

Hill, M. J., and Aries, V. C., 1971, Faecal steroid composition and its relationship to cancer of the large bowel, *J. Pathol.* **104:**129–139.

Hofmann, A. F., 1980, The medical treatment of gallstones: A clinical application of the new biology of bile acids, in: *The Harvey Lectures, Vol. 74,* Academic Press, New York, pp. 23–48.

Kay, R. M., and Truswell, A. S., 1977, Effect of citrus pectin on blood lipids and fecal steroid excretion in man, *Am. J. Clin. Nutr.* **30:**171–175.

Kay, R. M., and Truswell, A. S., 1980, Dietary fiber: Effects on plasma and biliary lipids in man, in: *Medical Aspects of Dietary Fiber* (G. A. Spiller and R. M. Kay, eds.), Plenum Medical, New York, pp. 153–173.

Kay, R. M., Wayman, M., and Strasberg, S. M., 1979, Effect of autohydrolyzed lignin and lactulose on gall bladder bile composition in the hamster, *Gastroenterology* **76:**1167.

Kelley, M. J., Thomas, J. N., and Story, J. A., 1981, Modification of spectrum of fecal bile acids in rats by dietary fiber, *Fed. Proc.* **40:**845.

Kritchevsky, D., and Story, J. A., 1974, Binding of bile salts *in vitro* by non-nutritive fiber, *J. Nutr.* **104:**458–462.

Miettinen, T. A., and Tarpila, S., 1977, Effect of pectin on serum cholesterol, fecal bile acids and biliary lipids in normolipidemic and hyperlipidemic individuals, *Clin. Chim. Acta* **79:**471–447.

Pomare, E. W., and Heaton, K. W., 1973, Alteration of bile salt metabolism by dietary fibre (bran), *Br. Med. J.* **4**:262–264.

Pomare, E. W., Heaton, K. W., Low-Beer, T. S., and Espiner, H. J., 1976, The effect of wheat bran upon bile salt metabolism and upon the lipid composition of bile in gallstone patients, *Dig. Dis.* **21**:521–526.

Ponz de Leon, M., Carulli, N., Loria, P., Iori, R., and Zironi, F., 1979, The effect of chenodeoxycholic acid (CDCA) on cholesterol absorption, *Gastroenterology* **77**:223–230.

Story, J. A., and Kritchevsky, D., 1976, Comparison of the binding of various bile acids and bile salts *in vitro* by several types of fiber, *J. Nutr.* **106**:1292–1294.

Thomas, J. N., Kelley, M. J., and Story, J. A., 1980, Regression of lipid accumulation in rats: Effect of pectin, *Fed. Proc.* **39**:784.

Watts, J. M., Jablonski, P., and Toouli, J., 1978, The effect of added bran to the diet on the saturation of bile in people without gallstones, *Am. J. Surg.* **135**:321–324.

Wilson, J. D., 1972, The role of bile acids in the overall regulation of steroid metabolism, *Arch. Intern. Med.* **130**:493–505.

19

Dietary Fibers and Intestinal Absorption of Lipids

GEORGE V. VAHOUNY

1. INTRODUCTION

Based on extensive epidemiological, clinical, and experimental trials, there is accumulating evidence that dietary fibers may influence one or more risk factors in coronary disease (including plasma lipids and lipoproteins), diabetes, and obesity. Considering the physicochemical complexity of intact dietary fibers and the diverse nature of the disorders contributing to risk, it has not been reasonable to develop a unified hypothesis for the mechanism of action for the dietary fibers in general, nor for specific fiber isolates. Some fiber materials, notably the mucilaginous and gelling fibers, demonstrate acute influences on plasma lipid levels and on fecal steroid excretion, while the effect of the particulate fibers tends to be less pronounced, even in prolonged studies (Spiller and Amen, 1976; Kelsay, 1978; Kay and Strasberg, 1978; Inglett and Falkehag, 1979; Zilversmit, 1979; Kay and Truswell, 1980; Kritchevsky *et al.*, 1980; Kay, 1982).

Since, by definition, dietary fibers are undigestible by pancreatic and intestinal amyloltic enzymes, most studies on major effects of dietary fibers on metabolism of lipids, and of cholesterol in particular, have been limited to intraluminal aspects. The available literature suggests that even at this isolated site, there are numerous responses to dietary fiber intake, any or all of which may influence, to varying degrees, the rate of nutrient absorption in general (Table I). These include: effects on gastric emptying; overall rates of intestinal

GEORGE V. VAHOUNY • Department of Biochemistry, The George Washington University, School of Medicine and Health Sciences, Washington, D. C. 20037.

TABLE I. Possible Intraluminal Effects of Dietary Fibers

Gastric emptying
Altered gastrointestinal transit times
Nutrient interactions (adsorption and ion exchange)
Effects on hydration and phase distribution of nutrients
Bulk-phase interference with diffusion
Modified cell turnover and villar structure

transit; fiber adsorption of nutrients and other luminal materials; hydratability of the fiber, which influences distribution of water-soluble materials between the hydrated fiber and intraluminal water; bulk phase interference with free diffusion to the intestinal surface; and a multitude of additional effects, resulting from changes in intestinal structure and cell turnover, which undoubtedly occur in response to the other actions described. The extent to which each of these contributes to overall cholesterol absorption and metabolism has not been fully described, and is largely dependent on the physicochemical nature of the fiber source under investigation.

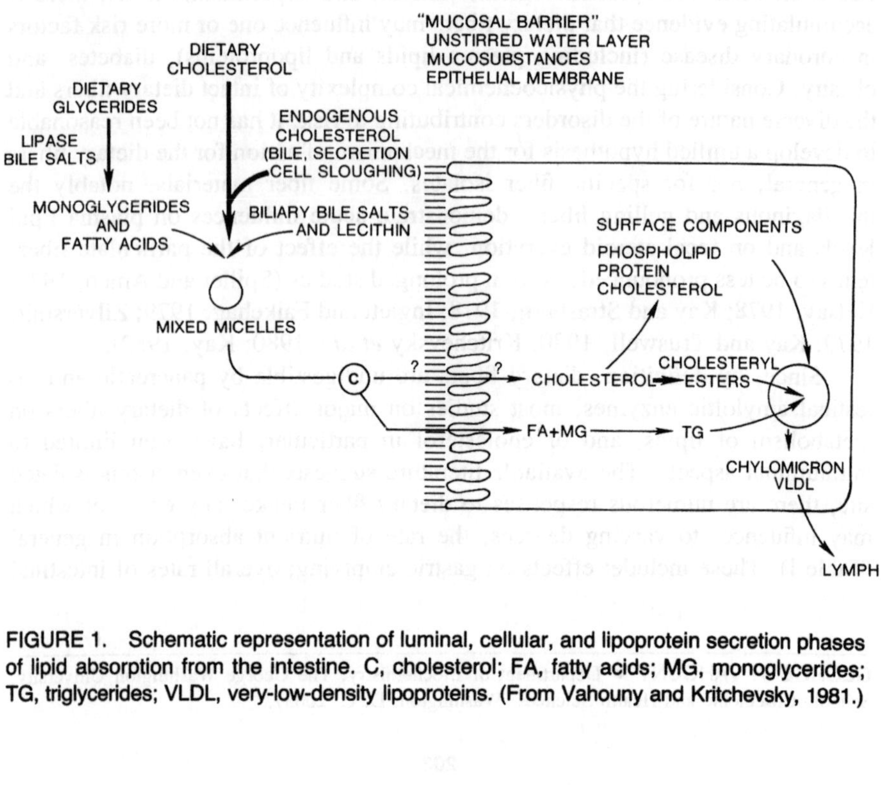

FIGURE 1. Schematic representation of luminal, cellular, and lipoprotein secretion phases of lipid absorption from the intestine. C, cholesterol; FA, fatty acids; MG, monoglycerides; TG, triglycerides; VLDL, very-low-density lipoproteins. (From Vahouny and Kritchevsky, 1981.)

2. MECHANISMS OF LIPID ABSORPTION

Several aspects of lipid absorption in general and cholesterol absorption in particular are well recognized and are schematically depicted in Fig. 1. The absorption of endogenous (biliary, secreted, or from sloughed cells) and of dietary cholesterol involves several complex physicochemical and enzymatic processes prior to appearance of cholesterol in the intestinal lymph. These relationships are treated in detail elsewhere (Treadwell and Vahouny, 1968; Dietschy and Wilson, 1970; Vahouny and Treadwell, 1976; Vahouny and Kritchevsky, 1981; Gallo, 1982) and are summarized as follows.

2.1. Luminal Phase

Only unesterified cholesterol is absorbed to any significant extent (Treadwell and Vahouny, 1968), and solubilization of this sterol in mixed micelles composed of bile components (the amphipaths, bile salts, and phospholipids) and products of triglyceride digestion (monoglycerides and fatty acids) is a requisite for efficient cellular transport. The distribution of cholesterol between the oil and micellar phases is influenced by the content and composition of dietary fat (Treadwell and Vahouny, 1968; Sylven and Borgström, 1969, Jandacek *et al.*, 1976). Thus, dietary fat is of importance in both luminal solubilization of cholesterol and in later steps, by providing significant levels of core lipids associated with formation of chylomicrons and very-low-density lipoproteins (VLDL).

There is a membrane barrier to lipid absorption, which is composed of at least three components: the unstirred water layer (UWL), a mucopolysaccharide "coat," and the highly differentiated membrane of the intestinal epithelial cell. Although the exact mechanism of transmembrane transport of lipids is not known, it has been demonstrated that the UWL presents the major resistance to transport of fatty acids and cholesterol into the cell (Salee *et al.*, 1972). This resistance is diminished by micellar bile salts, allowing more efficient transport of monomolecular lipids across this transport barrier (Westergaard and Dietschy, 1976). Furthermore, limited studies, presented below, suggest that mucosal surface-associated sialomucopolysaccharides also limit sterol transport and that this "barrier" is also overcome by micellar concentrations of bile salts and phospholipids.

2.2. Cellular Phase

The cellular phase of cholesterol absorption has been largely neglected due to technical difficulties of sterol exchanges and other problems inherent to broken cell preparations. It has been postulated that cholesterol exchanges rapidly between lipoproteins and organelles of the epithelial cell (Glover and Green, 1957;

Glover *et al.*, 1959). Sterol carrier proteins have been described in a variety of tissues having rapid cholesterol turnover (e.g., Kan and Ungar, 1974; Noland *et al.*, 1980), but to date, these types of carriers have not been evaluated in intestinal mucosa.

Cholesterol is absorbed exclusively into the intestinal lymphatic system (Biggs *et al.*, 1951; Hyun *et al.*, 1967), in contrast to absorption of fatty acids. The route of transport of these latter lipids is dependent on the extent of reesterification into triglycerides. Even with long-chain fatty acids, a significant portion is transported into portal blood associated with albumin, and this is increased under conditions which depress intestinal triglyceride synthesis (Hyun *et al.*, 1967). This aspect of fatty acid transport has not received sufficient attention, and may be of considerable importance during conditions in which dietary triglyceride hydrolysis is disturbed, or during rapid transit of lipids into the lower small intestine. In the upper intestine, chylomicron and VLDL synthesis and secretion are efficient and transport most of the absorbed lipid, following reesterification into triglycerides. However, in the lower small intestine, lipoprotein synthesis and secretion are less efficient (Sabesin *et al.*, 1975). Fat is still absorbed at this site, as evidenced by its lack of appearance as fecal neutral lipid. Thus, the absorbed fatty acids, which are unable to be transported efficiently as triglyceride in lipoprotein transport particles, likely appear in portal blood as albumin-associated unesterified fatty acids (Hyun *et al.*, 1967).

Following intragastric or intraduodenal administration of labeled cholesterol, radioactivity appears in intestinal lymph for periods up to 96 hr. The peak of lipid absorption, however, occurs within the first 8 hr, suggesting the distribution of labeled cholesterol among more than one cellular "pool" of cholesterol. At least one of these pools is rapidly exchangeable, as evidenced by the relatively rapid appearance of cholesterol label and mass into lymph. One or more others, however, are slowly exchangeable and continue to release cholesterol, which is largely unesterified, in a manner comparable to that of the slowly exchangeable "pools" demonstrated in whole-body turnover studies (Goodman and Noble, 1968).

During the net transfer of cholesterol from lumen to intestinal lymph, 70–90% becomes esterified to fatty acids in the mucosa (Treadwell and Vahouny, 1968). The fatty acids tend to reflect those in the diet, and the newly formed esters are incorporated into the core lipids of intestinal transport particles. Small amounts of unesterified cholesterol also appear in lymph, usually associated with the lipoprotein coat or membrane components.

The cellular esterification of cholesterol has been considered a rate-limiting step in absorption based on a variety of physiological and pharmacological evidence (Dietschy and Wilson, 1970; Treadwell and Vahouny, 1968; Vahouny and Treadwell, 1976; Vahouny and Kritchevsky, 1981; Gallo, 1982). This step can be attributed to either or both of two described enzyme systems, sterol ester hydrolase (SEH) and acyl CoA:cholesterol:*o*-acyl transferase (ACAT). SEH of pancreas and intestine have been extensively investigated in our laboratories (Hyun *et al.*, 1969; Calame *et al.*, 1975; Gallo *et al.*, 1977, 1980). The large majority of physiological evidence suggests that this mucosal esterifying enzyme is derived largely, if not entirely, from the enzyme in the exocrine pancreatic secretions. This transfer of pancreatic luminal enzyme into mucosal cells has been verified by biochemical (Gallo *et al.*, 1977) and immunological (Gallo *et al.*, 1980) techniques. The homogeneous enzyme preparation reversibly catalyzes both synthesis and hydrolysis of cholesterol esters depending on pH, and does not require prior activation of substrate (Hyun *et al.*, 1969). Regulatory aspects are implied by the fact that the enzyme protein is elaborated as a 65,000-MW monomer which has enzymatic activity against water-soluble esters but cannot synthesize or hydrolyze cholesterol esters (Calame *et al.*, 1975). Six of these subunits are aggregated by specific binding of cholic acid into the active 400,000-MW enzyme, which, unlike the monomeric protein, is heat and detergent stable, and insensitive to proteolytic enzymes and sulfhydryl reagents (Hyun *et al.*, 1969; Calame *et al.*, 1975).

The second esterifying activity is associated with microsomes and has therefore not been well characterized (Haugen and Norum, 1976; Norun *et al.*, 1977). In contrast to the stimulatory effects of bile salts on sterol ester hydrolase activity and on cholesterol absorption (Treadwell and Vahouny, 1968), the ACAT activity is inhibited by bile salts (Norum *et al.*, 1977). Further studies are required to elucidate the relative roles of these two esterifying systems in absorption of luminal cholesterol.

During absorption and esterification of endogenous and exogenous cholesterol, the newly derived esters do not accumulate in significant amounts in the intestinal mucosa (Treadwell and Vahouny, 1968); this suggests rapid incorporation and subsequent transport of this molecule in lipoprotein transport particles.

Because of the complex requirements and sites of lipoprotein component synthesis and organization, and difficulties in assessing cellular aspects of this process, most studies in this area have been directed at analyses of composition of intestinal and thoracic duct lymph lipoproteins. Furthermore, the majority of detailed studies on intestinal lipoprotein synthesis and release have dealt with transfer of the triglyceride components of luminal lipids, and considerably less attention has been given to cholesterol transport.

2.3. Secretory Phase—Lipoprotein Synthesis and Secretion

In contrast to triglyceride transport, which is largely associated with chylomicron synthesis and secretion (Vahouny *et al.*, 1980a), the transport of newly esterified cholesterol in lymph may be associated with both the chylomicron and nonchylomicron lipoprotein fractions of lymph (Vahouny *et al.*, 1980a). After fasting or feeding, rat intestinal lymph contains lipid transport particles comparable to VLDL of plasma with respect to flotation, composition, and electrophoretic properties. This lipoprotein fraction contains about half of the triglyceride and cholesterol of fasting lymph and has been shown to be of intestinal origin. During absorption and increased secretion of chylomicrons, the distribution of cholesterol transiently shifts toward chylomicron transport and is proportionately less in the VLDL fraction.

However, even during fat transport, when 70–90% of lymph triglycerides are associated with the chylomicron fraction and only 10–25% with the VLDL fraction, the "endogenous" cholesterol of lymph is almost equally distributed among these lipoproteins. Furthermore, this distribution appears to be dependent on the fat load and on the degree of unsaturation of the dietary fat, both of which markedly influence chylomicron size (Fraser *et al.*, 1968). The distribution of lipids between these lipoproteins may also be influenced by sex hormones (Vahouny *et al.*, 1980a).

The formation and release of intestinal lipoprotein transport particles are dependent on protein and phospholipid synthesis since these represent two important surface components of lipoproteins. Protein synthesis inhibitors, such as cycloheximide and puromycin, interrupt the assembly of intestinal transport particles and their subsequent release into lymph (Sabesin and Isselbacher, 1965). This has been largely attributed to interference with synthesis of important surface apolipoproteins, although these inhibitors have also been shown to alter intestinal phospholipid metabolism (Vahouny *et al.*, 1977; O'Doherty *et al.*, 1973). Under these conditions lipid generally accumulates in the epithelial cells, mimicking the genetic lipid transport abnormality, abetalipoproteinemia (Sabesin and Isselbacher, 1965).

Among the apolipoproteins associated with lymph chylomicrons and VLDL, apolipoprotein (apo) B, apo A-I, and apo A-IV are synthesized in the intestine in significant amounts, while other apoproteins (apo E and apo C), which become associated with the native particles, are of hepatic origin, and are "acquired" upon secretion of the nascent transport particles into lymph. It is generally held that apo B is a major determinant of chylomicron and VLDL transport by the intestine (Kostner, 1976).

Despite the fact that intestinal chylomicrons and VLDL have similar apolipoprotein compositions, recent evidence (Vahouny *et al.*, 1977, 1980) suggests that protein synthesis inhibition results in differential effects on triglyceride and cholesterol absorption in lymph, and that this effect, which is sex dependent, may be a response to differential inhibition of chylomicron and VLDL formation and/or secretion. This and other evidence suggesting a major role for VLDL in cholesterol absorption imply differences between factors regulating triglyceride and cholesterol transport from the intestine.

Thus, several steps in the overall process of cholesterol absorption from the intestinal tract may have a profound influence on the efficiency of absorption. These include the physical state of cholesterol in the intestinal lumen, the crucial role of fats and fatty acids in this process, the rates of transfer across the epithelial barrier, the activity of the cholesterol esterifying system(s), and the formation and release of lipoproteins in the intestinal cells.

2.4. Overview of Cholesterol Metabolism

As summarized in Fig. 2, overall cholesterol balance in the body is influenced by several factors. Cholesterol input is largely from two sources: intestinal absorption of dietary and/or endogenous cholesterol, and cholesterogenesis, which is largely a function of the liver and intestine in the adult. The extent of absorption, albeit inefficient at best, nevertheless plays an important role in regulation of cholesterogenesis by a feedback mechanism, and this appears to be a function of the rate-limiting enzyme hydroxymethylglutaryl coenzyme A reductase (Treadwell and Vahouny, 1968; Dietschy and Wilson, 1970; Vahouny and Treadwell, 1976). As a rough estimate, about two-thirds of the body cholesterol is derived by cholesterogenesis and one-third via absorption of endogenous and dietary cholesterol.

Output of cholesterol from the body occurs almost entirely from the bile as the neutral sterol, cholesterol, and as the major cholesterol metabolites, the bile acids. Of the unabsorbed cholesterol, the majority is hydrogenated by colonic bacteria and excreted as coprostanol in the neutral sterol fraction of feces. The secreted bile acids in bile are composed largely of conjugates of the primary bile acids, cholic and chenodeoxycholic acids, and the secondary bile acid, deoxy-

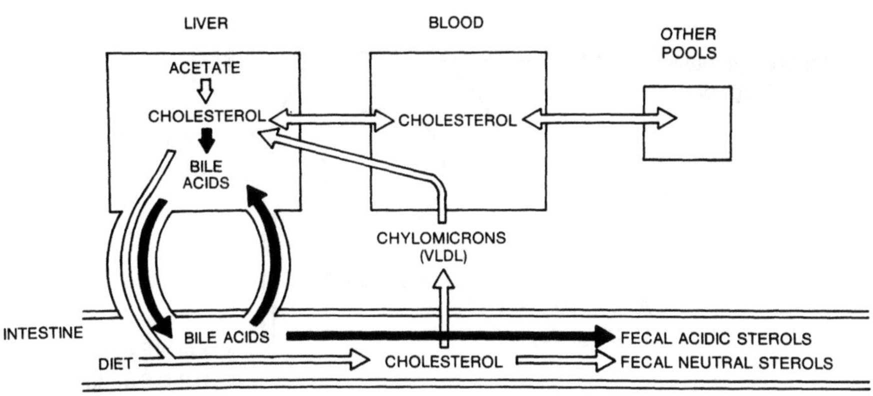

FIGURE 2. Overview of the metabolism of cholesterol in the body. VLDL, very-low-density lipoprotein. (Adapted from Dietschy and Wilson, 1970.)

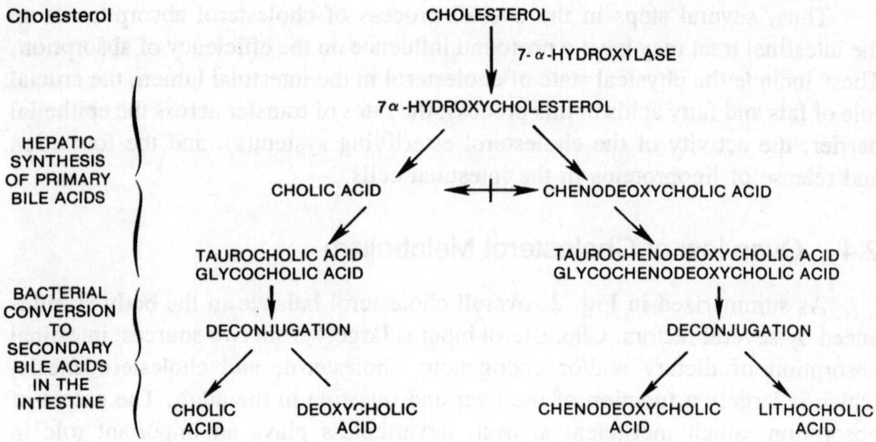

FIGURE 3. Pathways for the formation of primary and secondary bile acids, showing only major intermediates. (Adapted from Vahouny and Treadwell, 1976.)

cholic acid (see Fig. 3). These can be passively reabsorbed along the entire length of the small intestine and actively absorbed in the lower ileum. As shown in Fig. 4, resorption of bile acids exceeds 95%, contributing to efficient maintenance of the total enterohepatic pool of these cholesterol metabolites. Nevertheless, daily fecal excretion of 0.2–0.6 g of bile acids as fecal acidic steroids represents about 40% of the total cholesterol loss from the body. This occurs largely as the bacterial 7α-dehydroxylation products of cholic and chenodeoxycholic acids, namely deoxy- and lithocholic acid, respectively.

As is evident from Fig. 2, intestinal influences decreasing cholesterol ab-

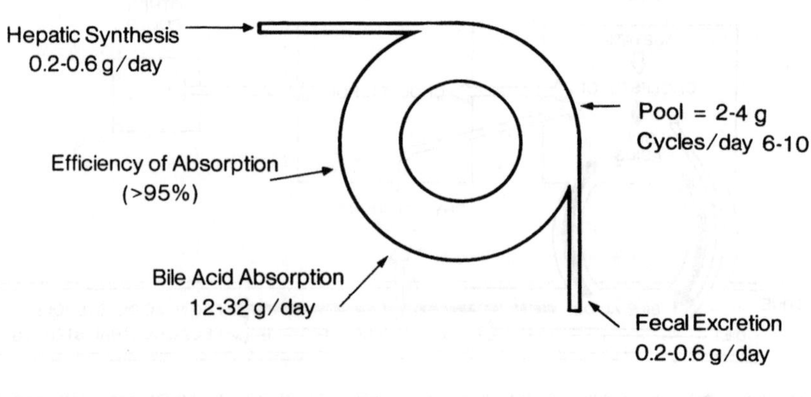

FIGURE 4. Schematic representation of the enterohepatic circulation of cholesterol in man. (Adapted from Hofmann and Mekhyian, 1973.)

sorption or increasing excretion of fecal neutral or acidic steroids can alter the balance of input and output and represent major therapeutic approaches to reducing plasma cholesterol levels.

3. DIETARY FIBERS AND CHOLESTEROL ABSORPTION

3.1. Background

Portman and Murphy (1958) initially demonstrated that rats maintained on a commercial stock diet, which contains alfalfa, excreted significantly more bile acids than did rats fed a semisynthetic diet. Similar observations have been made by others (Balmer and Zilversmit, 1974; Gustafsson and Norman, 1969; Kritchevsky *et al.*, 1973), and it has been shown that this effect is independent of plant sterols or fats in the stock ration (Portman, 1960).

Lin and co-workers (1957) reported that the inclusion of pectin in the diet resulted in an increased excretion of fecal saponifiable and nonsaponifiable lipids. By the balance technique, it was determined that "apparent cholesterol absorption" was decreased by 16–88% by feeding 500 mg pectin daily for 6 days.

Keys *et al.* (1960, 1961) reported that diets containing indigestible polysaccharides reduced blood cholesterol levels in humans, and Wells and Ershoff (1961) demonstrated that addition of pectin to a cholesterol-containing diet of rats largely counteracted the increase in plasma and liver cholesterol due to cholesterol feeding.

3.2. Acute Studies

The reports cited above prompted us to investigate the direct effects of citrus pectin on the absorption and lymphatic transport of cholesterol in the rat (Hyun *et al.*, 1963). Comparative studies were also carried out with cholestyramine, a commercial ion-exchange resin, which had been reported to effectively sequester bile acids *in vitro* (Tennant *et al.*, 1959), to reduce hypercholesteremia in humans (Bergen *et al.*, 1959), and to depress cholesterol-induced atherosclerosis in cockerels (Tennant *et al.*, 1960). These studies are described in some detail (Table II), since they typify the data obtained in studies using rats with indwelling catheters in the intestinal or thoracic duct lymphatic channels (Treadwell and Vahouny, 1968).

The direct intragastric administration of an aqueous emulsion containing cholestyramine at a level of 50 mg (group 2) resulted in a reduction in endogenous cholesterol in lymph, and this was further exaggerated by increasing the dose to 238 mg (group 3). This latter level was 2 molar equivalents with respect to the bile salt administered, and the inhibition observed was presumably due to

TABLE II. Acute Effects of Pectin and Cholestyramine on Absorption and Lymphatic Transport of Cholesterol[a]

Group[b]	Lymph cholesterol, mg Free	Lymph cholesterol, mg Esterified	[4-^{14}C]Cholesterol absorption, percent	Percent inhibition
1. Control	5.0 ± 0.1	9.9 ± 0.8	—	
2. Cholestyramine (CT) (50 mg)	5.1 ± 0.2	8.2 ± 0.3	—	17.2
3. Cholestyramine (238 mg)	3.5 ± 0.4	5.2 ± 1.0	—	47.5
4. Pectin (250 mg)	5.1 ± 0.4	9.8 ± 1.1		1.0
5. Complete emulsion	6.8 ± 0.3	24.0 ± 1.3	37.7 ± 1.8	—
6. Complete emulsion + CT (50 mg)	5.8 ± 0.4	19.9 ± 1.9	27.7 ± 4.0	26.5
7. Complete emulsion + CT (238 mg)	6.0 ± 0.3	9.0 ± 1.7	11.0 ± 2.0	70.9
8. Complete emulsion + pectin (250 mg)	5.9 ± 0.2	20.2 ± 0.7	30.7 ± 2.4	18.6

[a] Data derived from Hyun *et al.* (1963).
[b] All groups received an intragastric emulsion (3 ml) containing 50 mg albumin, 292 mg oleic acid, and 288 mg sodium taurocholate in addition to the indicated supplements.

interference with normal enterohepatic circulation of the expanded bile acid pool. Pectin, however, also administered at a level of 250 mg (in two separate doses, 1 hr apart), had no influence on 24-hr levels of endogenous unesterified or esterified cholesterol (group 4).

The administration of 50 mg [4-^{14}C]cholesterol in the complete emulsion containing albumin, oleic acid, and taurocholate resulted in a substantial increase in total lymph cholesterol, of which 89% of the newly appearing cholesterol in lymph was esterified. From tracer analysis, almost 38% of the fed cholesterol dose was absorbed over the 24-hr test period. The addition of 50 mg cholestyramine to the test dose significantly reduced lymph cholesterol and resulted in a 26% inhibition of absorption of the administered cholesterol. Again, with the higher dose of cholestyramine (group 7), a dramatic reduction in total lymph cholesterol was observed and absorption of the labeled cholesterol was reduced to only 30% of control.

Addition of pectin to the complete emulsion (125 mg in the emulsion and 125 mg in 3 ml saline, 1 hr later) also reduced lymph cholesterol and absorption of labeled cholesterol, albeit to a lesser extent than with the higher dose of cholestyramine. These studies provided the first direct demonstration that at least one mechanism by which pectin (and perhaps other dietary fiber components) exhibits a hypocholesteremic response in humans and experimental animals occurs by direct interference with intestinal absorption of cholesterol. They did not, however, address the issue of the mechanism by which pectin interferes with absorption of administered cholesterol.

3.3. Chronic Feeding Studies

Leveille and Sauberlich (1966) fed rats for 3–4 weeks on a defined diet containing 1% cholesterol or 1% cholesterol and 5% pectin. Using [4-^{14}C]cholesterol, it was found that less isotope accumulated in the liver, and a greater percentage of radioactivity was recovered in stools of animals on the pectin diet, supporting the results of earlier studies (Lin *et al.*, 1957; Hyun *et al.*, 1963). Although there was no difference in the fecal neutral sterol fraction between the two groups, rats fed 5% pectin excreted significantly more (132%) acidic steroids or bile acids. These results suggested that the effect of pectin on cholesterol absorption might be mediated by interference with the enterohepatic circulation of bile acids in a manner comparable to that observed with cholestyramine (Tennant *et al.*, 1959). Further evidence for bile acid sequestration by pectin was obtained in *in vitro* studies on bile acid transport in isolated intestinal segments from the lower ileum. In these trials, the presence of 0.18% pectin in the mucosal medium reduced bile acid transport by about 50% as judged by either serosal to mucosal ratios or by the final serosal concentration. Finally, the possibility that the effect of pectin on bile acid recirculation was mediated by altered intestinal microflora was largely eliminated in studies comparing the effects of pectin on plasma and liver cholesterol levels in control animals and rats treated with 1% succinylsulfathiazole. These studies represent the first definitive data on a mechanism of action of specific dietary fiber components on aspects of cholesterol absorption in the rat.

In most subsequent studies with other dietary fibers or fiber components in various species, decreases in plasma cholesterol levels or increases in fecal acidic and/or neutral sterols have been used as indicators of effects on cholesterol absorption. These approaches, albeit indirect, suggest that the gelling and mucilaginous fibers show the most consistent effects, while the effects of particulate fibers are variable. Thus, increased fecal output of acidic steroids have been reported for fiber isolates such as pectin (Leveille and Sauberlich, 1966; Jenkins *et al.*, 1976b; Kay and Truswell, 1977; Miettinen and Tarpila, 1977; Forman and Schneeman, 1980) and guar gum (Jenkins *et al.*, 1976b). In one of the more definitive studies on the effect of pectin feeding in humans, Kay and Truswell (1977) reported a reduction in plasma cholesterol of 13%, an increase in fecal fat of 44%, increased fecal neutral sterols (17%), and increased fecal acidic steroids (33%).

Other fiber sources have been reported to affect similar parameters of cholesterol metabolism, including bagasse (Walters *et al.*, 1975), cellulose (Stanley *et al.*, 1973), psyllium seed colloid or Metamucil® (which contains a large proportion of hemicellulose) (Stanley *et al.*, 1973; Beher and Casazza, 1971; Forman *et al.*, 1968), and various food-derived fibers (Antonis and Bersohn, 1962). The extensive bibliography of positive and negative effects of various dietary fibers and fiber isolates on these parameters of cholesterol metabolism is summarized elsewhere (Spiller and Amen, 1976; Kelsay, 1978; Kay and Strasberg, 1978; Inglett and Falkehag, 1979; Zilversmit, 1979; Kay and Truswell, 1980; Kritchevsky *et al.*, 1980).

With the exception of the early reports of Lin *et al.* (1957) and of Leveille and Sauberlich (1966) cited above, critical studies on the effects of prolonged feeding of dietary fibers on cholesterol absorption have been limited.

Kiriyama *et al.* (1968), using the balance technique to determine absorption, reported that both pectin and carboxymethylcellulose interfered with the cholesterol-induced increase in plasma and hepatic cholesterol in rats, while only pectin showed a significant effect on cholesterol absorption per se.

Kritchevsky *et al.* (1974) fed rats for 21 days on defined diets containing either cellulose (14%) or alfalfa meal (24.8–57.2%) and with variations in the levels of dietary carbohydrate, protein, or fat. Labeled cholesterol was given orally 24 hr prior to sacrifice. In all groups given alfalfa supplement, serum radioactivity was significantly less than in cellulose-fed groups, and excretion of labeled fecal neutral steroids was higher by 167–182%.

However, Balmer and Zilversmit (1974) used both the plasma isotope ratio method (Zilversmit, 1972) and the fecal isotope ratio method (Borgström, 1968) to determine effects of dietary fiber on cholesterol absorption in the rat. For these studies, diets were fed for one week prior to study and consisted of a semisynthetic mixture containing no dietary fiber and a commercial mouse stock diet. It was concluded that the nondigestible components of the stock diet did not inhibit cholesterol absorption despite a major influence of plasma cholesterol concentration and turnover on fecal excretion of both acidic and neutral steroids.

Using a different approach, Kelly and Tsai (1978) compared the effects of pectin, gum arabic, and agar on absorption, synthesis, and turnover in the rat. Animals trained to eat during a 2-hr period were fed defined diets containing 0.2% cholesterol and 5% levels of the fiber materials for 14 days prior to study. Twelve hours after an 8-g meal containing [4-^{14}C]cholesterol, absorption was determined by analysis of serum, liver, and carcass radioactivity. Serum cholesterol levels were lower in the pectin- and agar-fed groups, but liver cholesterol was less than controls only in animals given pectin. Cholesterol absorption was determined to be 49% of the administered dose in the control group, and was reduced to 35, 41, and 41% in the groups given pectin, gum arabic, and agar, respectively. Furthermore, it was demonstrated that pectin also increased the turnover of labeled cholesterol compared to controls.

As in the acute studies, absorption of cholesterol in the chronic studies cited above was determined from diets containing the fiber component under investigation. Thus, any effects of fiber feeding on cellular or hormonal aspects of intestinal function were largely overshadowed by a variety of intraluminal effects currently recognized for dietary fiber action (e.g., gastric emptying, altered transit, bulk-phase interference).

3.4. Direct versus Indirect Effects of Dietary Fibers on Cholesterol Absorption

In order to circumvent direct interactions of the diet or dietary fiber component on the test dose of cholesterol during absorption studies, we have fed defined diets to rats for 3 days (Vahouny *et al.*, 1978) or 5–6 weeks (Vahouny *et al.*, 1980b) prior to determination of lymphatic absorption of [4-^{14}C]cholesterol. Animals were fasted 24 hr prior to intraduodenal administration of 0.8 ml of a mixture containing a tracer dose of [1,2-^{3}H]cholesterol, 70 μg triolein, and 6.8% nonfat dry milk. Thus, effects of the diet per se on direct interactions with the cholesterol or on differential rates of gastric emptying were avoided.

In the short-term feeding studies (Vahouny *et al.*, 1978), inclusions of alfalfa meal or a yeast cell wall polyglycan (Robbins and Seeley, 1977) at a level of 15% of the defined diet had little effect on the subsequent rate or extent of appearance of labeled cholesterol into thoracic duct lymph. Prior feeding of a diet containing 15% wheat bran or 15% cellulose resulted in reduced cholesterol absorption to about 48% and 65% of control, respectively. The effects compared favorably to those obtained following feeding of diets containing 2% cholestyramine for three days. In all test groups, there were no differences in intestinal transit times, discounting this effect of fiber feeding as a major influence on sterol absorption. The possible retention of fiber in the intestine, perhaps by interaction with the intestinal surface, was, however, not discounted.

The longer-term feeding studies (Table III) gave results not entirely consistent with those obtained following 3-day feeding. The administration of cholestyramine as 2% of the diet again resulted in a dramatic effect on subsequent absorption of tracer cholesterol into lymph. Of the whole-fiber isolates, wheat bran and the yeast cell wall glycan showed minimal effects on absorption (79 and 82% of control, respectively), while feeding of 15% alfalfa-containing diets resulted in a 50% reduction in cholesterol absorption. Animals on all three test materials showed a tendency toward reduced transit times, although these were significant only in animals fed wheat bran.

Of the fiber components, feeding of 15% pectin-containing diets for 6 weeks resulted in a subsequent reduction in lymphatic absorption of cholesterol to 70% of control. These data suggest that the effects of pectin on cholesterol absorption

TABLE III. Effects of Prolonged Feeding of Dietary Fibers or Fiber Components on Intestinal Transit Times and Lymphatic Absorption of Cholesterol[a]

Dietary inclusion[b]	Intestinal transit times, hr	Lymph flow rates, ml/hr	Lymphatic cholesterol absorption, percent of administered dose/24 hr
Control	15.6 ± 1.5	5.8 ± 1.8	52.3 ± 3.0
Wheat bran	11.1 ± 2.4[c]	3.7 ± 0.8	41.4 ± 3.8
Alfalfa meal	12.6 ± 1.1	4.7 ± 0.8	26.9 ± 3.4[c]
Yeast cell wall	13.0 ± 4.9	2.6 ± 1.0	42.7 ± 7.6
Cellulose	11.5 ± 0.4	3.2 ± 0.7	12.5 ± 3.0[c]
Pectin	12.5 ± 0.7	2.8 ± 0.3[c]	36.4 ± 3.5[c]
Cholestyramine	11.1 ± 2.1	3.1 ± 0.9	19.9 ± 5.1[c]

[a] Vahouny *et al.* (1980b).
[b] Animals were fed *ad libitum* for 5–6 weeks on defined diets containing 15% levels of the fibers indicated at 2% cholestyramine. Transit times were determined following a single intragastric dose of 1 ml 10% carmine red dye. Following catheterization of the left thoracic duct, animals were fasted 24 hr prior to receiving a single intraduodenal test meal (0.8 ml) containing [1,2-^{3}H]cholesterol, 70 μg triolein, and 6.8% nonfat dry milk.
[c] $p < 0.05$.

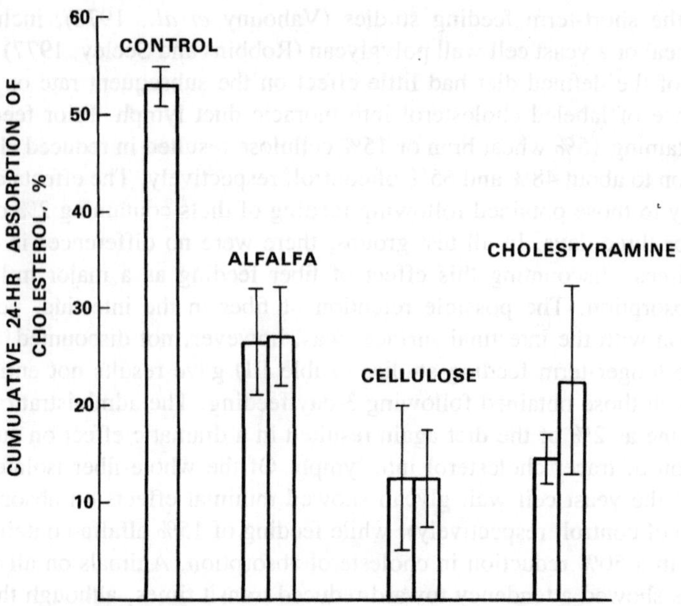

FIGURE 5. Reproducibility of the effects of feeding defined diets containing 2% cholestyramine or 15% levels of alfalfa meal or cellulose on subsequent lymphatic absorption of cholesterol administered as a single intraduodenal emulsion. Data represent means ± SEM for 4–5 animals. (From Vahouny *et al.*, 1980b.)

observed in earlier studies is not entirely dependent on simultaneous administration of the fiber with the lipid. As in the short-term studies, feeding of diets containing 15% cellulose showed the most dramatic effects on cholesterol absorption (Table III), and, as shown in Fig. 5, these results were reproducible. As in the case of wheat bran diets, the cellulose-containing diet resulted in a significant decrease in intestinal transit times, as well as a decrease in variance in transit times between animals.

The effects, or lack thereof, of the various dietary inclusions are not unique for cholesterol absorption. The test emulsion containing 70 μg triolein was labeled with glyceryl tri[1-^{14}C]oleate in order to determine effects on triglyceride digestion and absorption. Since many of the intraluminal and cellular factors regulating cholesterol and fatty acid or monoglyceride absorption are similar, direct intraluminal or cellular transport effects of the fiber inclusions could be expected to influence absorbability of both lipid classes. As shown in Fig. 6, each dietary inclusion affected the lymphatic absorption of triglyceride digestion products to the same relative extent as their effects on cholesterol absorption. These data suggest a generalized effect, when observed, on aspects of lipid absorption common to both cholesterol and triglycerides. However, as discussed in Section 4, this generalization may be inaccurate, based on the known differ-

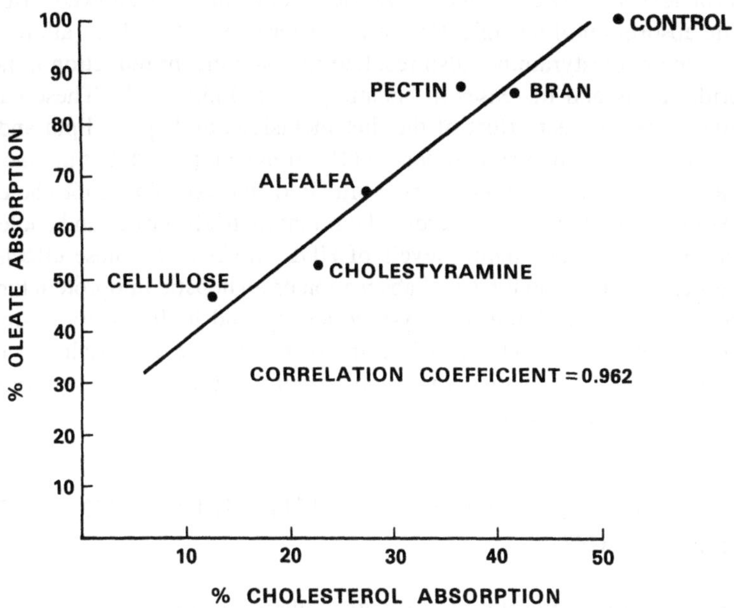

FIGURE 6. Correlation of effects of dietary fibers and cholestyramine on lymphatic absorption of cholesterol and triolein. (From Vahouny *et al.*, 1980b.)

TABLE IV. Serum and Hepatic Lipids following Prolonged Feeding of Defined Diets Containing Dietary Fibers or Cholestyramine[a]

Dietary inclusion[b]	Serum cholesterol, mg/dl	Hepatic lipid, mg/100 g wet weight		
		Cholesterol	Triglycerides	Phospholipids
Control	53 ± 1	427 ± 23	428 ± 43	2150 ± 160
Wheat bran	73 ± 11	400 ± 19	396 ± 36	2530 ± 120
Alfalfa meal	44 ± 12	459 ± 18	310 ± 26[c]	2750 ± 90[c]
Yeast cell wall	45 ± 3	464 ± 34	460 ± 44	2200 ± 350
Cellulose	54 ± 4	386 ± 7	280 ± 10[c]	2660 ± 90[c]
Cholestyramine	46 ± 5	398 ± 15	300 ± 28[c]	2490 ± 50[c]

[a] Vahouny *et al.* (1980b).
[b] See Table III.
[c] $p < 0.05$.

ences in absorbability of fatty acids and cholesterol directly into the portal system (Hyun *et al.*, 1967).

As shown in Table IV, the prolonged feeding of defined diets containing the test materials discussed above had no effect on plasma cholesterol levels. This was expected since these levels were already low and cholesterol was not included in the diets. Similarly, there were no significant effects on hepatic cholesterol levels. However, the same diet inclusions that showed effects of lymphatic absorption of a single dose of cholesterol and of triolein (alfalfa meal, cellulose, and cholestyramine) also resulted in a significant reduction in hepatic triglyceride levels and increases in hepatic phospholipid levels. These data are suggestive of an indirect effect of the diet inclusions on hepatic lipid and lipoprotein metabolism. Chen and Anderson (1979) have reported that the inclusion of 10% levels of pectin, guar gum, or oat bran into diets containing 1% cholesterol and 0.2% cholic acid results in decreased circulating triglycerides and cholesterol and to an increase in circulating levels of HDL cholesterol. These effects may in part represent effects on intestinal absorption or on hepatic lipoprotein turnover since both tissues contribute to very-low-density and high-density lipoprotein levels in the circulation. This possible indirect influence of certain fibers and fiber components on circulating cholesterol-containing lipoprotein fractions will be addressed in a later section.

4. MECHANISMS OF DIETARY FIBER EFFECTS ON LIPID ABSORPTION

Because of the indigestibility of dietary fibers and fiber components in the stomach and upper intestine, the effects of these materials on lipid (and other nutrient) metabolism have simplistically been considered to be largely assigned

TABLE V. Possible Mechanisms of Dietary Fiber Influences on Lipid Absorption

Direct effects
Gastric emptying
Altered transit times
Interference with bulk phase diffusion and availability to intestinal surface
Binding of fiber to intestinal surface coat
Sequestration of bile acids and other micellar components
Indirect effects
Effects on bile acid pool size and composition
a. Increased fecal excretion of acidic and neutral steroids
b. Increased 7α-hydroxylation of cholesterol
Altered responses of gut glucagon and pancreatic insulin
Adaptive changes in intestinal structure and function

to intraluminal sites. Even so, as summarized in Table V, the potential mechanisms by which fiber may influence lipid absorption are complex, and may all have more or less importance.

4.1. Direct Effects of Dietary Fibers and Components

It has been well established that acute ingestion of certain intact dietary fibers and fiber components may effect an improvement in glucose tolerance (Jenkins *et al.*, 1976a, 1977, 1978). This effect requires the presence of the fiber or fiber component in the test meal or ingestion immediately before the test meal, suggesting a direct influence on glucose availability for rapid absorption. This effect has been attributed to a decreased rate of gastric emptying (Jenkins *et al.*, 1977, 1978) and/or an effect on limiting intraluminal diffusion of glucose to the absorptive surface, thereby slowing absorption (Jenkins *et al.*, 1978). Thus, acute ingestion of gelling or viscous fiber components, such as pectin or guar gum, has been shown to delay gastric emptying and improve glucose tolerance. That this is related to the viscosity of the fiber has also been demonstrated, since the effect is lost when hydrolyzed, nonviscous guar is substituted for the viscous fiber component (Jenkins *et al.*, 1978).

Other nonviscous fiber components may, however, influence the rate but not the extent of glucose absorption by a different mechanism. Wheat bran and cellulose increase the rate of gastric emptying of the transit marker polyethylene glycol and reduce overall gastrointestinal transit time (G. V. Vahouny and T. Roy, unpublished results). Although the overall disappearance of an oral dose tracer glucose given with the marker is not affected under these dietary conditions, absorption must occur throughout a greater length of intestine and ultimately influence glucose appearance rates in plasma.

Despite differences in solubility characteristics of lipids, these same types of physiological responses to specific dietary fibers and fiber components should also be considered in the complex mechanisms of lipid digestion and/or absorption. Delayed lipid absorption from the upper small intestine, whether by limiting diffusibility of micellar lipids or by increasing the rate of transit, will, unlike the case of glucose, ultimately result in decreased overall absorption of lipids. This is due, in part, to at least two major influences on lipid absorption. The first involves the continued efficiency of micellar solubilization, and the second involves intestinal lipoprotein production. The phospholipid component of micelles is rapidly hydrolyzed by pancreatic phospholipase A, resulting in rapid absorption of the products of this reaction and loss of phospholipid from the micelle. Bile acids are also absorbed passively throughout the small intestine. Loss of either or both amphipaths during delayed absorption of lipids will differentially alter solubility and equibrium characteristics of cholesterol, fatty acids, and monoglycerides in the small bowel.

With respect to intestinal lipoprotein formation and secretion, it has been shown (Sabesin *et al.*, 1975) that although fatty acids in micellar form can be taken up and esterified in the ileum, subsequent transport from the cells is not efficient, and large chylomicronlike particles accumulate in the epithelial cells. Thus, delayed absorption of lipids in the upper intestine can be predicted to influence the overall efficiency of absorption of lipids in general, and may also alter the normal route of absorption of long-chain fatty acids (Hyun *et al.*, 1967). These possibilities have yet to be tested with respect to dietary fiber influences.

Finally there is recent evidence (G. V. Vahouny and T. Roy, unpublished results) that fiber components such as pectin can interact directly with the mucinous coat of the intestinal epithelium. This type of interaction is also likely to influence the rate and efficiency of absorption of various nutrients, including lipids.

4.2. Sequestration of Bile Acids and Other Micellar Components

The interaction of dietary fiber with bile acids was first reported by Eastwood and Boyd (1967), who observed appreciable quantities of bile acids associated with the insoluble material of small intestinal contents in the rat. The binding of bile acids to various grains, food fiber sources, and isolated fiber components has now been repeatedly demonstrated by various *in vitro* techniques (Eastwood and Hamilton, 1968; Balmer and Zilversmit, 1974; Kritchevsky and Story, 1974; Birkner and Kern, 1974; Story and Kritchevsky, 1975, 1976; Eastwood and Mowbray, 1976; Eastwood *et al.*, 1976; Vahouny *et al.*, 1980c). Adsorption of bile acids to dietary fibers is influenced by pH and osmolality, the structure of the bile acid, the nature of the micelles, and the physical and chemical form of the fiber (Eastwood and Hamilton, 1968). Thus, the binding of bile acids is greater at a relatively low pH (where acidic groups are un-ionized), is probably hydrophobic in nature (Eastwood and Hamilton, 1968), and appears to be reversible (Eastwood *et al.*, 1976). Certain fiber sources, such as alfalfa, show a preference for binding cholate and chenodeoxycholate compared to deoxycholate (Kritchevsky and Story, 1975), but this phenomenon appears not to be general for all fibers.

Among the fibers and fiber components, the mucilaginous and gel-forming polysaccharides and the lignins appear to be most effective in bile acid sequestration (Kay *et al.*, 1979; Kay and Truswell, 1980; Kay, 1982). The possible mechanism by which these interactions occur has received considerable attention (Eastwood and Hamilton, 1968; Kay *et al.*, 1979), but is not, as yet, well defined.

This sequestration ability of commercial bile acid "binding" resins and certain types of fibers and fiber components is, however, not merely confined to bile acids. The ability of cholestyramine, cereal fibers, and alfalfa to bind cholesterol from bile acid micelles has also been shown and this is roughly in proportion to bile acid binding (Balmer and Zilversmit, 1974; Eastwood and Mowbray, 1976; Vahouny *et al.*, 1980c). The studies of Eastwood and Mowbray (1976) suggested that the presence of additional micellar components, such as monoglycerides and fatty acids, reduced adsorption of the bile acids and cholesterol by cereal fiber, and that the extent of this reduction was dependent on the type of fatty acid included. However, our more recent studies (Vahouny *et al.*, 1980c, 1981) indicate that all components of mixed micelles are sequestered in roughly proportional amounts. As shown in Table VI, various commercial ion-exchange resins, dietary fiber sources, and fiber components were tested for their ability to sequester the individual amphipaths and amphiphiles comprising mixed micelles, characteristic of those occurring in the intestinal lumen. The

TABLE VI. Binding of Micellar Components by Ion-Exchange Resins and Dietary Fiber Sources[a]

	Percent bound				
Test material	Bile salt[b]	Lecithin	Cholesterol	Monoolein	Fatty acids[c]
Cholestyramine	82	99	95	96	96
DEAE–Sephadex	49	99	100	99	98
Guar gum[d]	36	22	23	23	33
Lignin	20	9	5	13	13
Alfalfa	7	4	1	19	18
Wheat bran	4	6	0	11	12
Cellulose	1.5	0.5	8	4	4

[a] Data from Vahouny *et al.* (1980c, 1981).
[b] Sodium taurocholate.
[c] Equimolar mixture of palmitic, oleic, and linoleic acids.
[d] Low viscosity.

commercial resins were most effective in sequestering all components of the micelle despite differences in the ability to bind the bile acid per se. Other studies (Vahouny *et al.*, 1980c) suggested this binding to be largely reversible. Among the fiber components and isolates tested, guar and lignin were the most effective sequestrants, while cellulose was without significant binding capacity.

These findings suggest that certain types of dietary fibers and isolated components may either interfere with the formation of intraluminal micelles or effectively alter the normal diffusion and accessibility of micellar lipids to the absorptive surface. If similar sequestration effects occur *in vivo,* the absorption of lipids might be compromised at several levels. These include the following:

1. Reduced accessibility of fatty acids, monoglycerides, and cholesterol for absorption in the upper intestine. Depending on the extent of micellar sequestration, this might be reflected as an increase in fecal fat and cholesterol or its bacterial metabolites.
2. Reduced accessibility of biliary lecithin for resorption, which has major implications for the ability of the upper intestine to synthesize and secrete the transport lipoproteins.
3. Reduced bile acid resorption, which could impact on several additional aspects of cholesterol metabolism (see Section 2). These include possible effects on cholesterol esterification required for efficient absorption, alterations in the resistances of the mucosal barrier (UWL), and changes in the level and composition of the enterohepatic pool of bile acids (e.g., Kay and Truswell, 1980), as well as a multitude of other responses. These effects might be expected to occur, even though the ultimate change in fecal output of bile acid metabolites (fecal acidic steroids) appears too small to account for the measured hypocholesteremic response to specific dietary fiber components.

4.3. Indirect Effects

The sequalae of effects resulting from bile acid pool size and composition with respect to cholesterol and lipoprotein metabolism are treated in detail elsewhere (Kay and Truswell, 1980), and require more serious consideration where assessing hypolipidemic effects of dietary fibers. Some of these responses, as well as those resulting from alterations in circulating insulin and glucagon levels, are also discussed in other chapters of this treatise. At present, it is not yet possible to evaluate the effects of these changes, particularly in hormone responses, on intestinal transport of lipids.

Another area which requires more careful assessment is the influence of altered dietary components, including fiber, on the nutrition of the gut itself. It is recognized that factors such as alterations in luminal nutrients and alterations in luminal bile acid levels may effect intestinal cytokinetics and morphology (e.g., Sprintz, 1971). Recent studies from our laboratories (Cassidy *et al.*, 1980, 1981) suggest that these parameters are also affected by commercial resins and by certain dietary fibers and fiber isolates. Such adaptive changes in structural and cytokinetic parameters of the small intestine might also be expected to be accompanied by changes in functional characteristics which have yet to be assessed. These types of indirect and adaptive effects of dietary fibers may ultimately prove to be of greater significance than several of the direct effects currently under investigation.

ACKNOWLEDGMENTS. This work was supported in part by the following grants: USPHS HL-02033, USDA 7900497, and NSF 7918561.

REFERENCES

Antonis, A., and Bersohn, I., 1962, The influence of diet on fecal lipids in South African white and Bantu prisoners, *Am. J. Clin. Nutr.* **11**:142–155.

Balmer, J., and Zilversmit, D. B., 1974, Effects of dietary roughage on cholesterol absorption, cholesterol turnover and steroid secretion in the rat, *J. Nutr.* **104**:1319–1328.

Beher, W. T., and Casazza, K. K., 1971, Effects of psyllium hydrocolloid on bile acid metabolism in normal and hypophysectomized rats, *Proc. Soc. Exp. Biol. Med.* **136**:253–256.

Bergen, S. S., Jr., Van Itallie, T. B., Tennant, D. M., and Sebrell, W. H., 1959, Effect of an anion exchange resin on serum cholesterol in man, *Proc. Soc. Exp. Biol. Med.* **102**:676–679.

Biggs, M. W., Friedman, M., and Byers, S. L., 1951, Intestinal lymphatic transport of absorbed cholesterol, *Proc. Soc. Exp. Biol. Med.* **78**:641–643.

Birkner, N. J., and Kern, F., Jr., 1974, *In vitro* adsorption of bile salts to food residues, salicylazosulfapyridine and hemicellulose, *Gastroenterology* **67**:237–244.

Borgström, B., 1968, Quantitative aspects of the intestinal absorption and metabolism of cholesterol and β-sitosterol in the rat, *J. Lipid Res.* **9**:473–381.

Calame, K. B., Gallo, L. L., Cheriathundam, E., Vahouny, G. V., and Treadwell, C. R., 1975, Purification and properties of subunits of sterol ester hydrolase from rat pancreas, *Arch. Biochem. Biophys.* **168**:57–65.

Cassidy, M. M., Lightfoot, F. G., Grau, L. E., Roy, T., Story, J. A., Kritchevsky, D., and Vahouny, G. V., 1980, Effect of bile acid-binding resins on the morphology of rat jejunum and colon, *Dig. Dis. Sci.* **25**:504–512.

Cassidy, M. M., Lightfoot, F. G., Grau, L. E., Story, J. A., Kritchevsky, D., and Vahouny, G. V., 1981, Effect of chronic intake of dietary fibers on ultrastructural topography of rat jejunum and colon, *Am. J. Clin. Nutr.* **34**:218–228.

Chen, W.-J. L., and Anderson, J. W., 1979, Effects of plant fibers in decreasing plasma total cholesterol and increasing high density lipoprotein cholesterol, *Proc. Soc. Exp. Biol. Med.* **162**:310–313.

Dietschy, J. M., and Wilson, J. D., 1970, Regulation of cholesterol metabolism, *N. Engl. J. Med.* **282**:1128–1138, 1179–1183, 1241–1249.

Eastwood, M. A., and Boyd, G. S., 1967, The distribution of bile salts along the small intestine of rats, *Biochim. Biophys. Acta* **137**:393–396.

Eastwood, M. A., and Hamilton, D., 1968, Studies on the adsorption of bile salts to non-absorbed components of the diet, *Biochim. Biophys. Acta* **152**:165–173.

Eastwood, M., and Mowbray, L., 1976, The binding of components of mixed micelles to dietary fiber, *Am. J. Clin. Nutr.* **29**:1461–1467.

Eastwood, M. A., Anderson, R., Mitchell, W. D., Robertson, J., and Pocock, S., 1976, A method to measure the adsorption of bile salts to vegetable fiber of different water holding capacity. *J. Nutr.* **106**:1429–1432.

Forman, L. P., and Schneeman, B. O., 1980, Effects of dietary pectin and fat on the small intestinal contents and exocrine pancreas of fats, *J. Nutr.* **110**:1992–1999.

Forman, D. T., Garvin, J. E., Forestner, J. E., and Taylor, C. B., 1968, Increased excretion of fecal bile acids by an oral hydrophilic colloid, *Proc. Soc. Exp. Biol. Med.* **127**:1060–1063.

Fraser, R., Cliff, W. J., and Courtice, F. C., 1968, The effect of dietary fat load on the size and composition of chylomicrons in thoracic duct lymph, *Q. J. Exp. Physiol.* **53**:390–398.

Gallo, L. L., 1982, Cholesterol absorption and transport, in: *CRC Handbook Series in Nutrition and Food* (M. Rechcigl, Jr., ed.), CRC Press, Boca Raton, Florida (in press).

Gallo, L. L., Newbill, T., Hyun, J., Vahouny, G. V., and Treadwell, C. R., 1977, Role of pancreatic cholesterol esterase in the uptake and esterification of cholesterol by isolated intestinal cells, *Proc. Soc. Exp. Biol. Med.* **156**:277–281.

Gallo, L. L., Chiang, Y., Vahouny, G. V., and Treadwell, C. R., 1980, Localization and origin of rat intestinal cholesterol esterase determined by immunocytochemistry, *J. Lipid Res.* **21**:537–545.

Glover, J. and Green, C., 1957, Sterol metabolism. 3. The distribution and transport of sterols across the intestinal mucosa of the guinea pig, *Biochem. J.* **67**:308–316.

Glover, J., Green, C., and Stainer, D. W., 1959, Sterol metabolism, 5. The uptake of sterols by organelles and intestinal mucosa ánd the site of their esterification during absorption, *Biochem. J.* **72**:82–87.

Goodman, D. S., and Noble, R. P., 1968, Turnover of plasma cholesterol in man, *J. Clin. Invest.* **47**:231–239.

Gustafsson, B. E., and Norman, A., 1969, Influence of the diet on the turnover of bile acids in germ-free and conventional rats, *Br. J. Nutr.* **73**:429–442.

Haugen, R., and Norum, K. R., 1976, Coenzyme A-dependent esterification of cholesterol in the rat intestinal mucosa, *Scand. J. Gastroenterol.* **11**:615–621.

Hofmann, A. F., and Mekhyian, H. S., 1973, Bile acids and the intestinal absorption of fats and electrolytes in health and disease, in: *The Bile Acids,* Vol. 2 (P. P. Nair and D. Kritchevsky, eds.), Plenum Press, New York, pp. 103–152.

Hyun, J., Vahouny, G. V., Kothari, H., Herm, E., Mortenson, J., and Treadwell, C. R., 1969, Purification and properties of pancreatic juice cholesterol esterase, *J. Biol. Chem.* **244**:1937–1945.

Hyun, S. A., Vahouny, G. V., and Treadwell, C. R., 1963, Effect of hypocholesterolemic agents on intestinal cholesterol absorption, *Proc. Soc. Exp. Biol. Med.* **112**:496–501.

Hyun, S. A., Vahouny, G. V., and Treadwell, C. R., 1967, Portal absorption of long-chain fatty acids in the rat, *Biochim. Biophys. Acta* **137**:296–305.

Inglett, G., and Falkehag, I. (eds.), 1979, *Dietary Fibers: Chemistry and Nutrition,* Academic Press, New York.

Jandacek, R. J., Webb, M. R., and Mattson, F. H., 1976, The effect of an aqueous phase on the solubility of cholesterol in an oil phase, *J. Lipid Res.* **8**:203–210.

Jenkins, D. J. A., Goff, D. V., Leeds, A. R., Alberti, K., Gill, M., Wolever, T. M. S., Gassull, M. A., and Hockaday, T. D. R., 1976a, Unabsorbable carbohydrates and diabetes: Decreased postprandial hyperglycemia, *Lancet* **2**:172–174.

Jenkins, D. J. A., Leeds, A. R., Gassull, M. A., Houston, H., Goff, D. V. and Hill, M. J., 1976b, The cholesterol lowering properties of guar and pectin, *Clin. Sci. Mol. Med.* **51**:8–15.

Jenkins, D. J. A., Leeds, A. R., Gassull, M. A., Cochet, B., and Alberti, K. G., 1977, Decrease in postprandial insulin and glucose concentrations by guar and pectin, *Ann. Intern. Med.* **86**:20–23.

Jenkins, D. J. A., Wolever, T. M. S., Leeds, A. R., Gassull, M. A., Haisman, P., Dilawari, J., Goff, D. V., Metz, G. L., and Alberti, K. G., 1978, Dietary fibres, fiber analogues and glucose tolerance: Importance of viscosity, *Br. Med. J.* **1**:1392–1394.

Kan, K. W., and Ungar, F., 1974, Stimulation factors for cholesterol side-chain cleavage in the corpus luteum and adrenal gland, *J. Steroid Biochem.* **5**:817–725.

Kay, R. M., 1982, Origin, functions and physiological significance of dietary fiber, *J. Lipid Res.* (in press).

Kay, R. M., and Strasberg, S. M., 1978, Origin, chemistry, physiological effects and clinical importance of dietary fibre, *Clin. Invest. Med.* **1**:9–24.

Kay, R. M., and Truswell, A. S., 1977, Effect of citrus pectin on blood lipids and fecal excretion in man, *Am. J. Clin. Nutr.* **30**:171–175.

Kay, R. M., and Truswell, A. S., 1980, Dietary fiber: Effects on plasma and biliary lipids in man, in: *Medical Aspects of Dietary Fiber* (G. A. Spiller and R. M. Kay, eds.), Plenum Medical, New York, pp. 153–174.

Kay, R. M., Strasberg, S. M., Petruka, C. N., and Wayman, M., 1979, Differential adsorption of bile acids by lignins, in: *Dietary Fibers: Chemistry and Nutrition* (G. E. Inglett and S. I. Falkehag, eds.), Academic Press, New York, pp. 57–66.

Kelly, J. J., and Tsai, A. C., 1978, Effect of pectin, gum arabic and agar on cholesterol absorption, synthesis, and turnover in rats, *J. Nutr.* **108**:630–639.

Kelsay, J. L., 1978, A review of research on effects of fiber intakes on man, *Am. J. Clin. Nutr.* **31**:142–159.

Keys, A., Anderson, J. T., and Grande, F., 1960, Diet type (fats constant) and blood lipids in man, *J. Nutr.* **70**:257–266.

Keys, A., Grande, F., and Anderson, J. T., 1961, Fiber and pectin in the diet and serum cholesterol concentration in man, *Proc. Soc. Exp. Biol. Med.* **106**:555–558.

Kiriyama, S., Okazaki, Y., and Yoshida, A., 1968, Hypocholesterolemic effect of polysaccharides and polysaccharide-rich foodstuffs in cholesterol-fed rats, *J. Nutr.* **97**:382–388.

Kostner, G.M., 1976, Apo B-deficiency (abetalipoproteinemia): A model for studying lipoprotein metabolism, in: *Lipid Absorption: Biochemical and Clinical Aspects* (K. Rommel, H. Goebell, and R. Bohner, eds.), University Park Press, Baltimore, pp. 203–236.

Kritchevsky, D., and Story, J. A., 1974, Binding of bile salts *in vitro* by non-nutritive fiber, *J. Nutr.* **104**:458–462.

Kritchevsky, D., and Story, J. A., 1975, *In vitro* binding of bile acids and bile salts, *Am. J. Clin. Nutr.* **28**:305–306.

Kritchevsky, D., Casey, R. P., and Tepper, S. A., 1973, Isocaloric, isogravic diets in rats. II. Effect on cholesterol absorption and excretion, *Nutr. Rep. Int.* **7**:61–69.

Kritchevsky, D., Tepper, S. A., and Story, J. A., 1974, Isocaloric osogravic diets in rats. III. Effects of nonnutritive fiber (alfalfa or cellulose) on cholesterol metabolism. *Nutr. Rep. Int.* **9**:301–308.

Kritchevsky, D., Story, J. A., and Vahouny, G. V., 1980, Influence of fiber on lipid metabolism, in: *Nutrition and Food Sciences,* Vol. 3 (W. Santos, N. Lopes, J. J. Barbosa, D. Chaves, and J. C. Valente, eds.), Plenum Press, New York, pp. 461–471.

Leveille, G. A., and Sauberlich, H. E., 1966, Mechanism of the cholesterol-depressing effect of pectin in the cholesterol-fed rat, *J. Nutr.* **88**:209–214.

Lin, T. M., Kim, K. S., Karvinen, E., and Ivy, A. C., 1957, Effect of dietary pectin, "protopectin" and gum arabic on cholesterol excretion in rats, *Am. J. Physiol.* **188**:66–70.

Miettinen, T. A., and Tarpila, S., 1977, Effect of pectin on serum cholesterol, fecal bile acids and biliary lipids in normolipidemic and hyperlipidemic individuals, *Clin. Chem. Acta* **79:**471–477.

Noland, B. J., Arebalo, R. E., Hansbury, E., and Scallen, T. J., 1980, Purification and properties of sterol carrier protein$_2$, *J. Biol. Chem.* **255:**4282–4289.

Norum, K. R., Lilljeqvist, A. C., and Drevon, C. D., 1977, Coenzyme A-dependent esterification of cholesterol in intestinal mucosa from guinea pig. Influence of diet on enzyme activity, *Scand. J. Gastroenterol.* **12:**281–288.

O'Doherty, P. J. A., Yousef, I. M., and Kuksis, A., 1973, Effect of puromycin on protein and glycerolipid biosynthesis in isolated mucosal cells, *Arch. Biochem. Biophys.* **156:**586–594.

Portman, O. W., 1960, Nutritional influences on the metabolism of bile acids, *Am. J. Clin. Nutr.* **8:**462–470.

Portman, O. W., and Murphy, P., 1958, Excretion of bile acids and β-hydroxy-sterols by rats, *Arch. Biochem. Biophys.* **76:**367–376.

Robbins, E. A., and Seeley, R. D., 1977, Cholesterol lowering effect of dietary yeast and yeast fractions, *J. Food Sci.* **42:**694–700.

Sabesin, S. M., and Isselbacher, K. J., 1965, Protein synthesis inhibition: mechanism for the production of impaired fat absorption, *Science* **147:**1149–1151.

Sabesin, S. M., Holt, P. R., and Clark, S. B., 1975, Intestinal lipid absorption: Evidence for an intrinsic defect of chylomicron secretion by normal rat distil intestine, *Lipids* **10:**840–846.

Salee, V. L., Wilson, F. A., and Dietschy, J. M., 1972, Determination of unidirectional rates for lipids across the intestinal brush border, *J. Lipid Res.* **13:**184–192.

Spiller, G. A., and Amen, R. J. (eds.), 1976, *Fiber in Human Nutrition,* Plenum Press, New York.

Sprintz, H., 1971, Factors influencing intestinal cell renewal, *Cancer* **28:**71–74.

Stanley, M. M., Paul, D., Gacke, D., and Murphy, J., 1973, Effect of cholestyramine, metamucil and cellulose on fecal bile acid excretion in man, *Gastroenterology* **65:**889–894.

Story, J. A., and Kritchevsky, D., 1975, Binding of sodium taurocholate by various foodstuffs, *Nutr. Rep. Int.* **11:**161–163.

Story, J. A., and Kritchevsky, D., 1976, Comparison of the binding of various bile acids and bile salts *in vitro* by several types of fiber, *J. Nutr.* **106:**1292–1294.

Sylven, C., and Borgström, B., 1969, Intestinal absorption and lymphatic transport of cholesterol in the rat: Influence of fatty acid chain length of the carrier triglyceride, *J. Lipid Res.* **10:**351–355.

Tennant, D. M., Siegel, H., Zanetti, M. E., Kuron, G. W., Ott, W. H., and Wolf, F. J., 1959, Reduction of plasma cholesterol in animals with bile acid sequestrants, *Circulation* **20:**969–970.

Tennant, D. M., Siegel, H., Zanetti, M. E., Kuron, G. W., Ott, W. H., and Wolf, F. J., 1960, Plasma cholesterol lowering action of bile acid binding polymers in experimental animals, *J. Lipid Res.* **1:**469–473.

Treadwell, C. R., and Vahouny, G. V., 1968, Cholesterol absorption, in: *Handbook of Physiology—The Alimentary Canal,* American Physiological Society, Bethesda, Maryland, pp. 1407–1438.

Vahouny, G. V., and Kritchevsky, D., 1981, Plant and marine sterols and cholesterol metabolism, in: *Nutritional Pharmacology* (G. A. Spiller, ed.), Alan R. Liss, New York, pp. 31–72.

Vahouny, G. V., and Treadwell, C. R., 1976, Cholesterol—Some aspects of its metabolism and balance in man, in: *Nutrition and Cardiovascular Disease* (E. B. Feldman, ed.), Appleton-Century-Crofts, New York, pp. 85–116.

Vahouny, G. V., Ito, M., Blendermann, E. M., Gallo, L. L., and Treadwell, C. R., 1977, Puromycin inhibition of cholesterol absorption in the rat, *J. Lipid Res.* **18:**745–752.

Vahouny, G. V., Roy, T., Gallo, L. L., Story, J. A., Kritchevsky, D., Cassidy, M., Grund, B. M., and Treadwell, C. R., 1978, Dietary fiber and lymphatic absorption of cholesterol in the rat, *Am. J. Clin. Nutr.* **31:**S208–212.

Vahouny, G. V., Blendermann, E. M., Gallo, L. L., and Treadwell, C. R., 1980a, Differential transport of cholesterol and oleic acid in lymph lipoproteins: sex differences in puromycin sensitivity, *J. Lipid Res.* **21**:415–424.

Vahouny, G. V., Roy, T., Gallo, L. L., Story, J. A., Kritchevsky, D., and Cassidy, M. M., 1980b, Dietary fibers. III. Effects of chronic intake on cholesterol absorption and metabolism in the rat, *Am. J. Clin. Nutr.* **33**:2182–2191.

Vahouny, G. V., Tombes, R., Cassidy, M. M., Kritchevsky, D., and Gallo, L. L., 1980c, Dietary fibers. V. Binding of bile salts, phospholipids and cholesterol from mixed micelles by bile acid sequestrants and dietary fibers, *Lipids* **15**:1012–1018.

Vahouny, G. V., Tombes, R., Cassidy, M. M., Kritchevsky, D., and Gallo, L. L., 1981, Dietary fibers. VI. Binding of fatty acids and monolein from mixed micelles containing bile salts and lecithin, *Proc. Soc. Exp. Biol. Med.* **166**:12–16.

Walters, R. L., McLean-Baird, I., Davies, P. S., Hill, M. J. K., Draser, B. S., Southgate, D. A. T., and Morgan, B., 1975, Effects of two types of dietary fiber on faecal steroid and lipid excretion, *Br. Med. J.* **2**:536–538.

Wells, A. F., and Ershoff, B. H., 1961, Beneficial effects of pectin in prevention of hypercholesterolemia and increase in liver cholesterol in cholesterol-fed rats, *J. Nutr.* **74**:87–92.

Westergaard, H., and Dietschy, J. M., 1976, The mechanism whereby bile acid micelles increase the rate of fatty acid and cholesterol uptake into the intestinal mucosal cell, *J. Clin. Invest.* **58**:97–108.

Zilversmit, D. B., 1972, A single blood sample dual isotope method for the measurement of cholesterol absorption in rats, *Proc. Soc. Exp. Biol. Med.* **140**:862–865.

Zilversmit, D. B., 1979, Dietary fiber, in: *Nutrition, Lipids and Coronary Heart Disease* (R. Levy, B. Rifkind, B. Dennis, and N. Ernst, eds.), Raven Press, New York, pp. 149–174.

20

Dietary Fiber and Lipoproteins

JON A. STORY and MICHAEL J. KELLEY

1. INTRODUCTION

Early investigations into the mechanism by which some sources of dietary fiber exert their hypocholestermic effects centered on changes in steroid excretion brought about by adsorption of bile salts and bile acids by some component(s) of dietary fiber (Story and Kritchevsky, 1976; Story, 1980). Subsequent examination of the relationship between the magnitude of change in steroid excretion and in serum cholesterol indicated that this single phenomenon could not explain the range of effects observed with various dietary fiber-related materials (Kay and Truswell, 1980; Story, 1980). Change in lipoprotein metabolism has emerged as an additional explanation for the involvement of dietary fiber in regulation of lipid metabolism. The association of the levels of low-density lipoproteins (LDL) with increased risk of atherosclerotic heart disease and high-density lipoproteins (HDL) with reduced risk, and the importance of these lipoproteins in regulation of all phases of lipid metabolism, have made examination of the effects of dietary fiber on these variables especially interesting.

In this review we will concern ourselves with recent observations on the effects of some sources of dietary fiber and some isolated polysaccharides on lipoprotein metabolism, since Kay and Truswell (1980) have recently reviewed this aspect of dietary fiber.

JON A. STORY and MICHAEL J. KELLEY • Department of Foods and Nutrition, Purdue University, West Lafayette, Indiana 47907.

2. LIPOPROTEIN METABOLISM

2.1. Animal Studies

Chen and Anderson (1979a,b) have investigated the effects of various dietary fiber sources and isolated polysaccharide components of dietary fiber on plasma lipoprotein cholesterol levels in rats fed cholesterol–cholic acid. Inclusion of cholesterol and cholic acid in the diets of rats causes increases in plasma cholesterol of sufficient magnitude to allow measurement of differences caused by dietary manipulation. As can be seen in Table I, addition of wheat bran (20%) to a semipurified diet which contained 1% cholesterol and 0.2% cholic acid resulted in an increase in plasma total cholesterol, while guar gum (7%) reduced total cholesterol levels below those observed in the basal (4% cellulose) group. All three diets with cholesterol and cholic acid added had plasma cholesterol levels significantly higher than the control, cholesterol-free dietary group. HDL cholesterol levels were reduced by cholesterol and cholic acid feeding in all groups, but were maintained at a higher level in guar gum-fed rats. HDL cholesterol as a percent of total cholesterol was reduced to the same extent in both cellulose- and wheat bran-fed groups, but was maintained at a higher level when guar gum was included in the diet.

In a second experiment (Chen and Anderson, 1979b), guar gum was compared with pectin, another fairly soluble polysaccharide component of dietary fiber, and oat bran, a source of dietary fiber rich in soluble polysaccharides. As can be seen in Table II, plasma cholesterol levels were higher in all groups fed cholesterol–cholic acid. Pectin, guar gum, and oat bran prevented the increase caused by cholesterol feeding, with pectin being the most efficient. HDL cholesterol levels were lower in all groups fed cholesterol–cholic acid, but pectin, guar gum, and oat bran reduced the magnitude of this change. All three materials

TABLE I. Effects of Wheat Bran and Guar Gum on Lipoprotein Cholesterol in Rats[a]

	B	BC[b]	BC + wheat bran[c]	BC + guar gum[c]
Plasma cholesterol				
Total, mg/dl	79	157	171	112
HDL, mg/dl	30	16	18	24
HDL/total	0.38	0.10	0.11	0.21
Liver cholesterol, mg/g	3.30	42.6	49.6	20.6

[a] Chen and Anderson (1979a). B, basal; BC, basal + cholesterol.
[b] 1.0% Cholesterol and 0.2% cholic acid.
[c] Wheat bran added at 20% and guar gum at 7% weight.

TABLE II. Modification of Lipoprotein Cholesterol by Pectin, Guar Gum, and Oat Bran[a]

	B[b]	BC[c]	BC + pectin[d]	BC + guar gum[d]	BC + oat bran[d]
Plasma cholesterol					
Total, mg/dl	74	133	84	105	113
HDL, mg/dl	34	18	26	31	25
HDL/total	0.46	0.14	0.31	0.30	0.22
Liver cholesterol, mg/g	3.5	51.1	13.5	22.9	27.1

[a] Chen and Anderson (1979b).
[b] Basal diet contained 46.5% sucrose, 16.6% starch, 15.6% casein, 6% cottonseed oil, and 10% cellulose and salt and vitamin mixes.
[c] Cholesterol (1%) and cholic acid (0.2%) were added at the expense of carbohydrate.
[d] Pectin and guar gum added at 10% of diet weight. Oat bran added at 36.5% by weight to provide 10% "plant fiber."

maintained HDL cholesterol at a higher percent of total cholesterol than the cellulose of the basal diet. Liver cholesterol levels were also reduced by pectin, guar gum, and oat bran in comparison to cellulose in rats fed cholesterol–cholic acid.

In obese Zucker rats (van Beresteyn *et al.*, 1979) fed 0.1% cholesterol, added wheat bran and cellulose resulted in higher serum cholesterol levels than those in animals fed the basal diet (Table III). HDL cholesterol was lower in wheat bran-fed animals.

Results of several additional experiments which examined lipoprotein cholesterol levels in experimental animals are summarized in Table IV. Chen *et al.* (1981) reported decreased total cholesterol level in pectin-, oat bran-, and oat gum-fed rats (10%) when compared to cellulose-fed animals, with 1% cholesterol

TABLE III. Effects of Wheat Bran and Cellulose on Lipoprotein Cholesterol in Zucker Rats[a]

	B[b]	B + cellulose[c]	B + wheat bran[c]
Plasma cholesterol			
Total, mg/dl	259	274	265
HDL, mg/dl	148	—	126
HDL/Total	0.57	—	0.48
Liver cholesterol, mg/g	7.7	7.4	6.2

[a] Van Beresteyn *et al.* (1979).
[b] Semipurified diet: 22.2% lactose; 21.3% wheat flour; 8.2% wheat starch; 19.6% casein; 21.6% butter oil, salt, and vitamin mixes; and 0.1% cholesterol.
[c] Cellulose and wheat bran were added at 10% and 12.5% of the total diet.

TABLE IV. Modification of Blood Lipids by Dietary Fiber and Its Components

	Total cholesterol	HDL cholesterol	LDL cholesterol	HDL/LDL
Chen *et al.* (1981)				
Pectin (10%)[a]	↓	↑	↓	↑
Oat bran (10%)	↓	↑	↓	↑
Oat gum (10%)	↓	↑	↓	↑
O'Donnell *et al.* (1981)				
Cellulose (10%)	Control			3.2
Oat bran (10%)	↓			1.3
Corn bran (10%)	↓			5.6
Wheat bran (10%)	nc[b]			4.2
Guar gum (10%)	↓			9.1

[a] Compared to 10% cellulose control.
[b] nc, No change.

and 0.2% cholic acid added to the diet. These three materials also resulted in increased HDL and decreased LDL cholesterol levels, again compared to cellulose-fed controls.

O'Donnell *et al.* (1981) observed a reduction in total cholesterol levels in gerbils fed 0.1% cholesterol when oat bran, corn bran, or guar gum was added at 10% of the diet in place of cellulose. The ratio of HDL/LDL cholesterol was increased modestly in groups fed corn bran and wheat bran, increased almost threefold in those fed guar gum, and reduced in those fed oat bran as compared to cellulose-fed controls.

The results of these animal studies indicate a reduction in LDL cholesterol and an increase or no change in HDL cholesterol in response to certain materials in cholesterol-fed animals. The materials which yield this response are, in general, soluble, viscous isolated polysaccharides, such as guar gum or pectin, or are dietary fiber-containing foods high in soluble polysaccharide components, such as oat bran. Little evidence has come forth to suggest any mechanism for the observed changes in lipoprotein cholesterol levels.

2.2. Human Studies

Reductions of total cholesterol in humans in response to several isolated polysaccharide components of dietary fiber and foods high in dietary fiber have been well documented (Anderson and Chen, 1979; Kay and Truswell, 1980). The observation that changes in plasma cholesterol were not uniformly distributed among the various lipoproteins has led to further investigation of this phenomenon.

Burslem *et al.* (1978) examined differences in lipoprotein composition be-

TABLE V. Alteration of Lipoprotein Lipid in Vegetarians[a]

	Plasma cholesterol, mg/dl	LDL cholesterol, mg/dl	HDL cholesterol, mg/dl	HDL/LDL + VLDL
20- to 30-year-olds				
Control (41)	179	116	49	0.38
Vegetarians (60)	130	80	41	0.46
30- to 40-year-olds				
Control (15)	194	129	51	0.36
Vegetarians (18)	142	86	42	0.42

[a] Burslem *et al.* (1978).

tween groups consuming vegetarian or mixed diets (Table V). Plasma cholesterol levels were lower in vegetarians of both age groups studied. Both LDL and HDL cholesterol levels were reduced, but LDL to a greater extent, as evidenced by the increase in HDL/LDL + VLDL (very-low-density lipoprotein) ratio. Dietary differences included significant differences in cholesterol, fat, and carbohydrate intake and the P/S ratio of the lipid consumed in addition to differences in dietary fiber intake.

Using guar in various hydrated states or administered as crispbread, Jenkins *et al.* (1980) observed decreases in serum cholesterol with all forms of guar (Table VI). Guar crispbread and hydrated guar resulted in the largest decrease in cholesterol level, both total and LDL. The HDL cholesterol levels were increased when hydrated guar was fed and remained unchanged with crispbread and decreased with semihydrated guar. Changes in serum cholesterol were manifested primarily as selective decreases in LDL cholesterol.

The effects of increased dietary fiber intake resulting from increased consumption of fruits and vegetables, supplementation with wheat bran (42.5 g/day), or supplementation with citrus pectin (9 g/day) were examined by Stasse-

TABLE VI. Effects of Guar on Serum Lipoproteins in Humans[a]

	Guar crispbread	Guar, hydrated	Guar, semihydrated
Number of subjects[b]	17	8	4
Guar consumed, g/day	13	11	11
Change in serum cholesterol, %			
Total	−10	−8	−3
LDL	−11	−11	−5
HDL	+0.4	+4	−3

[a] Jenkins *et al.* (1980).
[b] Two-week feedings.

TABLE VII. Changes in Serum Lipoprotein Cholesterol Levels in Hypercholesterolemic Men[a]

	Serum cholesterol, mg/dl	LDL cholesterol, mg/dl	HDL cholesterol, mg/dl
Control diet	252	169	48
Oat-bran diet[b]	234	159	48

[a] Kirby *et al.* (1981).
[b] An average of 94 g oat bran added to diet.

Wolthuis *et al.* (1980). They found that citrus pectin reduced total serum cholesterol levels significantly during the test period, while increased consumption of fruits and vegetables caused a small but insignificant reduction. Wheat bran supplementation resulted in a significant increase in serum cholesterol level. HDL cholesterol levels were not significantly changed in any of the experimental groups, indicating that the changes which occurred were selectively in LDL cholesterol levels. Addition of wheat fiber to the diet of normal humans was found to cause little change in serum cholesterol or the fraction of that cholesterol carried by HDL (Flanagan *et al.*, 1980).

Kirby *et al.* (1981) have recently observed selective reductions in LDL cholesterol in response to addition of 94 g oat bran/day to the diets of hypercholesterolemic men (Table VII). These patients were fed mixed diets matched for distribution of calories with added oat bran over a 10-day test period. Serum cholesterol was reduced 13% and the reduction was accounted for entirely by a lowering of LDL cholesterol level.

3. CONCLUSION

Experiments to date would seem to indicate that the effect of dietary fiber involves a selective decrease in LDL cholesterol with little change in HDL cholesterol. Some evidence from cholesterol-fed experimental animals suggests that HDL cholesterol may, in fact, be increased.

Second, the characteristics of the dietary fiber-related material used have a marked effect on the results. Wheat bran does not appear to alter blood cholesterol levels, although some reports of reductions (Munoz *et al.*, 1979; van Berge-Henegouwen *et al.*, 1979) have appeared. Oat bran appears to be very effective in lowering LDL while maintaining HDL, and, as evidenced by a limited amount of work, corn bran may be equally useful. Among the isolated polysaccharides commonly considered components of dietary fiber, pectins, guar gum, and oat gum are most effective in eliciting the above response, while cellulose appears to have no effect on lipoprotein cholesterol. As has been observed by others, all these materials are highly viscous polysaccharides or are composed of large amounts of such materials (Anderson and Chen, 1979; Kay and Truswell, 1980).

A primary focus for investigation of the mechanism for the effects of these materials is absorption and disposition of lipids. Dietary fiber may alter the site of absorption of lipids along the intestine. Composition of lipoproteins synthesized by the intestine changes as the site of absorption is moved more distal in the small intestine (Wu *et al.*, 1980). Distal chylomicrons are larger, apparently due to an inability to use intestinal phospholipids in lipoprotein synthesis. The affinity of some sources of dietary fiber for micellar phospholipids (Vahouny *et al.*, 1981) may also contribute to alteration of chylomicron synthesis by limiting phospholipid availability all along the intestine. Disposition of these chylomicrons and their remnants has not been examined. In addition, the alteration of insulin secretion would also have an effect on the clearance of chylomicrons through insulin's role in activating lipoprotein lipase. Information regarding the role of dietary fiber on these facets of lipoprotein metabolism is not available, making evaluation of the mechanisms involved in the observed changes impossible at this time.

The importance of a dietary component which would selectively lower LDL cholesterol is obvious. Most changes in lipoprotein cholesterol uniformly reduce all density classes. With current evidence suggesting the protective effect of higher levels of HDL cholesterol, maintenance of these levels while reducing LDL, a positively correlated risk factor, with dietary manipulation would be extremely valuble. Most of the observations to date have involved pharmacological use of isolated polysaccharides or single sources of dietary fiber which would be of limited application in free-living populations, although the use of these materials in certain high-risk populations is exciting. Further work is required before these findings could be applied as dietary recommendations aimed at improving risk of disease.

ACKNOWLEDGMENTS. Supported in part by the Indiana Agricultural Experiment Station (paper No. 8600), the Northeast Chapter of the Indiana Affiliate of the American Heart Association, and the Showalter Trust.

REFERENCES

Anderson, J. W., and Chem, W.-J. L., 1979, Plant fiber: Carbohydrate and lipid metabolism, *Am. J. Clin. Nutr.* **32**:346–363.

Burslem, J., Schonfeld, G., Howald, M. A., Weidman, S. W., and Miller, J. P., 1978, Plasma apoprotein and lipoprotein lipid levels in vegetarians, *Metabolism* **27**:711–719.

Chen, W.-J. L., and Anderson, J. W., 1979a, Effects of guar gum and wheat bran on lipid metabolism of rats, *J. Nutr.* **109**:1028–1034.

Chen, W.-J. L., and Anderson, J. W., 1979b, Effects of plant fiber in decreasing plasma total cholesterol and increasing high-density lipoprotein cholesterol, *Proc. Soc. Exp. Biol. Med.* **162**:310–313.

Chen, W.-J. L., Anderson, J. W., and Gould, M. R., 1981, Cholesterol-lowering effects of oat bran and oat gum, *Fed. Proc.* **40**:853.

Flanagan, M., Little, C., Milliken, J., Wright, E., McGill, A. R., Weir, D. G., and O'Moore, R. R., 1980, The effects of diet on high density lipoprotein cholesterol, *J. Hum. Nutr.* **34**:43–45.

Jenkins, D. J. A., Reynolds, D., Slavin, B., Leeds, A. R., Jenkins, A. L., and Jepson, E. M., 1980, Dietary fiber and blood lipids: Treatment of hypercholesterolemia with guar crispbread, *Am. J. Clin. Nutr.* **33**:575–581.

Kay, R. M., and Truswell, A. S., 1980, Dietary fiber: Effects on plasma and biliary lipids in man, in: *Medical Aspects of Dietary Fiber* (G. A. Spiller and R. M. Kay, eds.), Plenum Medical, New York, pp. 153–173.

Kirby, R. W., Anderson, J. W., Sieling, B., Rees, E. D., Chen, W.-J. L., Miller, R. E., and Kay, R. M., 1981, Oat-bran intake selectively lowers serum low-density lipoprotein cholesterol concentrations of hypercholesterolemic men, *Am. J. Clin. Nutr.* **34**:824–829.

Munoz, J. M., Sanstead, H. H., Jacob, R. A., Logan, G. M., Reck, S. J., Klevay, L. M., Dintzis, F. R., Inglett, G. F., and Shuey, W. C., 1979, Effects of some cereal brans and textured vegetable protein on plasma lipids, *Am. J. Clin. Nutr.* **32**:580–592.

O'Donnell III, J. A., Lee, H. S., and Hurt, H. D., 1981, Effect of plant fibers on plasma and liver lipids of adult male gerbils, *Fed. Proc.* **40**:853.

Stasse-Wolthuis, M., Albers, H. F. F., van Jeveren, J. G. C., de Jong, J. W., Hautrast, G. A. J., Hermus, R. J. J., Katan, M. B., Brydon, W. G., and Eastwood, M. A., 1980, Influence of dietary fiber from vegetables and fruits, bran, citrus pectin on serum lipids, fecal lipids, and colonic functions, *Am. J. Clin. Nutr.* **33**:1745–1756.

Story, J. A., 1980, Dietary fiber and lipid metabolism: An update, in: *Medical Aspects of Dietary Fiber* (G. A. Spiller and R. M. Kay, eds.), Plenum Medical, New York, pp. 137–152.

Story, J. A., and Kritchevsky, D., 1976, Dietary fiber and lipid metabolism, in: *Fiber and Human Nutrition* (G. A. Spiller and R. J. Amen, eds.), Plenum Press, New York, pp. 171–184.

Vahouny, G. V., Tombes, R., Cassidy, M. M., Kritchevsky, D., and Gallo, L. L., 1981, Dietary fibers V: Binding of bile salts, phospholipids and cholesterol from mixed micelles by bile acid sequestrants and dietary fibers, *Lipids* **15**:1012–1018.

van Beresteyn, E. C. H., van Schaik, M., and Kerkhof Mogot, 1979, M. F., Effects of bran and cellulose on lipid metabolism in obese female Zucker rats, *J. Nutr.* **109**:2085–2097.

van Berge-Henegouwen, G. P., Huybregts, H. W., van de Werf, S., Demacker, P., and Schade, W., 1979, Effect of a standardized wheat-bran preparation on serum lipids in young, healthy males, *Am. J. Clin. Nutr.* **32**:794–798.

Wu, A.-L., Clark, S. B., and Holt, P. R., 1980, Composition of lymph chylomicrons from proximal or distal rat small intestine, *Am. J. Clin. Nutr.* **33**:582–589.

21

Colon Cancer and Dietary Fiber

An Overview

GENE A. SPILLER

It is appropriate that what is probably the most intricate and complex effect of dietary fiber in the colon is being discussed after more direct relationships have been covered in earlier chapters. Furthermore, since no term in the lexicon of medicine strikes more terror than the word *cancer* (Robbins and Cotran, 1979), should some of the plant fiber polymers help to prevent colon cancer, this could be their most important beneficial effect in the colon.

The chapters that follow will attempt to answer the basic question: have the hypotheses of Malothra (1967) and Burkitt (1971) been sufficiently proven? If so, then this would allow us to make some preliminary recommendations for possible dietary changes for people on typical Western diets. I shall leave the answer to others in this treatise, but some of the key problems in researching the effects of plant fibers in the colon need to be briefly summarized.

In animal studies a basic dilemma faces the investigators: should they use highly purified substances, such as cellulose and pectin, so that the function of *specific* polymers may be identified, or should they use natural high-fiber products such as wheat bran, carrot powder, and other nonpurified sources? The former methodology is scientifcally appealing, as long as we remember how much the physicochemical properties of these polymers might have been altered in the process of isolating them from associated polymers in the plant cell wall. The alternative, using natural products, introduces many additional variables. Not only do we have a mixture of polymers, but we find almost endless polymeric patterns in various foods. In addition, there are many associated substances, the

GENE A. SPILLER • Department of Biology, Mills College, Oakland, California 94613.

effect of which may confound that of the fibrous polymers; examples are found in the cutins, glycoproteins, and silica.

In my opinion, both methodologies are important, but we must be careful not to generalize the effects of highly purified fiber polymers.

In human studies, the key problem is a lack of sufficient published data on dietary fiber composition of foods in various countries. The problem is further complicated by the absence of agreement as to the way fiber analyses for foods should be presented and which method of analysis to follow. This paucity of analytical data makes the task of the epidemiologist a difficult and often impossible one. Progress in this analytical field and increased availability of data for various countries is of crucial importance to the student of human cancer.

On the positive side, it appears that if dietary fiber is protective, it might be the water-insoluble, fecal-bulking fraction that is important (Spiller and Freeman, 1981).

Another point to be carefully considered in both animal studies and human surveys is that plant fiber polymers are highly *interactive* with other exogenous and endogenous compounds and that the remainder of the diet could alter the effect of fiber in the colon. To carry this point further, attempting to isolate fiber conceptually as the protective factor might be unwise: the etiology of colon cancer is undoubtely due to various factors, such as high fat and insufficient antioxidants in the diet. It is difficult to do studies with many variables, but in this case it is crucial if we want to solve the colon cancer puzzle.

REFERENCES

Burkitt, D. P., 1971, Epidemiology of cancer of the colon and rectum, *Cancer* **28**:3.

Malothra, S. L., 1967, Geographical distribution of gastrointestinal cancer in India with special reference to causation, *Gut* **8**:361.

Robbins, S. L., and Cotran, R. S., 1979, *The Pathologic Basis of Disease*, Saunders, Philadelphia, p. 141.

Spiller, G. A., and Freeman, J. H., 1981, Recent advances in dietary fiber and colorectal diseases, *Am J. Clin. Nutr*. **31**:231.

22

Dietary Fiber, Bile Acids, and Intestinal Morphology

MARIE M. CASSIDY, FRED G. LIGHTFOOT, and GEORGE V. VAHOUNY

1. INTRODUCTION

The rates of occurrence and mortality from colorectal cancer have been positively correlated with total fat intake in several populations (Carroll and Khor, 1975; Correa and Haenszel, 1978; Wynder, 1979). The possible relationships between fat ingestion and the subsequent steps in lipid metabolism have been explored in studies aimed at determining the link between bowel cancer risk, dietary fats, products of fat metabolism, and their appearance in the stool composition. Of the populations with higher incidences of bowel cancer there is also evidence of higher fecal steroids, bile acids, and 7-dehydroxylase activity (Hill *et al.*, 1975, Reddy *et al.*, 1975a) compared to control studies. Subjects with higher levels of fat consumption also demonstrate a high fecal concentration of important lipid metabolites (Reddy *et al.*, 1976).

Studies with animal models have evinced additional evidence supporting a relationship between bile acids, luminal lipid metabolites, and intestinal carcinogenesis (Nigro *et al.*, 1976; Reddy, 1975b). In several studies a low incidence of bowel cancer is observed in conjunction with a diet moderately high in fats.

MARIE M. CASSIDY • Department of Physiology, The George Washington University, School of Medicine and Health Sciences, Washington, D.C. 20037. FRED G. LIGHTFOOT • Department of Anatomy, The George Washington University, School of Medicine and Health Sciences, Washington, D.C. 20037. GEORGE V. VAHOUNY • Department of Biochemistry, The George Washington University, School of Medicine and Health Sciences, Washington, D.C. 20037.

This difference has been attributed to a protective effect of dietary fibers as a component of the diet (Burkitt *et al.*, 1972). The mechanisms postulated to account for this amelioration in effect include transit time of fecal material, alterations in gut floral metabolism, alterations in stool composition, particularly with respect to bile acid metabolites, binding of carcinogenic or precarcinogenic materials to luminal materials such as nondigested fibers, and alterations in lipid absorption leading to hypocholesteremia. The mechanism of action of dietary fibers has generally been considered as being limited to their interactions with some component of the gut contents, whether such contents be ingested food substances, physiological secretions, or bacterial flora. However, it is also possible that fiber per se or bile salt derivatives interact directly by some mechanism with the mucosa to evoke a fundamental change in the intrinsic absorptive capacity of the intestinal tract. Long-term effects of dietary patterns may be associated with biochemical and morphological alterations in the absorptive barriers.

Fatty acids and bile acid administration both evoke intestinal salt and water secretion and cause substantial and reversible alterations in the ultrastructural histology of the mucosal surface (Philips and Gaginella, 1979). The secondary bile acids have been demonstrated to be tumor promoters in experimental animals (Barisawa *et al.*, 1974), as is pancreaticobiliary diversion to the mid small bowel (Williamson *et al.*, 1979). Bile acid accumulation has been proposed to be responsible for the diarrhea states associated with colonic cancer. In an extensive series of studies designed to probe the mechanisms underlying the putative antiatherogenic and anticarcinogenic effects of a semipurified chronic dietary fiber intake and that of bile salt sequestrants, we have studied the ultrastructure of the small and large bowel in considerable detail.

2. MATERIALS AND METHODS

2.1. Animals and Diets in the Chronic Feeding Studies

Male albino rats of the Wistar strain (Carsworth Farms), weighing 150–200 g, were maintained in individual cages and provided the diet and drinking water *ad libitum* for 6 weeks. They were housed in quarters maintained at 23°C and with a 12-hr dark–light cycle. The isocaloric, isogravic diets administered in these studies were comparable to those used earlier (Vahouny *et al.*, 1980) and consisted of the following ingredients in g/100 g diet: dextrose, 55; casein, 25; corn oil, 14; salt mix, USP XIV, 5; vitamin mix. The ability of these various materials to bind cholesterol, phospholipids, and bile salts *in vitro* was also determined and compared to the quantitative morphological observations derived from scanning electron microscopy.

2.2. Methods

2.2.1. Chronic Feeding Studies

The dietary fiber materials, which were fed for a period of 6 weeks at a level of 15 g/100 g diet, included alfalfa (Bio-Serv, Frenchtown, N.J.), white wheat bran (Bio-Serv, Frenchtown, N.J.), cellulose (Solka floc, Brown and Co., Berlin, N.H.), and pectin (Bio-Serv, Frenchtown, N.J.). In the resin-fed groups cholestyramine (Questran, from Dr. H. P. Sarrett, Mead Johnson, Evansville, Ind.), colestipol, or DEAE–Sephadex (Secholex, a gift from Dr. A. Howard, Cambridge, England) was added as 2% of the diet at the expense of dextrose in the control group. Control animals were maintained on Purina Rat Chow. Food consumption and final body weight were similar in control and treated groups.

2.2.2. Colonic Infusion Studies of Bile Acids and Bile Salt Sequestrants

Wistar rats weighing 200 g were fasted for 2 days, with water *ad libitum*. A polytechnic catheter (PE #200) was implanted luminally at the junction of the cecum and the colon. It was threaded under the skin and externalized at the dorsal surface behind the neck. The animals were allowed to recover in individual cages and had free access to chow and water for the remainder of the study. Twice daily at 9:00 a.m. and 3:00 p.m. for 5 days, 1.0 ml of the appropriate test substances in 0.9% saline was introduced into the colon via the indwelling catheter. The test substances included: (1) Saline control (0.9% NaCl). (2) Cholestyramine: 100 mg given twice daily, which corresponds to the amount ingested per diem by rats fed a purified diet containing 2% cholestyramine. (3) Mixed bile acids: 164 μmol mixture of cholate, chenodeoxycholate, and deoxycholate in the proportions of 1.1 : 1.3 : 1.0 in 1 ml saline. This is equivalent to the amount of total and mixed bile acids sequestered by the dose of cholestyramine employed. (4) Cholestyramine–bile acid mixture: 100 mg of cholestyramine and 164 μmol of the mixed bile acids. This mixture was prepared by incubation of the resin and bile acids and reisolation of the resin containing bound bile acids as decribed previously (Vahouny *et al.*, 1980). A total of five animals were used for each group in two sets of experiments. All animal procedures were in accordance with the National Research Council's guide for the care and use of laboratory animals.

2.2.3. Morphological Methods

Three rats in each dietary group and three in the control group were prepared for both light and scanning electron microscopic (SEM) examination. At the end of the 6-week feeding period the rats were anesthetized with sodium pentobarbital and subjected to laparatomy. The alimentary tract from the pyloric sphincter to the terminal colon was removed and the jejunum was identified as the middle fifth of the small intestine. Colon samples were derived from the middle 5-cm segment of that organ. Rectangular segments of both regions were pinned flat as mucosal surface uppermost and fixed in 3% phosphate-buffered glutaraldehyde. Efforts were made to keep the degree of tissue stretch during fixation comparable in all conditions.

During fixation the mucosal surface was brushed gently with a sable brush to remove loose debris. For SEM 1-cm^2 tissue samples were dehydrated and critical-point-dried (Samdri PVT-3) using CO_2. The samples were mounted on aluminum stubs with mucosal surface uppermost and coated with approximately 10 nm of gold/palladium using a Hummer I sputtering device. They were coded and observed in either an ARM 1000 or a JSM-35 scanning electron microscope using 20–25 kV accelerating voltage. The microscopists were unaware of the identity of the coded samples. A preliminary assessment was made by a single viewer. Two other microscopists reassessed those samples and then analyzed the results of a repeated experiment. All of the numerical values recorded for the previously agreed upon criteria (number of villi, severity of damage) were pooled and examined statistically (cf. Table I). A minimum of 300 jejunal villi and 300 colonic ridges from three animals per condition were examined, and the number of villi or ridges with abnormal structure was recorded.

The degree of deviation from normal was graded on the following scale: 1, apical swelling of cells, disordered microvillar array; 2, dimpling of swollen cell surface, partial denudation of microvilli; 3, loss of most microvilli, tears in apical membrane; 4, extrusion of cell contents and loss of cells from the epithelial layer. The extent of agreement between the viewers was $\pm 3.5\%$.

Light and transmission electron microscopy (TEM) preparations were obtained by postfixation of glutaraldehyde-fixed samples in phosphate-buffered 2% OsO_4 followed by dehydration and embedment in Epon resin. Sections approximately 0.5 μm thick were cut on a Sorvall MT2-B ultramicrotome, stained with toluidine blue, and examined in a Zeiss phase contrast microscope equipped with a Reichert automatic camera for photographic recording.

Thin sections for TEM were stained with uranyl acetate and lead citrate and examined in a JEOL 100B transmission electron microscope. Quantitative morphology of lipid density in the epithelial cell layer was determined by volume measurements of lipid : cell ratios using the MOP III Digitizing Systems (Baltimore Instrument Co., Baltimore, Md.).

2.2.4. Bile Acid and Lipid Binding by Fibers and Resins

Binding of bile salts and lipids was determined as previously described by this laboratory (Vahouny *et al.*, 1980). Forty milligrams of the appropriate resin or fiber source was added to 5 ml of each micellar solution in a stoppered tube and the mixture was shaken in a Dubnoff incubator at 37°C for 1 hr. The tubes were centrifuged at 30,000*g* for 10 min and the entire supernatant was unified by mechanical agitation on a Vortex mixer prior to assay. Aliquots (0.1 ml) of the supernatant were added to 10 ml liquid scintillant (Scintiverse, Yorktown Research, Elmhurst, Pa.), and radioactivity was determined in a Beckman LS 250 liquid scintillation spectrometer using external standardization. Controls of the appropriate micellar media without added binding substances were carried through the same procedure. Binding was determined as the difference between the radioactivity of each micellar component added and that recovered in the supernatant after incubation. All studies were carried out at least three times and all isotope analyses were carried out in duplicate. Figures represent the mean ± standard error of the mean by statistical analysis.

3. RESULTS

SEM of jejunal tissue from animals fed the control chow diet is exemplified by Fig. 1. The villi are leaf-shaped, with some folds along the lateral sides. Individual epithelial cells are faintly demarcated as hexagonal shapes. Focal indentations in the smooth contour (arrow) have been identified as mucin-secreting goblet cells by cross-correlation with TEM studies. Figure 2 is a TEM of a goblet cell with contiguous enterocytes on either side. The goblet cells (GC) contain mucin granules of varying electron density and usually show some degree of secretory activity. The epithelial microvilli constitute a tightly packed brush border (BB) with a fuzzy coat or glycocalyx on the outer tips. The ultrastructural topography of control rat colon is similar in many respects to that of the small intestine (Fig. 3). The microvilli are, however, somewhat less densely packed and the ratio of goblet cells to epithelial cells is greater than in the jejunum (arrows). The ingestion of 15% alfalfa for 6 weeks was associated with disruption of the uppermost mucosal cells in both small intestine (Fig. 4) and large bowel (Fig. 5). Swollen, distorted epithelial cells and loss of apical membrane integrity were more obvious in the colonic samples.

Figure 6 is representative of jejunal samples from the bran-fed rats. These were essentially comparable to control samples. Occasional exfoliation of intact cells was seen and with the standard preparative procedures more mucin was associated with the surface of the tissue. The colons of the same animals were normal in appearance (Fig. 7), but as in the small intestine, there was evidence

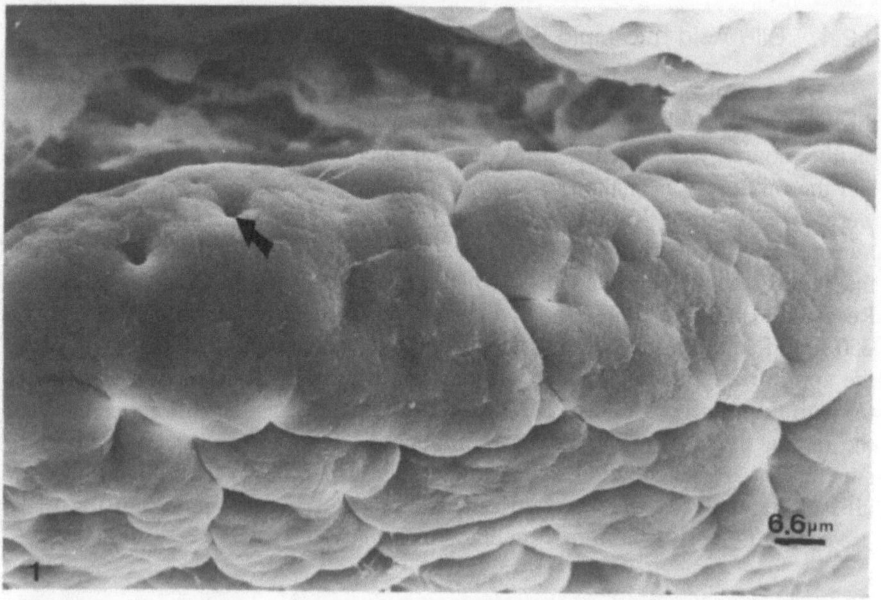

FIGURE 1. Rat jejunum, control sample. Individual villi are leaf-shaped with a smooth topographical appearance. Arrow points to apical indentation which is a goblet cell orifice.

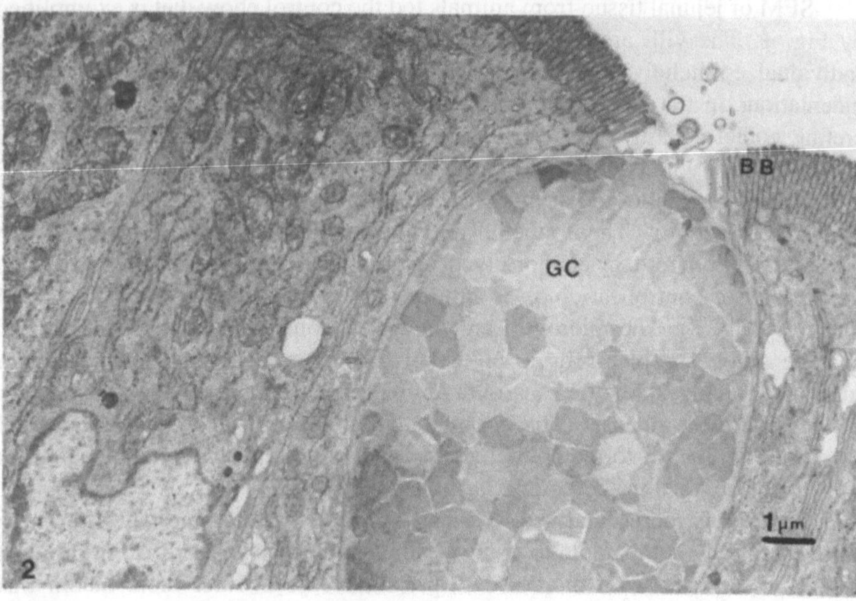

FIGURE 2. Rat jejunum, control TEM showing the typical intracellular structure of a goblet cell (GC) with adjacent brush border (BB) and glycocalyx material on the surface of the microvilli.

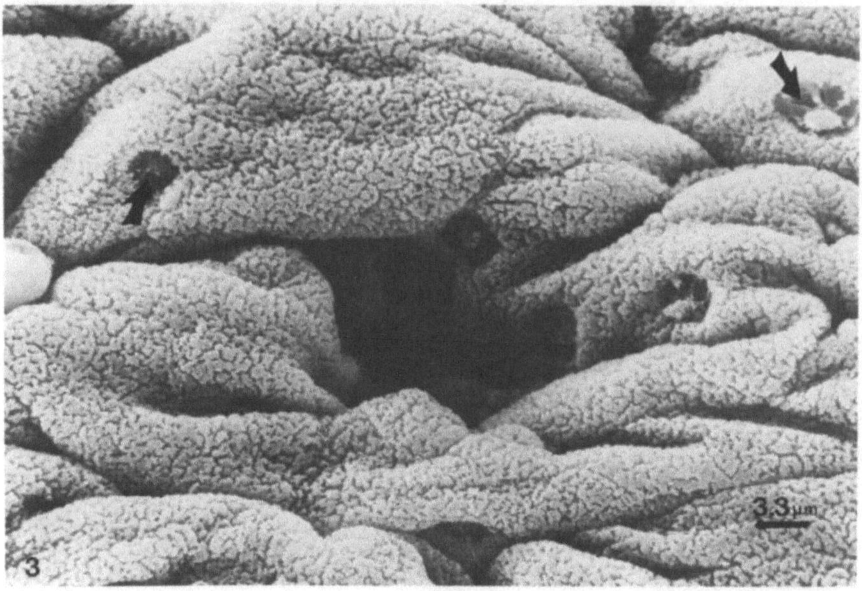

FIGURE 3. Rat colon, control SEM. The epithelial surface has whorled or folded appearance and the interspersed goblet cells show a wider orifice at the luminal surface than do those of the jejunum.

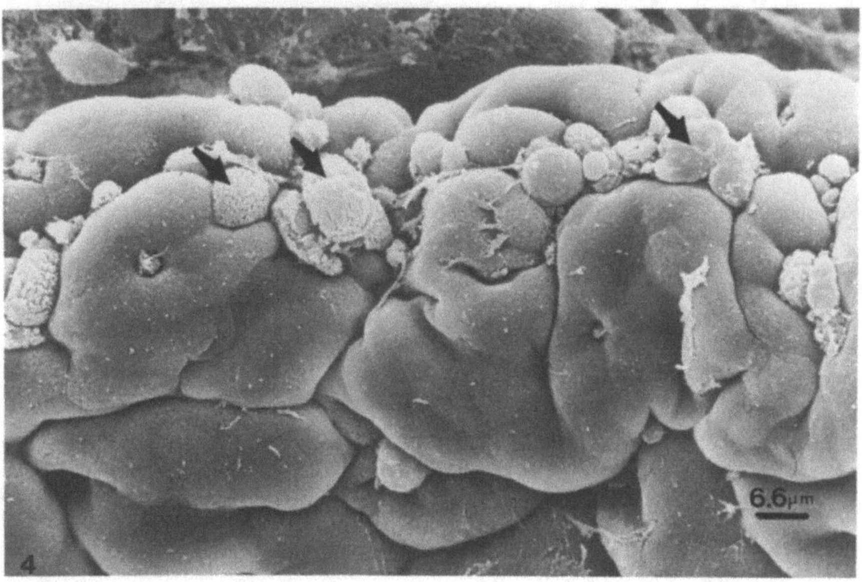

FIGURE 4. Rat jejunum from alfalfa-fed animals. The more apical cells exhibit cell swelling, loss of microvilli, and cell injury (arrows).

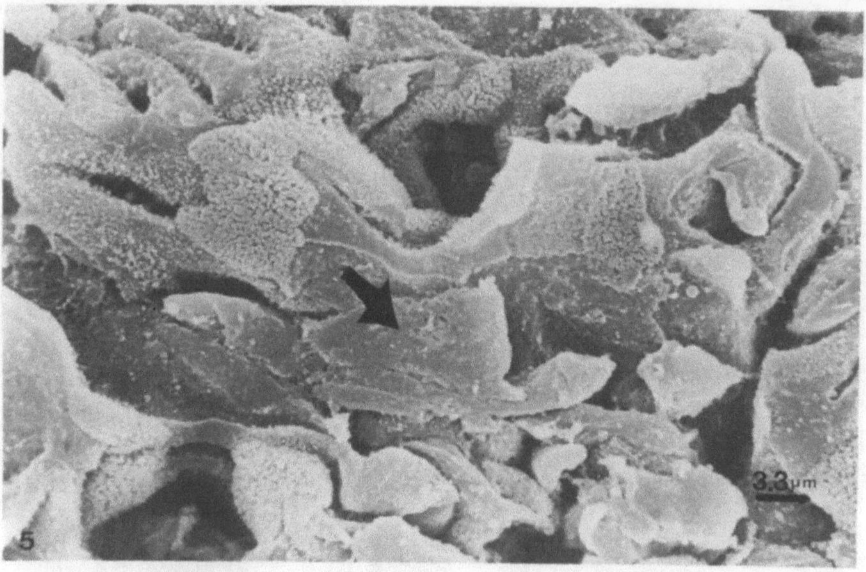

FIGURE 5. SEM of a colonic sample from alfalfa-fed rats. There is extensive loss of microvilli, and breaks in the apical membrane integrity of the epithelial cells (arrows). Many cells are swollen and distorted.

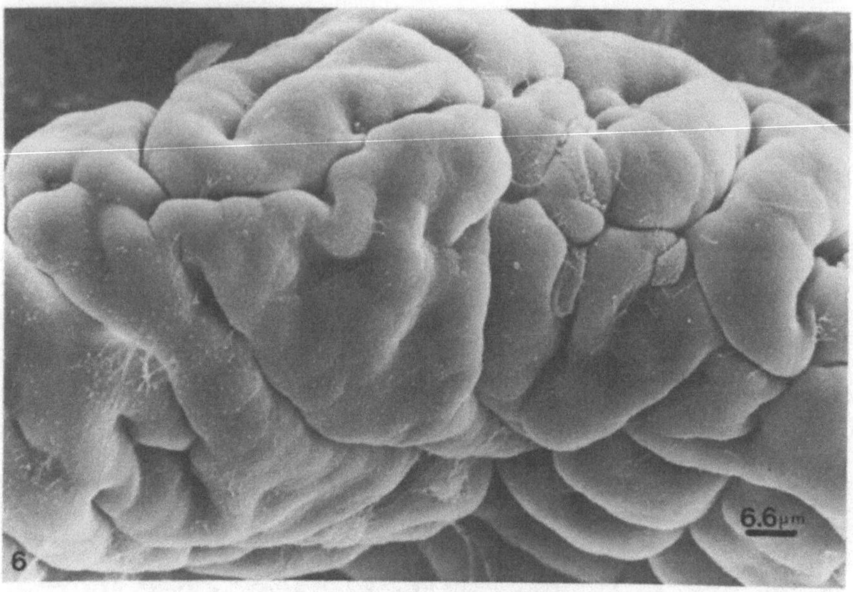

FIGURE 6. SEM of bran-fed animals. The appearance is essentially similar to normal control tissue. Visually, these animals seemed to have more mucin material associated with the tissue surface.

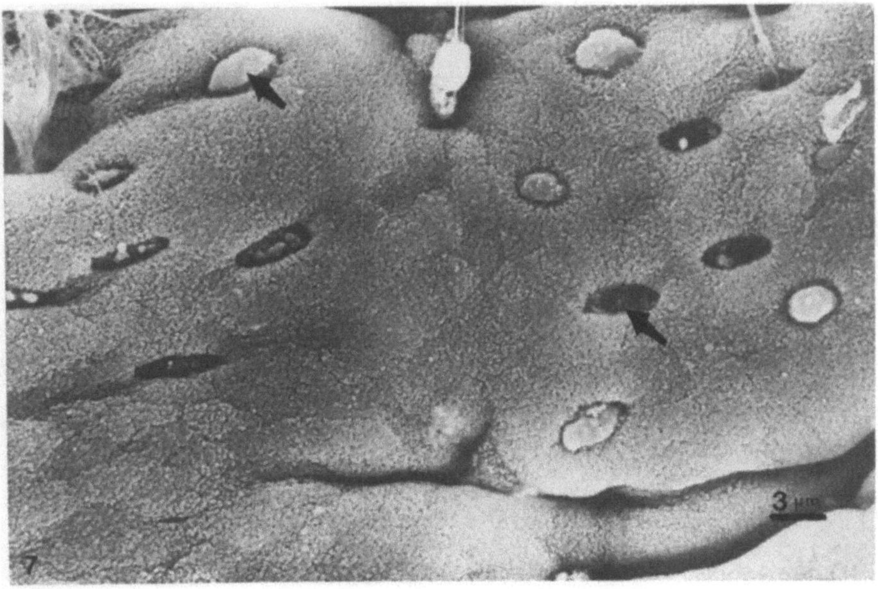

FIGURE 7. SEM of colonic surface in bran-fed rats. The epithelial cells are normal in shape and microvillar density. A larger proportion of the surface cells are goblet cells compared to controls and mucus secretion is apparent from many of these cells.

of enhanced goblet cell secretory activity (arrows). Cellulose feeding for 6 weeks had little apparent effect on the surface mucosal integrity of either small (Fig. 8) or large intestine (Fig. 9). More mucin was visually apparent on these samples, as with the bran-feeding regimen. There was some clumping and mild disarray of the microvillar processes in the colonic epithelial cells in this condition.

In the animals fed 2% cholestyramine, tissue damage at the villar apices was always observed (Fig. 10). Injured cells were frequently surrounded by normal cells and in several cases hemorrhagic debris was found on the tissue surface of both jejunum and colon. Complete loss of the microvillar covering was frequently found (Fig. 11) and a substantial portion of the colonic surface was thus affected by cholestyramine ingestion. Figure 12 is a micrograph from colestipol-treated animals and milder degrees of cell disruption were observed in this condition. Swollen, partially denuded jejunal cells surrounded by normal cells were apparent. The colonic samples were less smooth topographically than controls and showed evidence of cell necrosis, with several of the surface cells being completely stripped of microvilli (Fig. 13). Individual cell boundaries were sharply defined.

Chronic ingestion of 2% DEAE–Sephadex appeared to be associated with more convoluted jejunal villi than those characteristic of animals maintained on control chow. The jejunal surface was fairly normal, but some minor degree of

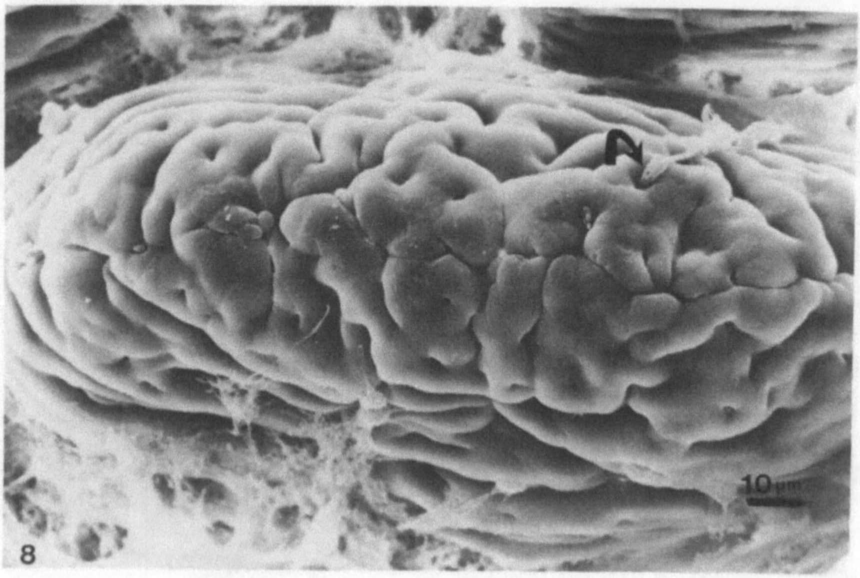

FIGURE 8. SEM of jejunum from cellulose-treated animals. Arrow points to goblet cell secreting a strand of mucus.

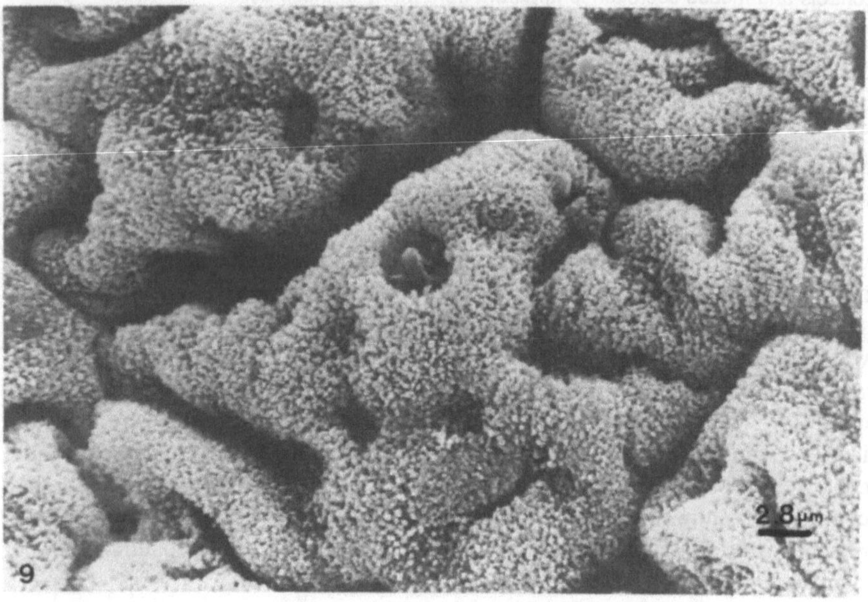

FIGURE 9. SEM of colonic surface from cellulose-fed rats. The microvilli are less densely packed than in control tissue and goblet cell activity is evident.

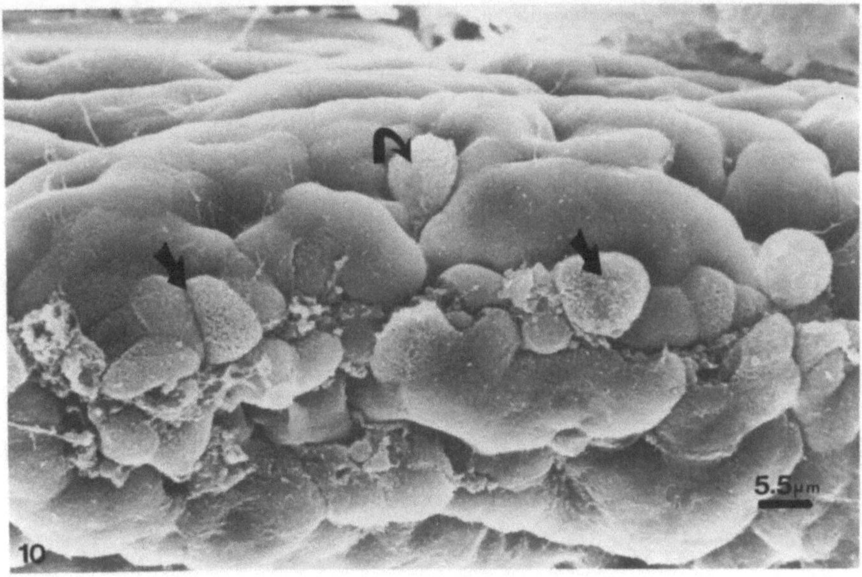

FIGURE 10. Jejunal villus from cholestyramine-treated animals. Cells at the villar apices exhibit swelling, microvillar denudation, and membrane damage.

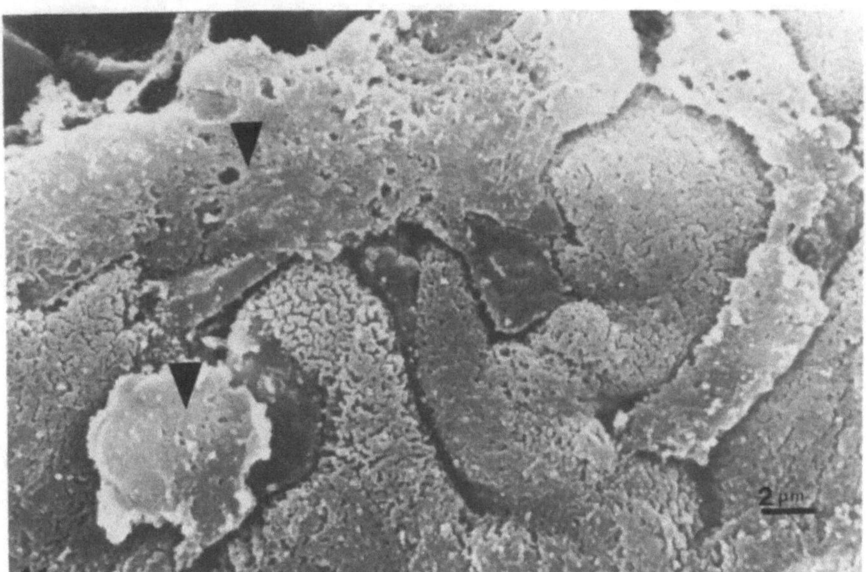

FIGURE 11. Colonic surface of cholestyramine-treated animals. Arrows point to areas of cell damage, apical membrane disruption, and epitheliolysis.

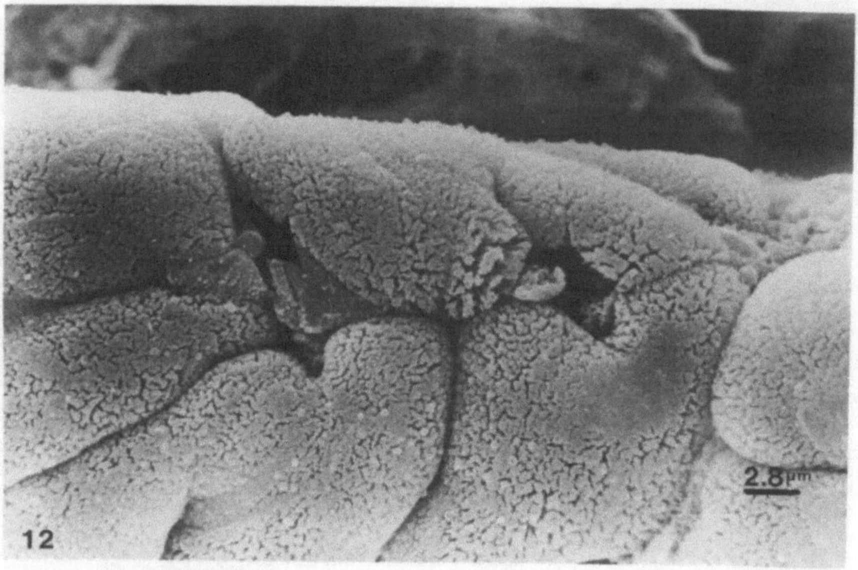

FIGURE 12. Jejunal villus from colestipol-treated rats. Occasional patches of microvillar denudation were observed.

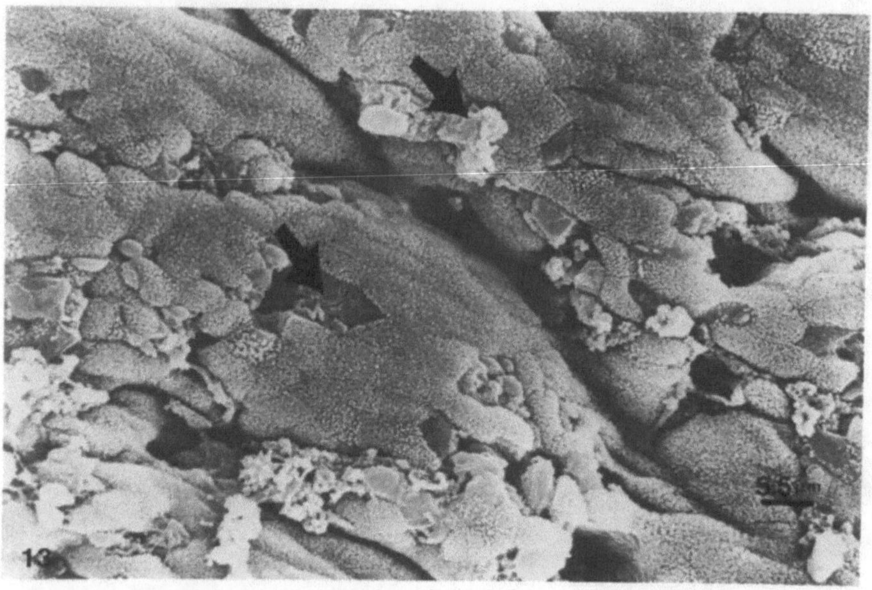

FIGURE 13. SEM of colonic surface in colestipol-fed animals. Many cells are devoid of the microvillar covering and others show progressive signs of cell injury.

cell swelling occurred (Fig. 14). Instead of the smooth ridges usually seen in control colonic tissue, the cells were arranged in deep whorls (Fig. 15) with random patches of cells devoid of microvilli. The alterations in ultrastructural intestinal morphology with chronic feeding semipurified fiber or bile-salt binding resins, as depicted in the preceding micrographs, have been quantitatively estimated. This was achieved by a blind-coding system of examination by three microscopists and the results are shown in Table I. Also included in this table is the *in vitro* bile salt-binding capability of the four fibers and the three resin substances used. It is apparent that a relationship does exist between the degree of bile salt binding *in vitro* of a specific dietary agent and the effect of chronic feeding of this agent on intestinal mucosal topography. The ingestion of 15% pectin or alfalfa or 2% cholestyramine or colestipol is associated with significant morphological deviations in both small and large intestine compared to control animals maintained on regular chow, bran, or cellulose. Figure 16 is a transmission electron micrograph of cholestyramine-treated jejunum showing intracelluar accumulation of lipid droplets.

In a series of experiments designed to test whether the resin per se or the resin–bile salt complex was the necrotic agent the effect of bile salt resin or resin–bile salt complex was probed by twice daily colonic infusion of these

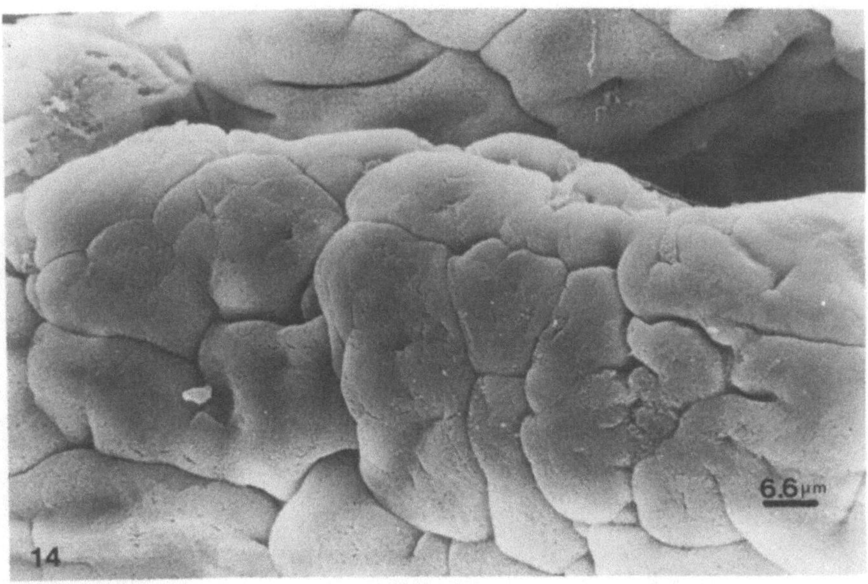

FIGURE 14. SEM of jejunal villi from animals treated with DEAE–Sephadex. Individual cell outlines are more clearly demarcated compared to controls and some loss of microvilli was apparent.

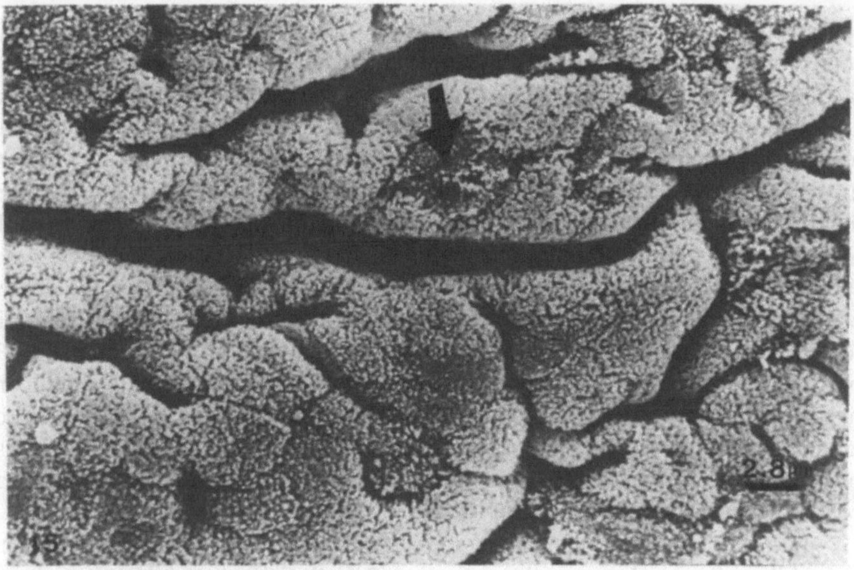

FIGURE 15. Colonic surface of Sephadex-treated rats. There is a mild disarray or clumping of the microvilli with occasional patches of denudation.

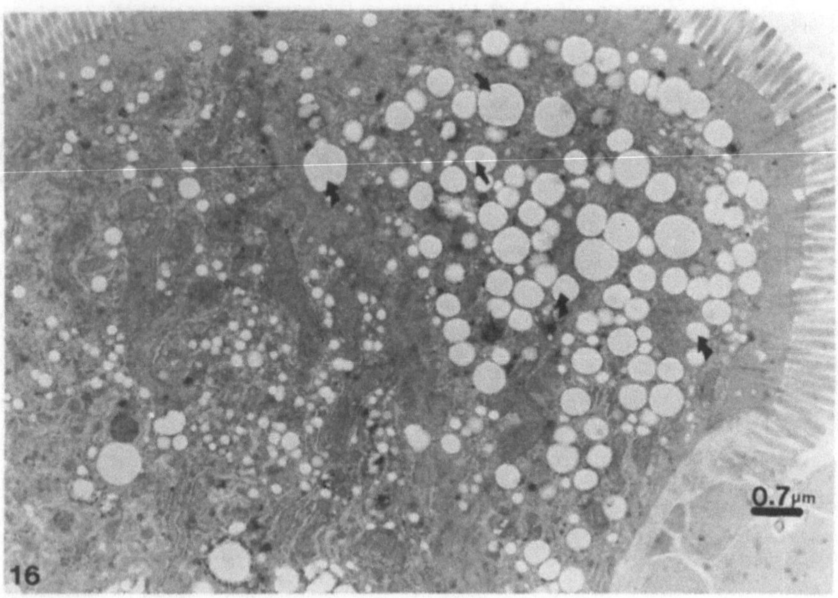

FIGURE 16. TEM of epithelial cells from cholestyramine-fed animals. In this condition the cells exhibit a high density of lipid droplets (arrows) compared to controls. The lipid globules are particularly prominent in the apical half of the enterocytes.

TABLE I. Effect of Fiber and Resin Feeding on Bile-Salt Binding and Morphology[a]

Dietary regimen	Average bile acid binding capacity *in vitro* (0–100%)	Percentage of intestinal villi with structural abnormality		Extent of deviation from control	
		Jejunum	Colon	Jejunum	Colon
Control chow	0	7.1 ± 2.3[d]	5.0 ± 0.8[d]	0.2 ± 0.1	0.1 ± 0.05
Bran	5[b,c]	5.0 ± 1.1[d]	15.0 ± 4.9[d]	0.12 ± 0.08	0.5 ± 0.03
Cellulose	0[c]	7.5 ± 2.1[d]	22.1 ± 4.1[d]	0.23 ± 0.07	0.7 ± 0.10
Pectin	7[b]	30.7 ± 4.7[e]	29.4 ± 8.0[d]	1.18 ± 0.1	1.32 ± 0.1
Alfalfa	15[b,c]	32.8 ± 7.2[e]	58.6 ± 4.3	0.9 ± 0.2	3.2 ± 0.2
DEAE–Sephadex	30–40	13.0 ± 3.6[b]	40.6 ± 9.1[e]	1.6 ± 0.7	2.5 ± 0.06
Colestipol	50–60	35.9 ± 12.6[f]	55.0 ± 10.1[e]	11.5 ± 0.4	2.9 ± 0.4
Cholestyramine	80–100[b,c]	64.2 ± 4.7	39.5 ± 10.5[e]	3.7 ± 0.2	3.2 ± 0.6

[a] Reprinted with permission from Cassidy *et al.* (1980, 1981). Values shown are means ± SE for a minimum of 300 intestinal villi or colonic folds.
[b] After Kritchevsky and Story (1975).
[c] After Vahouny *et al.* (1981).
[d] After Cassidy *et al.* (1981).
[e] Significantly different from control ($p < 0.01$).
[f] Significantly different from control and DEAE–Sephadex.

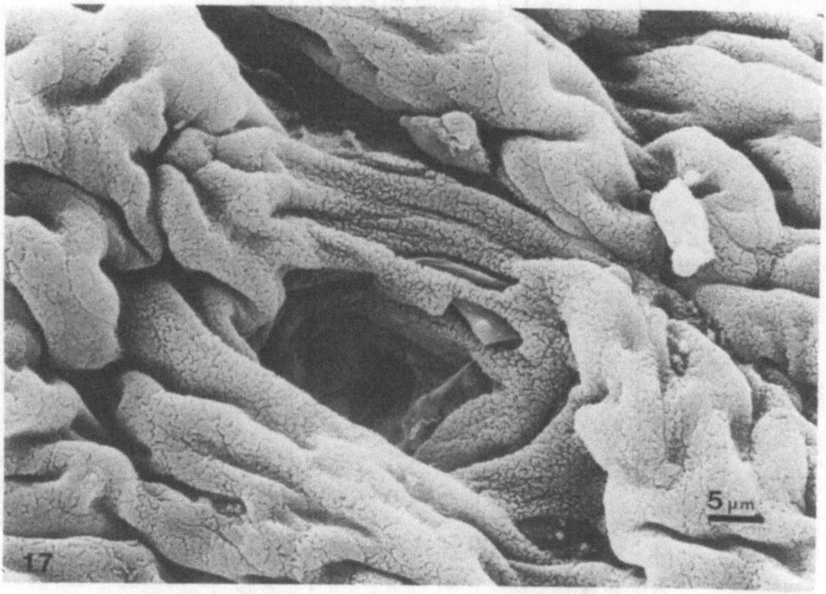

FIGURE 17. SEM of colonic surface from animals perfused twice daily for 5 days with the control colonic catheterization fluid. The tissue is normal in appearance.

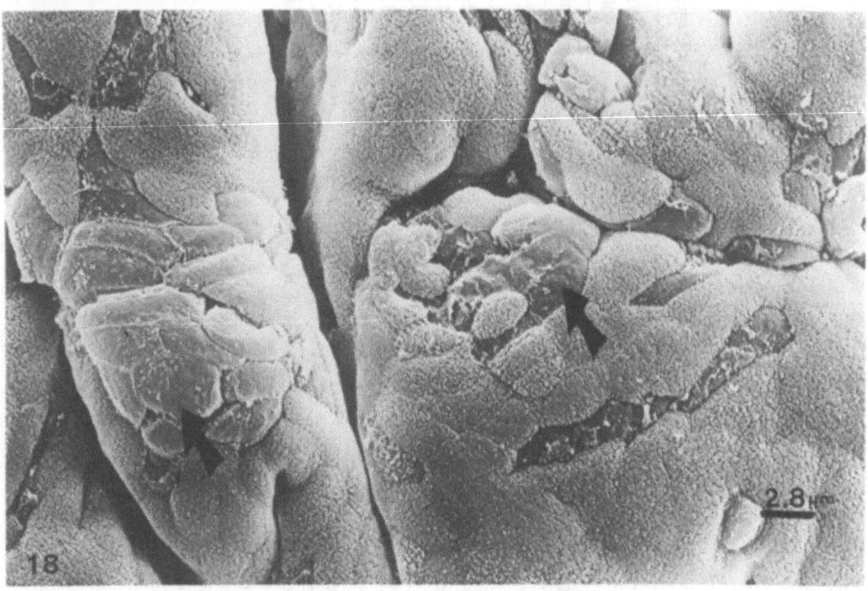

FIGURE 18. SEM of colonic surface in animals administered cholestyramine by colonic infusion for 5 days. Loss of microvilli, cell swelling, and membrane destruction are evident.

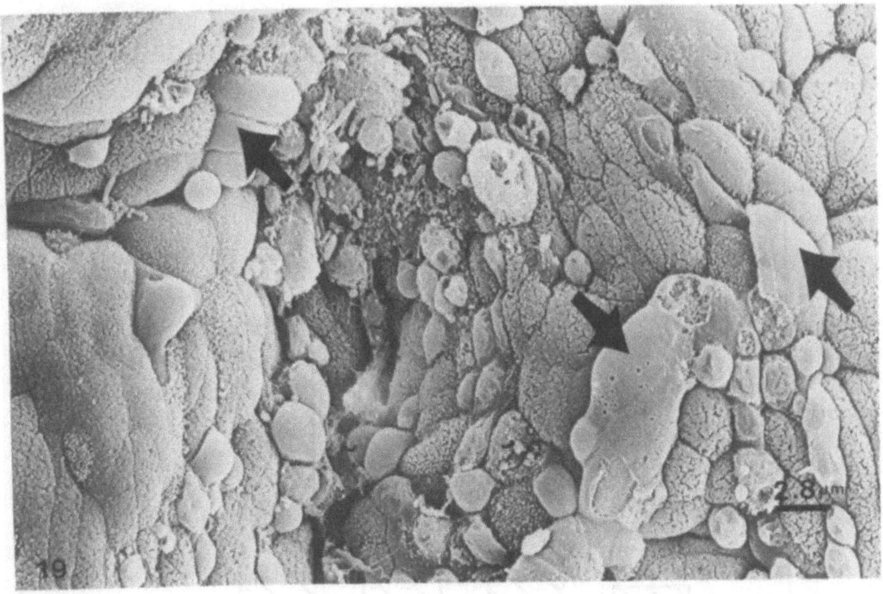

FIGURE 19. SEM of colonic tissue from rats exposed to the mixed bile salt mixture via colonic infusion. The sequence of cell swelling, denudation of microvilli, membrane destruction, and cell lysis is apparent.

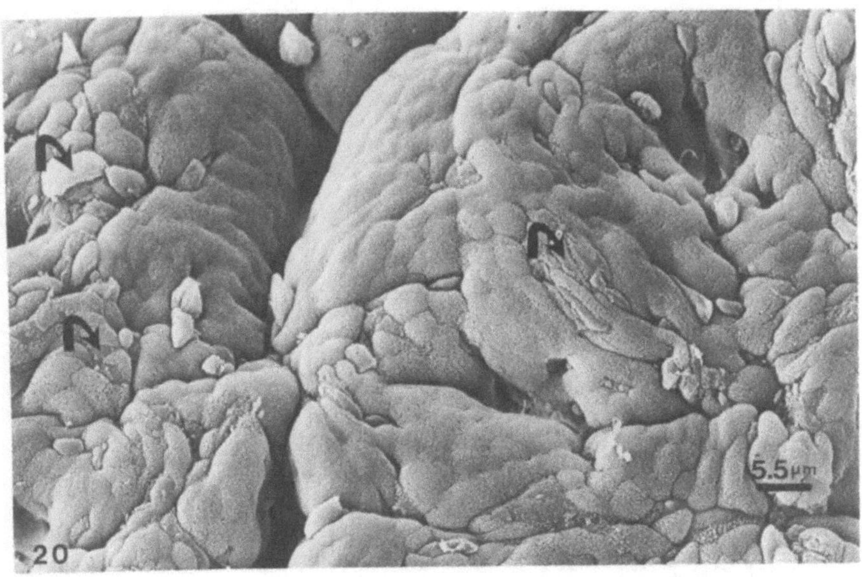

FIGURE 20. Colonic surface from rats infused with the bile salt–cholestyramine mixture for 5 days. There is a distortion of cell architecture (arrows), some loss of microvilli, and patches of cell injury.

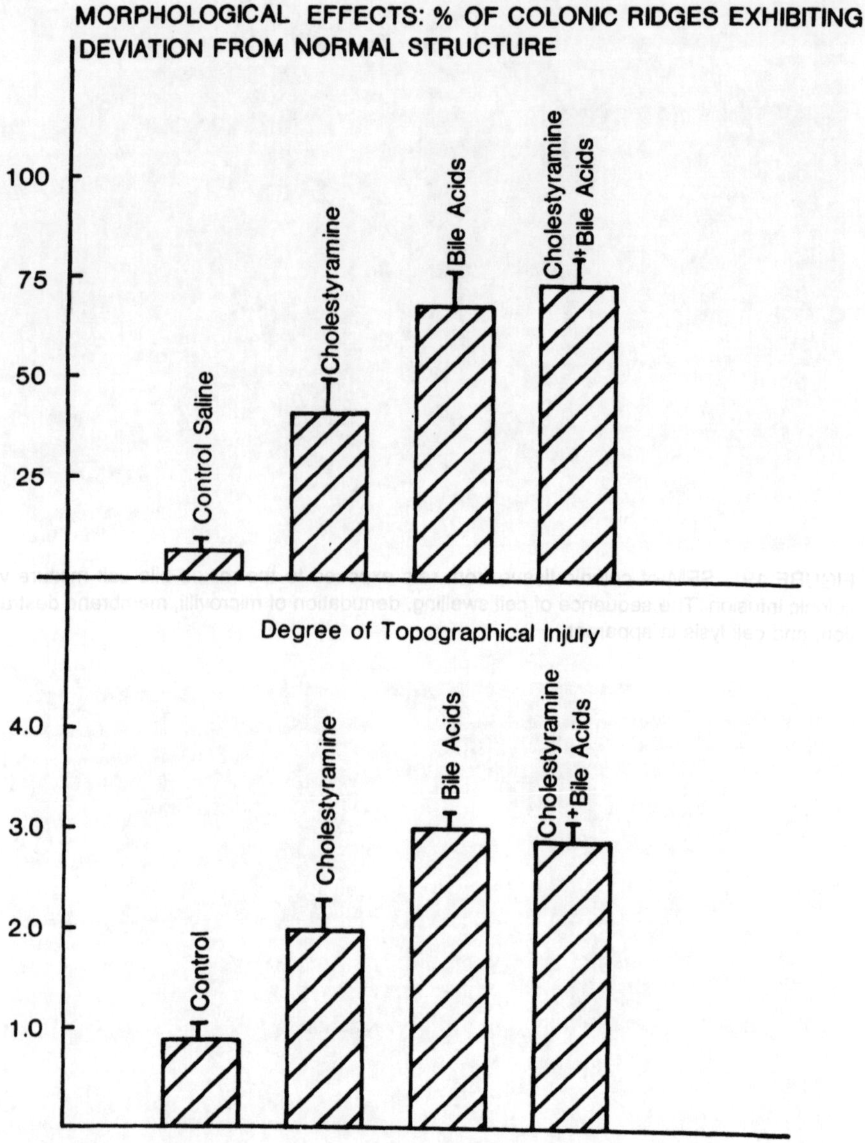

FIGURE 21. Histogram of frequency and degree of morphological damage in the colonic infusin studies. Both cholestyramine and the bile salt solutions evoked similiar deviations from normal histology, but the infusion of both materials together does not potentiate the individual effects.

mixtures. With administration of standard physiological saline the colonic topography did not differ from that of the animals fed the control diet (Fig. 17). With the addition of cholestyramine to the control medium the sequence of cell damage and epitheliolysis was clearly evident (Fig. 18). The mixed bile salt mixture without cholestyramine showed a very similar pattern of injury (Fig. 19). When cholestyramine bound to the bile salts was infused a similar pattern of tissue damage was evident (Fig. 20), although distortion of cellular architecture was more obvious (arrows). The extent and degree of morphological effect of these agents was quantitated and the results are expressed in Fig. 21 as a histogram. Both cholestyramine and the mixed bile salt mixture caused similar effects on colonic topography. Surprisingly, perhaps, the infusion of both ma-

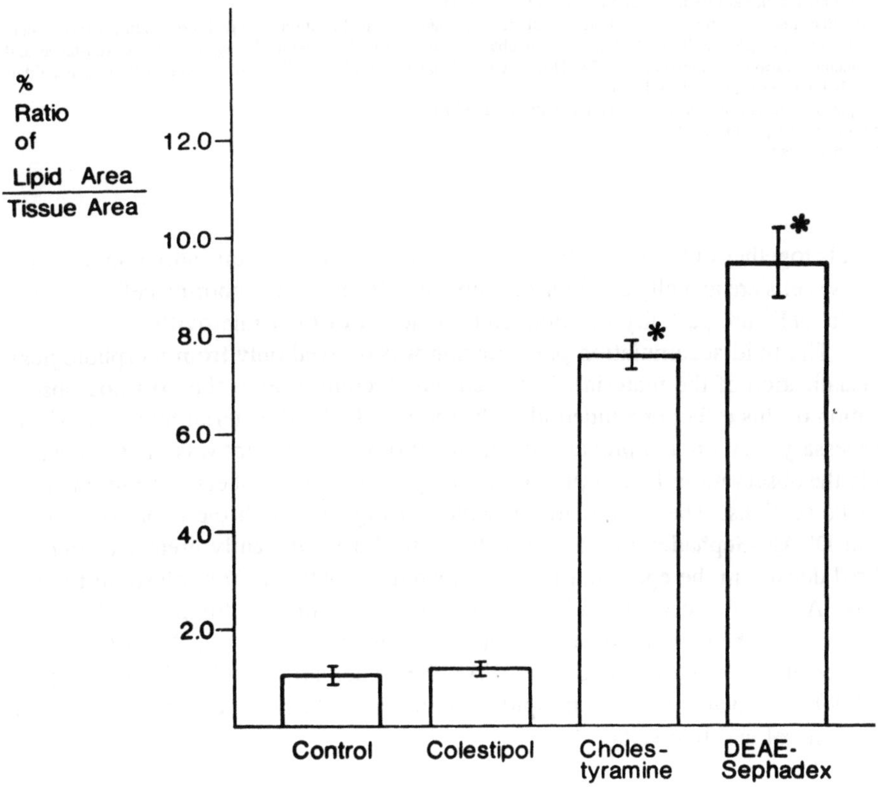

FIGURE 22. The morphometric lipid density of jejunal epithelial cells in groups of rats fed the control diet, DEAE–Sephadex, colestipol, or cholestyramine. Both cholestyramine and DEAE–Sephadex show significantly greater mucosal lipid accumulation when compared to control-fed animals and colestipol-treated rats. *, Significantly different from control ($p < 0.01$).

TABLE II. Binding of Components of Taurochenodeoxycholate–Phospholipid Micelles by Bile Acid Sequestrants and Dietary Fibers[a]

	Binding,[c] percent		
Test substance[b]	Bile salt	Phospholipid	Cholesterol
Cholestyramine	91.9 ± 0.6	99.0 ± 0.2	86.2 ± 4.8
DEAE–Sephadex	84.2 ± 0.3	99.4 ± 0.2	99.6 ± 0.1
Guar gum[d]	31.1 ± 2.7	21.7 ± 4.6	23.4 ± 0.7
Guar gum[e]	37.6 ± 3.0	33.6 ± 1.6	27.2 ± 0.5
Lignin	38.7 ± 1.4	1.6 ± 1.5	14.5 ± 0.7
Alfalfa	14.4 ± 1.1	1.5 ± 1.3	83.6 ± 0.6
Wheat bran	12.1 ± 0.7	0 ± 0	9.5 ± 0.8
Cellulose	3.5 ± 1.0	1.3 ± 1.0	4.7 ± 1.7

[a] Reprinted with permission from Vahouny *et al.* (1981).
[b] Micellar mixtures contained 5 mM taurochenodeoxycholate, 625 μM lecithin, 250 μM monoolein, 500 μM oleic acid, and 250 μM cholesterol. Triplicate incubations were carried out with 40 mg of each test substance and contained various combinations of [7α-^{3}H]taurochenodeoxycholate, [1-^{14}C]dioleylphosphatidylcholine, and either [1,2-^{3}H]cholesterol or [4-^{14}C]cholesterol.
[c] Figures represent means from 6–12 incubations ± SEM.
[d] Low viscosity, food grade.
[e] High viscosity.

terials together did not appear to evoke a potentiation of their individual effects. These cells are apically swollen and separated from the neighboring cells. Several of the cells are partially or completely denuded of their microvilli.

The lipid accumulation phenomenon was derived only from morphological visualization of the material. In the chronic feeding studies the post hoc observation of this effect precluded identification of the lipid material by virtue of the necessary fixation and preparative methodology. An attempt was made to quantify the observations by morphometric analyses of large numbers of representative tissue sections. The results are presented in Fig. 22. Both the cholestyramine- and DEAE–Sephadex-treated animals showed a significantly greater degree of lipid density in the epithelial layer compared to control-fed or colestipol-treated rats. A partial piece of evidence which may explain this finding is shown in Table II. When these resins are incubated *in vitro* with micelles of varying composition cholestyramine and DEAE–Sephadex are capable of binding up to 98–99% of available phospholipid in addition to their sequestration effects on cholesterol and bile salts.

4. DISCUSSION

SEM of the alimentary epithelial surface possesses certain advantages as a morphological technique. It avoids the tedious thin-sectioning and stereological reconstruction associated with transmission or light microscopy procedures. Rapid assessment of surface mucosal integrity, even in human biopsy material, is possible, and is a factor of considerable importance in utilizing this methodology in the determination of the etiologic factors and their progressive identification in a wide variety of gastrointestinal disease states. A disadvantage lies in the fact that only the upper third of the intestinal surface, by virtue of its architectural configuration, is accessed by this technique. It is generally considered, however, that the epithelial cells in this region constitute the more differentiated and hence functionally competent cells in the mucosal layer. The transport of important solutes, e.g., inorganic ions, water, glucose, and amino acids, has been ascribed to the cells in this particular region of the villi and hence their morphological characteristics deserve serious study.

In the studies outlined in this paper we have attempted to characterize the morphological consequences of certain dietary regimens in a rat model. The data thus far obtained would seem to indicate that there is a morphological component which should be added to the expanding list of intestinal mechanisms associated with modification of dietary patterns by fiber ingestion or the administration of bile salt sequestrant materials. A complete and detailed appraisal of normal variations ascribable to species, diet, developmental status, or of the normal ultrastructural topography of small and large intestine is still lacking. Hence, it must be emphasized that the information presented here represents an attempt to quantify readily observable morphological characteristics in groups of animals subjected to feeding regimens in which lipid absorption and metabolic characteristics are fairly well defined. Two primary questions arising from these findings are: (1) Are the differences which have been noted attributable to a degree of deviation which might be expected to be within the normal range? (2) Can such deviations in mucosal structure be interpreted in light of other known and relevant physiological and biochemical observations derived from these or other similar feeding studies?

In regard to the first question, it would seem reasonably well established that normal mucosal cell production, exfoliation, and replacement in the gastrointestinal tract is an unobvious phenomenon when viewed by ultrastructural techniques (Cassidy and Lightfoot, 1979; Creamer *et al.*, 1961). The kinetics of the phenomenon and its regulation have been documented by biochemical and autoradiographic techniques. Currently it seems as if cell expulsion at the villus tips occurs via loss of intact cells (Harding and Morris, 1977; Lightfoot and Cassidy, 1978) without any overt disruption of apical membrane integrity. The cell swelling, loss of microvilli, distortion, and injury seen with certain fibers and resins have not been reported in any control population of rats either by us or others (Gaginella *et al.*, 1977; Ivey *et al.*, 1978). The cell damage phenomenon appears to be limited to those cells of the small intestine or colon that are closest to the lumen. We have, therefore, concluded that the ultrastructural differences induced by the perturbant feeding patterns are mainly limited to the mature, senescent enterocytes in both small and large intestine.

It is accepted that various types of irritant agents that evoke similar stages of cell necrosis can stimulate crypt cell production, migration, and villar repair rates via a tip to crypt feedback mechanism (Sprinz, 1971). Specific experimental examples include the effect of aspirin in an ulceration model (Yeomans *et al.*, 1973) and bile salt influences on epithelial cell dynamics (Roy *et al.*, 1975). It is true that particular dietary regimens and hormonal agents stimulate jejunal and colonic hypertrophy (Dworkin *et al.*, 1976; Hageman and Stragand, 1977; Maudsley *et al.*, 1976; Mak and Chang, 1976).

These results would suggest that one possible intestinal adaptation to chronic feeding patterns or bile salt sequestrations may involve altered rates of cytokinetic loss and renewal. The limited focal lesions found in our studies may represent early mucosal irritation and injury which could lead to later hyperplastic and perhaps neoplastic responses. Further studies are necessary to demonstrate such a continum between mild epithelial cell damage and longitudinal development of carcinomatous lesions. The type of *in situ* cell degeneration noted here causes breaks in the continuity of the mucosal barrier allowing penetration of luminal materials to subepithelial layers. In the *normal* mature rat, 1.7% of intact antigenic protein is absorbed by the small intestine and 0.13% by the colon (Nigro *et al.*, 1973; Asano *et al.*, 1973; Warshaw *et al.*, 1977). The mechanism of necrosis may relate to membrane damage by lysolecithin or to extraction of membrane phospholipids, leading to epitheliolysis.

With respect to a putative association between other documented effects of these dietary materials and the observed structural changes, we have probed several other possible mechanisms implicated in the antiatherogenic and anticarcinogenic sequelae of high-fiber feeding. Particular studies carried out with the same groups of rats included intestinal transit times, the direct measurement of cholesterol absorption into the lymphatic system, certain parameters of hepatic lipid metabolism, and the bile salt-binding capabilities of these agents *in vitro*.

Only cellulose and bran significantly reduced intestinal transit time (Vahouny *et al.*, 1980). There was qualitative evidence of more goblet cell activity and mucin secretion in both of these conditions, although these preliminary observations have not been quantitatively documented. It is not yet established whether cells in the crypt region are predetermined to remain as such or whether some lability between columnar and goblet cells occurs during the villar maturation process. Cellulose feeding induced a significant depression of cholesterol absorption, while the bran diet mildly lowered lipid uptake. Neither of these semipurified diets bind bile acids *in vitro* and their effect on mucosal surface structure is indistinguishable from controls.

Feeding of pectin and alfalfa and the three resins, on the other hand, is associated with lowered cholesterol absorption, binding of bile salts from mixed micelles, and the microscopic appearance of distinctly abnormal mucosal ultrastructure. It would seem that the most clearly discernible correlation between the morphological and biochemical findings lies in the *in vitro* bile salt sequestration phenomenon. There is some evidence that bran and cellulose exert a protective effect against the induction of tumorigenesis with 1,2-dimethylhydrazine in the colon (Barbolt and Abraham, 1978; Fleiszer *et al.*, 1978; Freeman *et al.*, 1978). By contrast, bile acids, fed as such, enhance the mutagenic activity of several carcinogens, even in germ-free animals (Reddy *et al.*, 1978). A similar type of stimulation is observed with cholestyramine (Nigro *et al.*, 1973; Asano *et al.*, 1970). In the colonic perfusion studies we found similar damage to the colonic surface when either bile acids or cholestyramine or the bile acid–resin complex was presented twice per diem for 5 days. There was no evidence of an additive histological effect when bile salts and cholestyramine were used together. Chadwick *et al.* (1979) have recently emphasized the specific bile salt molecular structure required to cause various effects in the intestinal tract, including salt and water secretion and epitheliolysis. Ammon (1979) has concluded that biliary lecithin is critical for the protection of mucosal gallbladder from the potentially damaging effects of bile salts. The finding by Levine *et al.* (1980) that antral instillation of bile salts significantly elevates serum gastrin levels in man suggests that gastrin secretion may be modulated by these dietary regimens and, in consequence, trophic effects on intestinal epithelium could be predicted.

One other observation from our morphological appraisals is the significant accumulation of lipid droplets within the jejunal enterocytes with cholestyramine or DEAE–Sephadex administration. The nature of this material has not yet been identified, but was apparent in three out of four separate feeding studies. An explanation for such lipid-laden cells rests on the assumption of a greater blockage of lipid exit mechanisms at the basolateral membranes of the mucosa than that initially presumed to occur at the micellar entry step, when bile sequestrants are present in the lumen. An obligatory requirement for luminal phospholipid in the exit step via chylomicron packaging and exit has been postulated (O'Doherty *et al.*, 1973). Our observations on the capacity of the sequestrants to bind, not only cholesterol and bile acids, but also 98–99% of available phospholipids in mixed micelles is pertinent. With the feeding of cholestyramine or DEAE–Sephadex there may be a lack of intracellular phospholipid for this essential step. An analogous finding of intracellular lipid deposition has been reported with the use of hydrophobic detergents. Detergent and resin agents may prove to be useful probes of lipid serosal transfer mechanisms in the intestinal tract. It should be mentioned that alternatively (1) permeation of resin into the cell could conceivably bind intracellular phospholipid essential to chylomicron formation, or (2) the resin could also bind other essential luminal components of the lipid transfer process, e.g., pancreatic cholinesterase (Gallo *et al.*, 1980).

In summary, we have sought to investigate structural alterations in the alimentary organs in response to chronically ingested dietary materials. Whether the relationships we have consistently observed are causative or merely correlative epiphenomena remains to be determined in studies designed to test possible underlying mechanisms. Ultimately, any mechanism involved in the explanation of health status relating to diet must originate from some signals which are perceived by this organ and relayed to a host of other homeostatic systems, whether these be morphological, biochemical, physiological, or pathological in nature. Scanning electron microscopy offers a relatively easy and rapid approach to the assessment of morphological modifications and also provides novel clues in the pursuit of other important adaptive responses which may lie in the molecular or biochemical realm.

ACKNOWLEDGMENTS. This work was supported in part by USPHS grant HL-02033 (GVV), USDA grant 7900497 (GVV), Office of Naval Research Contract #N0014-79-C-603 (MMC), and the ITT Continental Baking Co. Research and Development Fund of the Biochemistry Department, George Washington University.

REFERENCES

Ammon, H. V., 1979, Effect of taurine conjugated bile salts with and without lecithin on water and electrolyte transport in the canine gallbladder *in vivo*, *Gastroenterology* **76**:778–783.

Asano, T. M., Pollard, M., and Madsen, D., 1970, Effect of cholestyramine on 1,2-dimethylhydrazine induced enteric carcinoma in germfree rats, *Proc. Soc. Exp. Biol. Med.* **150**:780–785.

Barbolt, T. A., and Abraham, R., 1978, The effect of bran on dimethylhydrazine-induced colon carcinogenesis in the rat, *Proc. Soc. Exp. Biol. Med.* **157**:656–659.

Barisawa, I., Magadia, N. E., Weisburger, J. H., and Wynder, E. L., 1974, Promoting effect of bile acids on colon carcinogenesis after intrarectal installation of *N*-methyl-*N*-nitrosoguanidine in rats, *J. Natl. Cancer Inst.* **55**:1093–1097.

Burkitt, D. P., Walker, A. R. P., and Painter, N. S., 1972, Effect of dietary fiber on stools and transit times and its role in the causation of disease, *Lancet* **2**:1408–1412.

Carroll, K. K., and Khor, H. T., 1975, Dietary fat in relation to tumorigenesis, *Prog. Biochem. Pharmacol.* **1975**:308–353.

Cassidy, M. M., and Lightfoot, F. G., 1979, Electron microscopic study of gastrointestinal response to acetylsalicylic acid, *J. Submicrosc. Cytol.* **11**:449–462.

Cassidy, M. M., Lightfoot, F. G., Grau, L. E., Roy, T., Kritchevsky, D., and Vahouny, G. V., 1980, Effect of bile salt-binding resins on the morphology of rat jejunum and colon: A scanning electron microscopy study, *Dig. Dis Sci.* **25**:504–512.

Cassidy, M. M., Lightfoot, F. G., Grau, L., Story, J., Kritchevsky, D., and Vahouny, G. V., 1981, Effect of chronic intake of dietary fibers on the ultrastructural topography of rat jejunum and colon: A scanning electron microscopy study, *Am. J. Clin. Nutr.* **34**:218–228.

Chadwick, V. S., Gaginella, T. S., Carlson, G. L., and Delongnie, J. C., 1979, Effects on molecular structure of bile-acid induced alterations in absorptive function, permeability and morphology, *J. Lab. Clin. Med.* **94**:661–664.

Correa, P., and Haenszel, W., 1978, The epidemology of large bowel cancer, *Adv. Canc. Res.* **26**(2):141.

Creamer, B., Shorter, R. J., and Banforth, J., 1961, The turnover and shedding of epithelial cells, *Gut* **196**(2):110–120.

Dworkin, L. D., Levine, G. M., Farber, M. J., and Spector, M. H., 1976, Small intestinal mass of the rat is partially determined by indirect effects on ultraluminal nutrition, *Gastroenterology* **71**:626–630.

Fleiszer, D., Murray, D., MacFarlane, J., and Brown, R. A., 1978, Protective effects of dietary fiber against chemically induced bowel tumors in rats, *Lancet* **2**:552–553.

Freeman, J. J., Spiller, G. A., and Kinn, Y. S., 1978, A double-blind study on the effects of purified cellulose dietary fiber on 1,2-dimethylhydrazine-induced rat colonic neoplasia, *Cancer Res.* **38**:2912–2917.

Gaginella, T. S., Lewis, J. C., and Phillips, S. F., 1977, Rabbit ileal mucosa exposed to fatty acids, bile acids and other secretatogues. Scanning electron microscopic appearances, *Dig. Dis. Sci.* **22**:781–789.

Gallo, L., Chiang, Y., Vahouny, G. V., and Treadwell, C. R., 1980, Localization and origin of rat intestinal cholesterol esterase determined by immunocytochemistry, *J. Lipid Res.* **21**:537–545.

Hageman, R. F., and Stragand, J., 1977, Fasting and refeeding cell kinetic response of jejunum, ileum and colon, *Cell Tissue Kinet.* **10**:3–14.

Harding, R. K., and Morris, G. P., 1977, Cell loss from normal and stressed gastric mucosae of the rat. An ultrastructural analysis, *Gastroenterology* **72**:857–863.

Hill, M. J., Drasar, B. S., Williams, R. E. O., Meade, T. W., Cox, A. G., Simson, J. E. P., and Morson, B. C., 1975, Fecal bile acids and clostridia in patients with cancer of the large bowel, *Lancet* **1**:535–539.

Ivey, K. J., Baskin, W. M., Krause, W. J., and Terry, B., 1978, Effect of aspirin and acid on human jejunal mucosa: An ultrastructural study, *Gastroenterology* **76**:50–75.

Kritchevsky, D., and Story, J., 1975, *In vitro* binding of bile acids and bile salts, *Am. J. Clin. Nutr.* **28**:305–306.

Levine, J. S., Connan, J. E., and Kern, M., 1980, Effect of antral instillation of bile salts on fasting serum gastrin levels, *Dig. Dis. Sci.* **25**:449–452.

Lightfoot, F. G., and Cassidy, M. M., 1978, The effect of aspirin and prostaglandin E_1 on intestinal mucus secretory patterns, *Scanning Electron Microscopy* **2**:719–729.

Mak, K. M., and Chang, W. W. L., 1976, Pentagastrin stimulates epithelial cell proliferation in duodenal and colonic crypts in fasted rats, *Gastroenterology* **71**:1117–1120.

Maudsley, D. V., Lief, J., and Kobayashi, Y., 1976, Ornithine decarboxylase in rat small intestine: Stimulation with food or insulin, *Am. J. Physiol.* **231**:1557–1561.

Nigro, N. E., Bhadrachari, N., and Chomachi, C., 1973, A rat model for studying colonic cancer: Effect of cholestyramine on induced tumors, *Dis. Colon Rectum* **16**:438–443.

Nigro, N. D., Campbell, R. L., Singh, D. V., and Lin, Y. N., 1976, Effect of diet high in beef fat on the composition of fecal bile acids during intestinal carcinogenesis in the rat, *J. Natl. Cancer Inst.* **57**:883–888.

O'Doherty, P. J. A., Kakis, G., and Kuksis, A., 1973, Role of luminal lecithin in intestinal fat absorption, *Lipids* **8**:249–255.

Philips, S. F., and Gaginella, T. S., 1979, Effects of fatty acids and bile acids on intestinal water and electrolyte transport, in: *Mechanisms of Intestinal Secretion* (H. Binder, ed.), Alan R. Liss, New York, pp. 287–294.

Reddy, B. S., Weisburger, J. H., and Wynder, E. L., 1975a, Effect of high risk and low risk diets for colon carcinogenesis of fecal microflora and steroid in man, *J. Nutr.* **105**:878–884.

Reddy, B. S., Mastromarino, A., and Wynder, E. L., 1975b, Further leads on metabolic bowel cancer, *Cancer Res.* **35**:3421–3426.

Reddy, B. S., Narisawa, T., Vukusich, D., Weisburger, J. H., and Wynder, E. L., 1976, Effect of quality and quantity of dietary fat and dimethylhydrazine in colon carcinogenesis in rats, *Proc. Soc. Exp. Biol. Med.* **151**:237–239.

Reddy, B. S., Weisburger, J. H., and Wynder, E. L., 1978, Colon cancer: Bile salts as tumor promoters, in: *Carcinogenesis,* Vol. 2 (T. J. Slaga, A. Swak, and R. K. Boutwell, eds.), Raven Press, New York, pp. 453–464.

Roy, C. C., Laurendeau, G., Doyon, G., Chartrand, L., and Rivest, M. R., 1975, The effect of bile and sodium taurocholate on the epithelial dynamics of the rat small intestine, *Proc. Soc. Exp. Biol. Med.* **149:**1000–1004.

Sprinz, H., 1971, Factors influencing intestinal cell renewal cell, *Cancer* **28:**71–74.

Vahouny, G. V., Roy, T., Gallo, L., Story, J., Kritchevsky, D., and Cassidy, M. M., 1980, Dietary fibers. III. Effects of chronic intake on cholesterol absorption and metabolism in the rat, *Am. J. Clin. Nutr.* **33:**2182–1291.

Vahouny, G. V., Tombes, R., Cassidy, M. M., Kritchevsky, D., and Gallo, L. L., 1981, Dietary fibers. V. Binding of bile salts, phospholipids and cholesterol from mixed micelles by bile acid sequestrants and dietary fibers, *Lipids* **15:**1012–1018.

Warshaw, A. L., Bellini, G. A., and Walker, W. A., 1977, The intestinal mucosal barrier to intact antigenic protein. Differences between colon and small intestine, *Am. J. Surg.* **133:**55–58.

Williamson, R. C., Bauer, F. L. R., Ross, J. S., Watkins, J. B., and Malt, R. A., 1979, Enhanced colonic carcinogenesis with azoxymethane in rats after pancreaticobiliary diversion to mid small bowel, *Gastroenterology* **76:**1386–1392.

Wynder, E. L., 1979, Dietary habits and cancer epidemiology, *Cancer* **43:**1955–1961.

Yeomans, N. D., St. John, D. J. B., and du Boer, W. G. R. M., 1973, Regeneration of the gastric mucosa after aspirin-induced injury in the rat, *Am. J. Dig. Dis.* **18:**773–780.

Dietary Fiber and Colon Carcinogenesis

A Critical Review

BANDARU S. REDDY

1. INTRODUCTION

Rapid progress has been made in the understanding of basic concepts concerning carcinogenesis (Griffin and Shaw, 1979; Emmelot and Kriek, 1979). Agents that are genotoxic by definition interact with the gene to yield an abnormal genetic material. In the classic Berenblum–Shubik terminology, such agents would be considered initiating agents (Slaga *et al.*, 1978). It is thought that cancer causation requires the action of such an agent, which alters the genetic material in specific ways. The second broad class of agents act by epigenetic mechanisms and tend to increase the development of lesions initiated by genotoxic carcinogens. Thus, this list includes co-carcinogens, which operate at the same time as genotoxic carcinogens and can alter the metabolism of a genotoxic agent with an increased ratio of activation/detoxification metabolites. Such agents can also act as more classic tumor promoters, which exhibit their effect after the action of a genotoxic carcinogen.

During the past two decades, epidemiologic studies have investigated the influence of environmental factors on the occurrence of cancer of the large bowel, breast, stomach, and upper alimentary tract. Researchers have compared patterns by occurrence between and within population groups, studying in particular the

BANDARU S. REDDY • Naylor Dana Institute for Disease Prevention, American Health Foundation, Valhalla, New York 10595.

difference in rates of the disease between the sexes (where appropriate), over time, and between groups categorized demographically, socioeconomically, and according to migratory and dietary habits. The consistency of their findings suggests that environmental factors in general, and dietary factors in particular, play a dominant role in the development of these cancers in humans.

Epidemiologic studies suggest that diets particularly high in total fat and low in fiber and in certain vegetables as well as high intake of beef are generally associated with an increased incidence of large bowel cancer in man (Armstrong and Doll, 1975; Jain *et al.*, 1980; Reddy *et al.*, 1980a; Wynder *et al.*, 1969; Graham and Mettlin, 1979; Correa and Haenszel, 1978; Burkitt, 1975). Dietary fat may be a risk factor in the absence of factors that are protective, such as use of highly fibrous foods and fiber (West, 1980; Reddy *et al.*, 1978a). As an example, in Finland, where the dietary intake of fat is similar to that of many of the Western countries and the fiber intake is higher, the incidence of colon cancer is lower than in all of the Western countries. On the basis of epidemiologic and laboratory evidence, new efforts are underway to delineate knowledge on other dietary components suspected of playing a role in carcinogenesis (Reddy *et al.*, 1980a).

This brief review evaluates current research on the relation between dietary fiber and large bowel cancer in humans, including the use of animal models to determine if implicated factors can be modified experimentally. This review also presents an evaluation of the mechanism whereby certain dietary fibers modify the risk for the development of large bowel cancer. This chapter, however, does not review the epidemiologic data, since they have been the subject of several reviews (Correa and Haenszel, 1978; Wynder, 1975).

2. ETIOLOGIC FACTORS: CORRELATION AND CASE-CONTROL STUDIES

Since the risk of large bowel cancer closely parallels a country's level of economic development, cross-national correlations between the incidence of colon cancer and dietary habits have been used to select hypotheses for testing in case control and cohort studies. These studies have shown that certain food preferences appear to be associated with either a high or a low risk for colon cancer. Such correlations may be spurious, but when they are supported by experimental evidence from animal studies and underlying mechanisms can be described, further studies seem worthwhile.

Burkitt (1971, 1975) first observed the rarity of large bowel cancer in most African populations and suggested that populations consuming a diet rich in fiber have a lower incidence of this type of cancer, while those eating refined carbohydrates and little fiber have a higher incidence. He argued that large bowel tumors are related to factors characteristic of modern Western society whereby intestinal transit time is slowed, small, firm stools are produced, and fecal bacterial flora are altered. Slower transit would allow more time for gut bacteria to degrade intraluminal components, produce carcinogens, and enable such carcinogens to act. There is, however, no support for the suggestion that faster transit results in increased degradation of substrates by gut bacteria (Walters *et al.*, 1975). A recent study comparing low-risk populations in Kuopio, Finland, with those at high risk in Copenhagen indicated that transit time and stool weight had few significant correlations with diet and defecation habits, but that stool weights were higher in the Kuopio population (IARC Intestinal Microecology Group, 1977). The Finnish population is unique in this respect because in this country the total dietary fat is similar to countries with high rate of colon cancer, although the incidence of large bowel cancer is different. The crucial difference in dietary intake between Finland and Denmark and other Western countries may relate to dietary fiber and meat. Our data also suggest that one of the factors contributing to the low risk of large bowel cancer in Kuopio appears to be that a high intake of dietary fiber (mainly cereal fiber) leads to increased stool bulk, in effect diluting tumorigenic compounds in the colon (Reddy *et al.*, 1978a). The results are consistent with a possible role for dietary fiber in the prevention of large bowel cancer in humans.

Case control studies have been conducted to study the possible relation of dietary fiber to large bowel cancer. Recently, Dales *et al.* (1979) found that among American blacks significantly more colon cancer patients than controls reported that their diet was high in saturated fat and low in fiber. Investigating many dietary constituents, Modan *et al.* (1975) discovered that those contributing less to the diets of patients with colon cancer than to the diets of controls were those containing fiber. Bjelke (1974), who interviewed hospitalized patients and controls in Minnesota and in Norway, learned that colorectal cancer patients less frequently ate vegetables; in particular, the Minnesota patients ate less cabbage. Similarly, Graham *et al.* (1978) found that individuals who ate vegetables such as cabbage, broccoli, and Brussels sprouts had a lower risk of colon cancer.

These studies indicte that diets with a high intake of total fat and beef and a low intake of certain fibers and certain vegetables are generally associated with an increased incidence of large bowel cancer in humans. Even in populations consuming a high amount of fat, high dietary fiber acts as a protective factor in colon carcinogenesis.

3. PROTECTIVE EFFECT OF DIETARY FIBER IN COLON CARCINOGENESIS: POSSIBLE MECHANISM

The possible mechanism of protective effect of dietary fiber against colon cancer has been the subject of a recent workshop (Talbot, 1980). Although the concept of fiber involvement in colon carcinogenesis is attractive, the data often appear contradictory and confusing. Discrepancies may have arisen from the general misuse of fiber teminology. As well, experimental design has failed to account for the possible subtle effect of inhibitors, especially in relation to the promoting process. Evaluations of the biologic function of dietary fiber have often lacked complete information on the nature of the fiber.

Dietary fiber has been defined as that part of ingested plant material that is resistant to digestion by the secretions of the gastrointestinal tract. It comprises a heterogeneous group of carbohydrates, including cellulose, hemicellulose, and pectin, and a noncarbohydrate substance, lignin (Trowell *et al.*, 1976). According to Van Soest (1978), fibers can be classified into three groups: vegetable fibers, which are highly fermentable and have little indigestible residue; brans, which are less fermentable; and chemically purified fibers, such as cellulose, which are relatively nonfermentable. Pectins and gums, soluble substances that are not true fibers, are considered part of the dietary fiber complex because of the similar effects they can elicit in the diet. Wheat bran and vegetable and fruit fibers have different percentages of cellulose, hemicellulose, and lignin.

The protective effect of dietary fiber may be due to adsorption, dilution, or metabolism of co-carcinogens, promoters, and yet-to-be-identified carcinogens by the components of the fiber (Table I) (Reddy *et al.*, 1978a; Nigro *et al.*, 1979; Cummings *et al.*, 1979). There is evidence that alfalfa, wheat straw, and some other fibers can bind considerable amounts of bile acids *in vitro* (Story and Kritchevsky, 1979). This indicates that the different types of nonnutritive fibers

TABLE I. Modifying Factors in Colon Cancer

Dietary fat[a]	Dietary fibers[a,b]	Micronutrients[c]
1. Increases bile acid secretion into gut	1. Certain fibers increase fecal bulk and dilute carcinogens and promoters	1. Modify carcinogenesis at activation and detoxification level
2. Increases metabolic activity of gut bacteria	2. Modify metabolic activity of gut bacteria	2. Act also at promotional phase of carcinogenesis
3. Increases secondary bile acids in colon	3. Certain fibers bind carcinogens and/or promoters	
4. Alters immune system	4. Modify the metabolism of carcinogens and/or promoters	
5. Stimulates mixed function oxidase system		

[a] Dietary factors, particularly high total dietary fat, and a relative lack of certain dietary fibers and vegetables have a role.
[b] High dietary fiber or fibrous foods may be a protective factor even in the high dietary fat intake.
[c] Include vitamins, minerals, and antioxidants.

possess specific binding properties. Dietary fiber could also affect the enterohepatic circulation of bile salts (Kern, 1978). Fiber not only influences bile acid metabolism (Reddy *et al.*, 1978a; Cummings *et al.*, 1979), thereby reducing the formation of potential tumor promoters in the colon, but also exerts a solventlike effect in that it dilutes potential carcinogens and co-carcinogens by its bulking effect (Reddy *et al.*, 1978a) and is able to bind bile acids and certain carcinogenic compounds (Story and Kritchevsky, 1979; Rubino *et al.*, 1979; Smith-Barbaro *et al.*, 1981).

Smith-Barbaro *et al.* (1981) in our laboratory determined the capacity of various fibers to bind the colon carcinogen 1,2-dimethylhydrazine (DMH) *in vitro* (Fig. 1). The percent of DMH bound to wheat bran, corn bran, alfalfa fiber, dehydrated citrus pulp, and citrus pectin was dependent on the pH of the medium as well as the type of fiber examined. Results from this study show that at colonic pH, a greater percent of DMH was bound by wheat bran than by citrus pulp or pectin. Therefore, it is possible that certain fibers bind carcinogen at colonic pH, thus making it unavailable for contact with the colonic mucosa. Other fibers, such as pectin (soluble fiber), do not bind DMH at colonic pH, but may modify the metabolism of carcinogen via activation/deactivation steps either in the liver and/or in the colonic mucosa.

Investigations have been carried out in several laboratories to determine whether there are differences in fecal constituents between populations at high and low risk of colon cancer, and whether changes in the fiber content of the

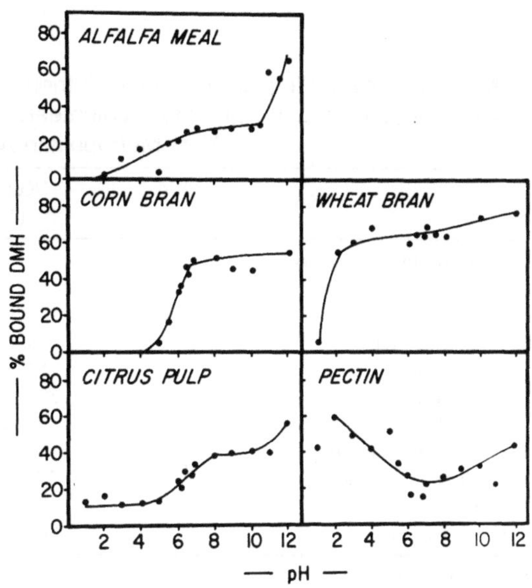

FIGURE 1. Effect of pH on *in vitro* binding of 1,2-dimethylhydrazine (DMH) to various dietary fibers.

diet would alter the concentration of fecal bile acids and the activity of fecal microflora. This chapter, however, does not review the effect of dietary fiber on the excretion of fecal bile acids and the activity of gut microflora, since these topics are adequately covered by Hill (Chapter 25) and Cummings and Branch (Chapter 26).

Recently, we studied healthy individuals in Kuopio, Finland, an area of low risk for the development of colon cancer (Reddy *et al.*, 1978a). Dietary histories indicated that the total fat consumption is similar to that in the United States but that the main source of fat is milk and other dairy products. The intake of cereal fiber in Finland is higher and the daily output of feces three times higher than that of healthy individuals in the United States (Table II). The concentration of fecal secondary bile acids, mainly deoxycholic acid and lithocholic acid, and the extent of fecal bacterial β-glucuronidase activity are less in Kuopio than in the United States, but the total daily output is the same in the two populations because of the threefold greater daily output of feces in Kuopio (Table III). This suggests that increased fecal bulk dilutes suspected carcinogens and promoters that may be in direct contact with the large bowel mucosa.

Recent studies also indicate that the dietary intake of fiber is significantly higher in Kuopio than in Copenhagen, a high-risk area for the development of colon cancer (IARC Intestinal Microecology Group, 1977). Transit times were not different, but stool weight was significantly higher in Kuopio, particularly in the autumn. This suggests that certain foods may have effects on transit time independent of their effects on stool weight. Cummings *et al.* (1978) demonstrated that fiber from carrot, cabbage, apple, bran, and guar gum produces

TABLE II. Dietary Intake of Various Nutrients and Fecal Excretion of Various Constituents in Middle-Aged Male Volunteers from Kuopio, Finland, and New York Metropolitan Area[a]

	Kuopio (15)	New York (40)
Dietary constituents, g/day		
Total protein	93 ± 4	89 ± 2
Total fat	110 ± 4	115 ± 3
Saturated fat	59 ± 3	49 ± 2
Other fats	51 ± 3	66 ± 2
Carbohydrates	320 ± 4	285 ± 4
Total fiber	32 ± 3	14 ± 2
Fecal constituents, g/day		
Fresh feces excreted	277 ± 20	76 ± 12
Fiber	26 ± 2	9 ± 1
Fecal dry matter	61 ± 8	22 ± 1

[a] Averages ± SEM.

TABLE III. Fecal Bile Acids of Healthy Male Subjects from Kuopio, Finland, and New York Metropolitan Area[a]

	mg/g		mg/day	
Bile acids	Kuopio (15)	New York (20)	Kuopio (15)	New York (20)
Cholic acid	0.20 ± 0.06	0.24 ± 0.04	12 ± 2.9	6 ± 1.4
Chenodeoxycholic acid	0.13 ± 0.03	0.23 ± 0.03	8 ± 1.3	5 ± 1.1
Deoxycholic acid	1.72 ± 0.16[b]	3.74 ± 0.26	104 ± 12	88 ± 5.1
Lithocholic acid	1.40 ± 0.16[b]	3.27 ± 0.15	84 ± 5	77 ± 4.5
Ursodeoxycholic acid	0.08 ± 0.02[b]	0.13 ± 0.01	5 ± 1.1	3 ± 0.3
3α,7β,12α-Trihydroxy-5β-cholanic acid	0.04 ± 0.01[b]	9.12 ± 0.01	2 ± 0.8	3 ± 0.3
12-Ketolithocholic acid	0.06 ± 0.02[b]	0.13 ± 0.01	4 ± 1.0	3 ± 0.2
Other bile acids	0.93 ± 0.08[b]	3.8 ± 0.26	56 ± 5.0[b]	89 ± 6.0
Total bile acids	4.59 ± 0.42[b]	11.7 ± 0.54	277 ± 22	275 ± 14

[a] Averages ± SEM.
[b] Significantly different from New York, $p < 0.05$, or better.

different responses in fecal weight in humans related to the intake of pentose-containing polysaccharides in the fiber. The fecal weight increased by 127% when bran was added to the diet and 20% when guar gum was added; carrot, cabbage, and apple produced intermediate changes. Adding fiber to the diet shortened the mean transit time through the gut and significantly diluted an inert marker in the feces. In another study, Cummings *et al.* (1976) reported that an increase in cereal fiber intake from 17 to 45 g/day increased the fecal weight from 79 to 228 g/day and diluted the fecal bile acids. Kay and Truswell (1977) showed that adding wheat fiber to the diet decreased the concentration of fecal bile acids and neutral steroids because of the bulking effect of fiber, whereas the addition of pectin to the diet increased the fecal steroid and bile acid output. These results suggest that the effect on fecal bile acid excretion may depend on the type of fiber consumed.

The effect of dietary wheat bran and alfalfa at 15% level on the composition of fecal bile acids was studied in rats fed a semipurified diet (Reddy *et al.*, 1980d). Diets containing wheat bran and alfalfa caused a significant increase in stool weight. The concentration of fecal bile acids, particularly hyodeoxycholic acid, β-muricholic acid, deoxycholic acid, and lithocholic acid, was lower in rats fed wheat bran compared to those fed a control diet, but the daily output of these bile acids was the same in both groups (Table IV). Alfalfa had no effect on the concentration of fecal bile acids, but the daily excretion of deoxycholic acid, lithocholic acid, and 12-ketolithocholic acid was increased compared to the control diet. It is apparent from this study that the fecal excretion of bile acids varies with the type and amount of dietary fiber.

TABLE IV. Fecal Bile Acid Excretion in Rats Fed Wheat Bran[a]

	mg/g		mg/day per kg body weight	
Bile acids	Control diet	Wheat bran	Control diet	Wheat bran
Hyodeoxycholic acid	0.90 ± 0.14	0.54 ± 0.04[b]	2.99 ± 0.43	2.62 ± 0.20
β-Muricholic acid	0.19 ± 0.03	0.05 ± 0.01[b]	0.66 ± 0.12	0.25 ± 0.09[b]
Cholic acid	0.25 ± 0.05	0.18 ± 0.02	0.89 ± 0.21	0.90 ± 0.15
Deoxycholic acid	1.79 ± 0.30	1.00 ± 0.04[b]	6.01 ± 1.03	4.92 ± 0.28
Chenodeoxycholic acid	0.03 ± 0.01	0.09 ± 0.02	0.11 ± 0.03	0.45 ± 0.14[b]
Lithocholic acid	0.44 ± 0.07	0.27 ± 0.02	1.46 ± 0.23	1.34 ± 0.14
Ursodeoxycholic acid	0.15 ± 0.03	0.07 ± 0.02	0.52 ± 0.12	0.36 ± 0.10
12-Ketolithocholic acid	0.08 ± 0.02	0.04 ± 0.01	0.28 ± 0.05	0.20 ± 0.08
Unknown bile acids	1.01 ± 0.04	0.63 ± 0.08[b]	3.81 ± 0.30	3.70 ± 0.40
Total bile acids	4.92 ± 0.47	2.90 ± 0.14[b]	16.7 ± 1.62	14.1 ± 0.77

[a] Mean ± SEM, $N = 8$.
[b] Significantly different from the control group, $p < 0.05$.

Until recently, not only was the nature of the carcinogens responsible for colon cancer obscure, but there were no real leads. Because of potential importance of fecal mutagens in the genesis of large bowel cancer and of possible role of dietary factors in the induction of colon cancer, the fecal mutagenic activity of various population groups with distinct dietary habits and varied colon cancer incidences was determined by several investigators. Ehrich *et al.* (1979) have demonstrated that the stools of South African urban whites, who consume a high-fat, low-fiber diet and who are at high risk for colon cancer development, were higher (17% of the individuals) in mutagenic activity with *Salmonella typhimurium* strains TA98 and TA100 without microsomal activation compared to South African urban and rural blacks (0–5% of the individuals), who consume a low-fat, high-fiber diet and who are at low risk. Bruce *et al.* (1977) were the first to show that the feces of some normal humans consuming a high-fat, low-fiber diet contained compounds that caused direct mutagenesis of TA98 and TA100 in the Ames assay and suggested that the mutagenicity is due to an *N*-nitroso compound. They have suggested that this compound is produced in the stomach or small intestine. They have also demonstrated that increased dietary fiber, α-tocopherol, or vitamin C reduced fecal mutagens. Kuhnlein *et al.* (1981) compared fecal mutagens (water extracts) from a group of vegetarians consuming high-fiber diets with those from persons on typical North American diets containing meat. On TA100 and TA98, ovolactovegetarians and strict vegetarians had lower levels of fecal mutagens than nonvegetarians. Correlation studies between the pH of the fecal homogenate and mutagenicity indicate the presence of several fecal mutagens.

Recently, we have studied fecal mutagens of three populations with distinct risk for the development of colon cancer: a high-risk population in New York

TABLE V. Mutagenic Activity of Fecal Samples Collected from Healthy Male Subjects from Kuopio, Finland, and New York Metropolitan Area[a]

Population group	Percent samples with mutagenic ratio greater than 3[b]					Percent samples showing mutagenic activity in at least one test system
	TA98		TA100		TA98 and TA100[c]	
	+S9	−S9	+S9	−S9		
New York non-Seventh-Day Adventists (18)	0	22	6	11	11	22
Kuopio (15)	13	0	0	0	0	13
New York Seventh-Day Adventists (11)	0	0	0	0	0	0

[a] Values represent samples which showed a mutagenic ratio of three or greater in *Salmonella*/mammalian microsomal assay system.
[b] Mutagenic ratio is the number of his^+ revertant colonies on the test plate divided by the number of his^+ spontaneous revertant colonies on control plates.
[c] Samples showing activity both in TA98 and TA100 tester systems.

(non-Seventh-Day Adventists) (SDA) consuming a high-fat, low-fiber, mixed Western diet, low-risk vegetarian SDA, and a low-risk population in Kuopio, Finland, consuming a high-fat, high-fiber diet (Reddy *et al.*, 1980b). Fecal samples of non-SDA were highly mutagenic in TA98 without microsomal activation, followed by TA100 without activation and TA100 with activation (Table V). None of the samples of SDA tested showed mutagenic activity in any of the tester systems, whereas Kuopio samples exhibited activity only in TA98 with microsomal activation. In general, the fecal extracts of non-SDA consuming a high-fat, low-fiber, mixed Western diet showed a higher mutagenic activity than did volunteers from Kuopio consuming a high-fat, high-fiber diet.

It has been reported that certain compounds are noncarcinogenic or mutagenic alone, but enhance the tumorigenic or mutagenic properties of carcinogens. For example, Rao *et al.* (1979) demonstrated that B[*e*]P enhanced the 2-AAF- and B[*a*]P-induced mutagenic activity in TA1538 and TA98. Nagao *et al.* (1977) and Matsumoto *et al.* (1972) reported comutagenic properties of harman and norharman for a variety of carcinogens and mutagens. We have studied two groups of volunteers, namely SDA and non-SDA who did not show any fecal mutagenic activity in our earlier study (Reddy *et al.*, 1980b). Samples collected from non-SDA exhibited a significantly higher comutagenic activity in both TA100 and TA98 with S9 activation than did SDA (Table VI) (Reddy *et al.*, 1980c). This is the first demonstration that the fecal samples contain comutagenic activity and that this activity differs in high- and low-risk populations for the development of colon cancer.

If fecal mutagens are involved in the genesis of colon cancer, it would be

TABLE VI. Co-mutagenic Activity of Fecal Extracts from Non-Seventh-Day Adventists and Seventh-Day Adventists in *Salmonella*/Mammalian Microsomal Assay System[a,b]

	Number of his$^+$ revertants/plate 230–250 mg dry feces			
	TA 98		TA 100	
Population group	$-$S9[c]	+S9[d]	$-$S9[c]	+S9[d]
Non-Seventh-Day Adventists (18)	8 ± 4	1832 ± 250[e]	218 ± 98	1157 ± 140
Seventh-Day Adventists (16)	7 ± 3	846 ± 106	104 ± 37	627 ± 104

[a] Non-Seventh-Day Adventists were consuming a high-fat, high-meat diet and Seventh-Day Adventists were consuming a vegetarian diet. These two groups were agematched volunteers from New York metropolitan area not exhibiting any fecal mutagenic activity in our previous screening.

[b] Co-mutagenic activity is the number of his$^+$ revertant colonies obtained with the combination of MNNG or 2-AAF and fecal extract minus the number of his$^+$ revertant colonies induced by MNNG or 2-AAF alone. Values given are mean ± SEM.

[c] MNNG-induced.

[d] 2-AAF-induced.

[e] Significantly different from Seventh-Day Adventists, $p < 0.01$.

of interest to extend these studies to populations at varied levels of risk for colon cancer development. It is possible that the fecal samples may contain co-mutagens and antimutagens that contribute to the overall mutagenic potential of the feces. Isolation and identification of these compounds could lead to a better understanding of the mutagenic load in the colon.

4. EXPERIMENTAL STUDIES IN COLON CARCINOGENESIS

Research on the mechanisms of cancer causation in the large bowel has been assisted by the discovery over the last 20 years of several animal models that mirror the type of lesions seen in man. These models are (1) induction of large bowel cancer in rats through chemicals such as 3-methyl-4-aminobiphenyl or 3-methyl-2-naphthylamine; (2) use of derivatives and analogs of cycasin and methylazoxymethanol (MAM), such as azoxymethane (AOM) and 1,2-dimethylhydrazine (DMH), which work well in rats and mice of selected strains; (3) intrarectal administration of direct-acting carcinogens of the type of alkylnitrosourea type, such as methylnitrosourea (MNU) or *N*-methyl-*N*′-nitro-*N*-nitrosoguanidine (MNNG), which lead to cancer of the descending large bowel in every species tested so far; and (4) the oral administration of large doses of 3-methylcholanthrene, which leads to large bowel cancer in selected strains of hamsters.

The relation between dietary fiber consumption and colon cancer has been studied in experimental animals. Wilson *et al.* (1977) found that Sprague–Dawley rats fed a diet containing 20% corn oil or beef fat and 20% wheat bran had fewer benign colon tumors induced by DMH given by gastric intubation than rats fed a control diet containing 20% fat and no bran. It is possible that with additional time, a number of tumors that were classified as benign might have developed a more invasive character. There was no difference in the incidence of colon cancer between the rats fed corn oil and those fed beef fat. In this study, the experimental diets containing high fat/no bran and high fat/high bran differed substantially in caloric density. Animals on a high-bran diet might have consumed quantitatively more vitamins and minerals than those fed a high-fat, no-bran diet because the rats adjust their food intake to compensate for nondigestible fiber in high-bran diet.

Recently, Freeman *et al.* (1978) compared the incidence of colon tumors induced by DMH in Sprague–Dawley rats fed either a fiber-free diet or a diet containing 4.5% purified cellulose. Among the animals ingesting cellulose, fewer had colonic neoplasms, and the total number of colon tumors in this group was lower. This protective effect appeared to be associated with a shift in tumor distribution from the proximal colon to a more distal site. Although the mechanism for this apparent redistribution of tumors within the colon remains obscure, some change in the luminal physiochemical environment or some inherent difference in the mucosa of the two areas may be responsible.

A recent study by Fleiszer *et al.* (1978) indicated that the incidence of colon tumors induced by DMH in rats decreases as the dietary intake of fiber increases. The number of animals used in this study was small. The diets in that study, namely high-bran diet (28% fiber), special chow (15% fiber), rat chow (5% fiber), Flexical diet (0% fiber), differed not only in consistency (that is, solid or liquid) but also in the proportions of protein and fats, which have been shown to have an independent effect on colon carcinogenesis induced by DMH. However, the colon cancer incidence in the fiber-free (Flexical) group was lower than in the rat chow group. Some reduction in tumor incidence in the rats ingesting a high-fiber diet might be expected on the basis of reduced energy intake. Although the study's findings suggest that reduced intake alone cannot account for the significant protective effect of dietary bran, a better experimental design might have strengthened the results.

In another study, Cruse *et al.* (1978) found that a diet containing 20% wheat bran had no effect on colon carcinogenesis induced by DMH in rats. Not only were the number of animals used in this study small (ten rats/group), but also the doses of the chemical in their experiment were so high (40 mg/kg body weight per week for 13 weeks) that any protective effect of bran might have been unobservable. In a study of the effect of diet on chemical carcinogenesis, it is important to avoid exposing the animal to an excessive level of carcinogen for a long period, as this may obscure more subtle changes induced by certain dietary modifications. In addition, differences in caloric density of experimental diets attributable to the dilutional effect of added fiber further complicated the interpretation of the data. In fact, the data presented by Cruse *et al.* (1978) suggest that a high-fiber diet reduces the frequency of death due to DMH in rats.

The effect of a diet containing 15% alfalfa, pectin, or wheat bran on colon carcinogenesis by MNU or AOM was studied in F344 rats by Watanabe *et al.* (1979). In this study, the experimental diets were not adjusted isocalorically. The addition of pectin or wheat bran to the diet greatly inhibited colon tumor incidence induced by AOM, a carcinogen requiring host-mediated metabolic activation (Table VII). However, the incidence of AOM-induced colon tumors was not influenced by the addition of alfalfa to the diet. The diets containing wheat bran and pectin did not protect against MNU-induced colon carcinogenesis. In this study, the rats fed the pectin diet showed about 25% less body weight compared to those fed a control diet or diets containing wheat bran or alfalfa. The question also arises whether the observed results on pectin are related to the dietary pectin per se or to the lower body weight gain of the rat as a result of pectin feeding. Inasmuch as calorie restriction and consequent weight loss could decrease the tumor incidence (Clayson, 1975), the effect of pectin on tumorigenesis may also be due to a depressed weight gain. If this were so, the rats fed

TABLE VII. Colon Tumor Incidence in Female F344 Rats Fed Diets Containing Pectin, Alfalfa, or Wheat Bran and Treated with Azoxymethane or Methylnitrosourea

	Percent animals with colon tumors	
Diet	Azoxymethane-treated	Methylnitrosourea-treated
Control	57	69
Pectin	10[a]	59
Alfalfa	53	83[b]
Wheat bran	33[a]	60

[a] Significantly different from the groups fed the control diet or alfalfa diet by χ^2 test, $p < 0.05$.
[b] Significantly different from the other groups, $p < 0.05$.

the wheat bran should have had the same tumor incidence rates as animals fed the control diet, because these two groups did not differ in body weight gain. In this study, this was not the case. Thus, modifying effect of pectin on AOM-induced colon carcinogenesis may not be ascribed to reduced food intake. It might be explained in terms of diet-dependent intestinal mucosal, as well as hepatic, microsomal carcinogen-metabolizing enzyme inhibitors or inducers in the diet that modify the capacity of the animal to metabolize the carcinogen (Wattenberg, 1978). However, this factor would presumably not play a role in MNU-induced carcinogenesis because MNU does not require metabolic activation either in the liver or in the intestine. These results thus indicate that the protective effect of fiber in colon carcinogenesis depends on the type of carcinogen and the source of fiber.

The effect of alfalfa, wheat bran, and cellulose on the incidence of intestinal tumors induced by AOM was further studied in Sprague–Dawley rats fed diets containing 10% alfalfa, wheat bran, or cellulose and 30% beef fat; 20% alfalfa, bran, or cellulose and 6% beef fat; or 30% alfalfa, bran, or cellulose and 6% beef fat (Nigro *et al.*, 1979). The presence of 10% fiber in the high-fat diet did not reduce the frequency of intestinal tumors. The diets were not adjusted isocalorically. Apparently, the effect of AOM plus the high dietary intake of fat was too great to be affected by the dietary fiber. In addition, at the time of purchase from commercial source, the animals were 4 months old and consuming laboratory chow. However, in the same study, presence of 20% bran or cellulose or 30% of any fiber in a diet containing 6% fat significantly reduced the frequency of intestinal tumors. All the groups, except those with a diet containing 20% alfalfa, had a lower frequency of tumors in the proximal half of the large bowel than the groups not ingesting fiber. The addition of 30% fiber to the diet increased daily fecal weight. The concentration, but not the total daily excretion, of fecal steroids was significantly lower in the groups with a lower tumor frequency.

Therefore, the inhibition of tumor formation by dietary fiber may be due to the dilution of bile steroids in the lumen of the intestine by the additional bulk (Nigro *et al.*, 1979). The investigators suggest that the risk of large bowel cancer may be lowered by a reduction in the amount of fat and an increase in fiber in the diet.

Bauer *et al.* (1979) have demonstrated that the protective effect of dietary fiber against colon carcinogenesis probably occurs at the promotional stage rather than in the initiating period. Rats were fed a fiber-free diet with 20% corn oil or diets containing 20% corn oil and 20% wheat bran, 20% carrot fiber, or 6.5% citrus pectin from 3 days before the first injection of DMH (15 mg/kg body weight/week for 12 weeks) until 14 days after the last injection. They were then transferred to a standard rat pellet diet for 10–12 weeks and autopsied. The fiber-containing diets were fed only for about 15 weeks out of a total experimental period of 25 weeks. There was no difference in the incidence of colorectal tumors between the groups fed a fiber-free diet and those fed a diet containing wheat bran or carrot fiber. The citrus pectin group had a higher incidence of colorectal tumors. However, it is possible that the high tumor yield resulting from large doses of the carcinogen in this study masked any protective effect of dietary fiber. In addition, these results and those of others reported above have suggested that the continual feeding of a high-fiber diet protects against colon carcinogenesis, while a switch from a high-fiber to a low-fiber diet after administration of the carcinogen has no observable effect. These observations imply that dietary fiber protects against tumorigenesis during the promotional phase.

The effect of dietary wheat bran and dehydrated citrus fiber at 15% level and 5% dietary fat on intestinal carcinogenesis induced by AOM and 3,2′-dimethhyl-4-aminobiophenyl (DMAB) was studied in male F344 rats (Reddy *et al.*, 1981; Reddy and Mori, 1981). Composition of diets was adjusted so that all animals in different experimental groups consumed approximately the same amount of protein, fat, minerals, and vitamins. The animals fed the wheat bran or citrus fiber and treated with AOM had a lower incidence (number of animals with tumors) and multiplicity (number of tumors/tumor-bearing rat) of colon tumors and tumors of the small intestine than did those fed the control diet and treated with AOM (Table VIII; Reddy *et al.*, 1981). Although 15% purified pectin in the diet (Watanabe *et al.*, 1979) inhibited the colon tumor incidence better than did 15% dehydrated citrus fiber, in this study the inhibition of colon tumor multiplicity was more pronounced with the dehydrated citrus fiber compared with purified pectin. Because dehydrated citrus fiber contains about 20% pectin, the pectin content of this diet was considerably lower than that of the diet used in the study by Watanabe *et al.* (1979). Therefore, the modifying effect of dehydrated citrus fiber in intestinal tumor incidence may be due to the pectin content of the citrus fiber, its fecal bulking effect, and its nonfiber components.

The animals fed the wheat bran and treated with DMAB had a lower

TABLE VIII. Colon Tumor Incidence in F344 Male Rats Fed Diets Containing Wheat Bran or Citrus Fiber and Treated with Azoxymethane

	Animals with colon tumors						Colon tumors per tumor-bearing rat[c]		
	Total[b]		Adenoma		Adenocarcinoma				
Diet[a]	No.	%	No.	%	No.	%	Total	Adenoma	Adenocarcinoma
Control (96)	86	90	83	86	60	63	3.45 ± 0.16	2.37 ± 0.16	1.08 ± 0.18
Wheat bran (51)	36	71[d]	24	47[d]	20	39[d]	155 ± 0.12[e]	0.94 ± 0.13[e]	0.61 ± 0.11[e]
Citrus pulp (51)	32	63[d]	21	41[d]	20	39[d]	1.78 ± 0.18[e]	0.90 ± 0.14[e]	0.88 ± 0.16

[a] Effective number of animals in each group is shown in parentheses.
[b] Total represents animals with adenomas and/or adenocarcinomas.
[c] Mean ± SEM.
[d] Significantly different from the group fed the control diet by χ^2 test ($p < 0.05$ or better).
[e] Significantly different from the group fed the control diet by Student's t test ($p < 0.05$ or better).

TABLE IX. Colon Tumor Incidence in F344 Male Rats Fed Diets Containing Wheat Bran or Citrus Fiber and Treated with 3,2′-dimethyl-4-aminobiphenyl

	Animals with colon tumors (tumor incidence)						Colon tumors/tumor-bearing rat (tumor multiplicity)[c]		
	Total[b]		Adenoma		Adenocarcinoma				
Diet[a]	No.	%	No.	%	No.	%	Total[b]	Adenoma	Adenocarcinoma
Control (94)	43	46	38	40	13	14	0.75 ± 0.10	0.58 ± 0.09	0.17 ± 0.05
Wheat bran (50)	13[d]	26	10[d]	20	4	8	0.28 ± 0.07[e]	0.20 ± 0.06[e]	0.08 ± 0.03
Citrus pulp (50)	22	44	18	36	7	14	0.61 ± 0.12	0.47 ± 0.10	0.14 ± 0.05

[a] Effective number of animals in each group is shown in parentheses.
[b] Total represents animals with adenomas and/or adenocarcinomas.
[c] Mean ± SEM.
[d] Significantly different from the group fed the control diet by χ^2 test, $p < 0.05$ or better.
[e] Significantly different from the group fed the control diet by Student's *t* test.

TABLE X. Small Intestinal, Ear Duct, and Salivary Gland Tumor Incidences in F344 Male Rats Fed Diets Containing Wheat Bran or Citrus Fiber and Treated with 3,2′-dimethyl-4-aminobiphenyl

Diet[a]	Animals with small intestinal tumors (tumor incidence)						Small intestinal tumors/tumor-bearing rat (tumor multiplicity)[c]			Tumors in other organs	
	Total[b]		Adenoma		Adenocarcinoma						
	No.	%	No.	%	No.	%	Total[b]	Adenoma	Adeno-carcinoma	Ear duct, %	Salivary gland, %
Control (94)	52	56	12	13	45	48	0.74 ± 0.09	0.12 ± 0.03	0.62 ± 0.09	24	29
Wheat bran (50)	17[d]	34	2	4	15[d]	30	0.46 ± 0.08[e]	0.02 ± 0.02[e]	0.44 ± 0.10	16	20
Citrus pulp (50)	14[d]	28	3	6	12[d]	24	0.34 ± 0.09[e]	0.05 ± 0.03	0.29 ± 0.08[e]	16	26

[a] Effective number of animals in each group is shown in parentheses.
[b] Total represents animals with adenomas and/or adenocarcinomas.
[c] Mean ± SEM.
[d] Significantly different from the group fed the control diet by χ^2 test, $p < 0.05$ or better.
[e] Significantly different from the group fed the control diet by Student's t test.

incidence and multiplicity of colon and small intestinal tumors compared to those fed the control diet and treated with DMAB (Tables IX and X; Reddy and Mori, 1981). Animals fed the diet containing citrus fiber developed fewer DMAB-induced small intestinal tumors than did the rats fed the control diet.

5. CONCLUSIONS

During the last decade, as this review indicates, some progress has been made in the understanding of the role played by the dietary constituents in general and specifically the role of fibers in cancer of the large bowel. The populations with high incidences of cancer of the breast and large bowel are characterized by consumption of a high level of dietary fat. Furthermore, dietary fat may be a risk factor for colon cancer in the absence of factors that are protective, such as use of certain highly fibrous foods and fiber. Thus, alteration of dietary habits leading to a higher intake of certain fibers would be indicated to decrease the risk of this important cancer. Beginning this dietary pattern early in life may prove most beneficial.

The demonstration of two stages in experimentally induced cancer in animal models suggests that there are two stages in environmentally induced cancer in humans (Diamond *et al.*, 1980). Most human cancers probably result from a complex interaction of carcinogens, co-carcinogens, and tumor promoters. Most of nutritional or dietary factors act at the promotional phase of carcinogenesis. Because promotion is a reversible process, in contrast to the rapid, irreversible process of initiation by carcinogens, manipulation of promotion would seem to be the best method of cancer prevention. Such studies have not been conducted to any great extent in animal systems and should be considered for future studies.

In addition to pinpointing harmful environmental agents and their elimination, which is also one of the practical methods of preventing cancer, there seems to be on the horizon the promise of select methods of prevention. Although there is much need for further research to understand the modification of carcinogenic processes, the successes thus achieved have expanded into some understanding of the numerous steps involved in the activation and detoxification of various chemical carcinogens and promoters by a variety of means. Application of a number of such basic tools, elimination of harmful agents from the environment, reduction of biochemical activation processes and a concomitant increase in detoxification reactions, as well as trapping of active intermediates by harmless nucleophilic reagents, point to a promising future for preventive efforts. Thus, there is hope that these important types of cancer can be controlled by modifying the environment with respect to not only the genotoxic carcinogens, but also epigenetic carcinogens, and thus that human cancer risk can be feasibly reduced (Reddy *et al.*, 1978b).

ACKNOWLEDGMENTS. This work was supported in part by grants CA-16382 through the National Large Bowel Cancer Project, CA-12376, and CA-17613, and contract CP-85659 from the National Cancer Institute. The author thanks Mrs. A. Banow for preparation of the manuscript.

REFERENCES

Armstrong, B., and Doll, R., 1975, Environmental factors and cancer incidence and mortality in different countries with special reference to dietary patterns, *Int. J. Cancer* **15**:617–631.

Bauer, H. G., Asp, N., Oste, R., Dahlquist, A., and Fredlund, P., 1979, Effect of dietary fiber on the induction of colorectal tumors and fecal β-glucuronidase activity in the rat, *Cancer Res.* **39**:3752–3756.

Bjelke, E., 1974, Epidemiologic studies of cancer of the stomach, colon and rectum; with special emphasis on the role of diet, *Scand. J. Gastroenterol.* **9**(31):1–253.

Bruce, W. R., Varghese, A. J., Furrer, R., and Land, P. C., 1977, A mutagen in the feces of normal humans, in: *Origins of Human Cancer* (H. H. Hiatt, J. D. Watson, and J. A. Winsten, eds.), Cold Spring Harbor Laboratory, Cold Spring Harbor, New York, pp. 1641–1646.

Burkitt, D. P., 1971, Epidemiology of cancer of the colon and rectum, *Cancer* **28**:3–13.

Burkitt, D. P., 1975, Large bowel carcinogenesis. An epidemiologic jigsaw puzzle, *J. Natl. Cancer Inst.* **54**:3–6.

Clayson, D. B., 1975, Nutritional and experimental carcinogenesis, *Cancer Res.* **35**:3292–3300.

Correa, P., and Haenszel, W., 1978, Epidemiology of large bowel cancer, in: *Advances in Cancer Research,* Vol. 26 (G. Klein and S. Weinhouse, eds.), Academic Press, New York, pp. 1–141.

Cruse, J. P., Lewin, M. R., and Clark, C. G., 1978, Failure of bran to protect against experimental colon cancer in rats, *Lancet* **2**:1278–1280.

Cummings, J. H., Hill, M. J., Jenkins, D. J. A., Pearson, J. R., and Wiggins, H. S., 1976, Changes in fecal composition and colonic function due to cereal fiber, *Am. J. Clin. Nutr.* **29**:1468–1473.

Cummings, J. H., Southgate, D. A. T., Branch, W., Houston, H., Jenkins, D. J. A., and James, W. P. T., 1978, Colonic response to dietary fibre from carrot, cabbage, apple, bran and guar gum, *Lancet* **1**:5–8.

Cummings, J. H., Hill, M. J., Jivraj, T., Houston, H., Branch, W. J., and Jenkins, D. J. A., 1979, The effect of meat protein and dietary fiber on colonic function and metabolism. 1. Changes in bowel habit, bile acid excretion and calcium absorption, *Am. J. Clin. Nutr.* **32**:2086–2093.

Dales, L. G., Friedman, G. D., Wry, H. K., Grossman, S., and Williams, S. R., 1979, A case control study of relationships of diet and other traits to colorectal cancer in American blacks, *Am. J. Epidemiol.* **109**:132–144.

Diamond, L., O'Brien, T. G., and Baird, W. M., 1980, Tumor promoters and the mechanism of tumor promotion, in: *Advances in Cancer Research,* Vol. 32 (G. Klein and S. Weinhouse, eds.), Academic Press, New York, pp. 1–74.

Ehrich, M., Ashell, J. E., van Tassell, R. L., *et al.*, 1979, Mutations in the feces of 3 South African populations at different levels of risk for colon cancer, *Mutat. Res.* **64**:231.

Emmelot, P., and Kriek, E. (eds.), 1979, *Environmental Carcinogenesis: Occurrence, Risk, Evaluation and Mechanism,* Elsevier, Amsterdam, Holland.

Fleiszer, D., MacFarlane, J., Murray, D., and Brown, L. A., 1978, Protective effect of dietary fibre against chemically induced bowel tumours in rats, *Lancet* **2**:552–553.

Freeman, H. J., Spiller, G. A., and Kim, Y. S., 1978, A double blind study on the effect of purified cellulose dietary fiber on 1,2-dimethylhydrazine-induced rat colonic neoplasia, *Cancer Res.* **38**:2912–2917.

Graham, S., and Mettlin, C., 1979, Diet and colon cancer, *Am. J. Epidemiol.* **109:**1–20.

Graham, S., Dayal, H., Swanson, M., Mittelman, A., and Wilkinson, G., 1978, Diet in the epidemiology of cancer of the colon and rectum, *J. Natl. Cancer Inst.* **61:**709–714.

Griffin, A. C., and Shaw, C. R. (eds.), 1979, *Carcinogenesis: Identification and Mechanism of Action,* Raven Press, New York.

IARC (International Agency for Research on Cancer) Intestinal Microecology Group, 1977, Dietary fiber, transit time, fecal bacteria, steroids and colon cancer in two Scandinavian populations, *Lancet* **2:**207–211.

Jain, M., Cook, G. M., Davis, F. G., Grace, M. G., Howe, G. R. and Miller, A. B., 1980, A case–control study of diet and colorectal cancer, *Int. J. Cancer* **26:**757–768.

Kay, R. M., and Truswell, A. S., 1977, Effect of wheat fibre on plasma lipids and faecal steroid excretion in man, *Br. J. Nutr.* **37:**227–235.

Kern, F. Jr., Birkner, H. J., and Ostrower, V. S., 1978, Binding of bile acids by dietary fiber, *Am. J. Clin. Nutr.* **31:**S175–S179.

Kuhnlein, U., Bergstrom, D., and Kuhnlein, H., 1981, Mutagens in feces from vegetarians and non-vegetarians, *Mutat. Res.* **85:**1–12.

Matsumoto, T., Yoshida, D., and Mizusaki, S., 1972, Enhancing effect of harman on mutagenicity in *Salmonella, Mutat. Res.* **56:**85–88.

Modan, B., Barell, V., Lubin, F., and Modan, M., 1975, Dietary factors and cancer in Israel, *Cancer Res.* **35:**3503–3506.

Nagao, M., Yahagi, T., Kawachi, T., Kosuge, T., Tsuji, K., Wakabayashi, K., Mizusaki, S., and Matsumoto, T., 1977, Co-mutagenic action of nonharman and harman, *Proc. Jpn. Acad.* **53:**95–98.

Nigro, N. D., Bull, A. W., Klopfer B. A., Pak, M. S., and Campbell, R. L., 1979, Effect of dietary fiber on azoxymethane-induced intestinal carcinogenesis in rats, *J. Natl. Cancer Inst.* **62:**1097–1102.

Rao, T. K., Young, J. A., Weeks, C. E., Slaga, T. J., and Epler, J. L., 1979, Effect of the co-carcinogen benzo(e)pyrene on microsome-mediated chemical mutagenesis in *Salmonella typhimurium, Environ. Mutagenesis* **1:**105–112.

Reddy, B. S., and Mori, H., 1981, Effect of dietary wheat bran and dehydrated citrus fiber on 3,2′-dimethyl-4-aminobiphenyl-induced intestinal carcinogenesis in F344 rats, *Carcinogenesis* **2:**21–25.

Reddy, B. S., Hedges, A. T., Laakso, K., and Wynder, E. L., 1978a, Metabolic epidemiology of large bowel cancer: Fecal bulk and constituents of high-risk North American and low-risk Finnish population, *Cancer* **42:**2832–2838.

Reddy, B. S., Weisburger, J. H., and Wynder, E. L., 1978b, Colon cancer: Bile salts as tumor promotors, in: *Carcinogenesis,* Vol. 2, *Mechanisms of Tumor Promotion and Co-Carcinogenesis* (T. J. Slaga, A. Sivak, and R. K. Boutwell, eds.), Raven Press, New York, pp. 453–464.

Reddy, B. S., Cohen, L., McCoy, G. D., Hill, P., Weisburger, J. H., and Wynder, E. L., 1980a, Nutrition and its relationship to cancer, in: *Advances in Cancer Research,* Vol. 32 (G. Klein and S. Weinhouse, eds.), Academic Press, New York, pp. 237–345.

Reddy, B. S., Sharma, C., Darby, L., Laakso, K., and Wynder, E. L., 1980b, Metabolic epidemiology of large bowel cancer: Fecal mutagens in high- and low-risk population for colon cancer, A preliminary report, *Mutat. Res.* **72:**511–522.

Reddy, B. S., Sharma, C., and Wynder, E. L., 1980c, Fecal factors which modify the formation of fecal co-mutagens in high- and low-risk population for colon cancer, *Cancer Lett.* **10:**123–132.

Reddy, B. S., Watanabe, K., and Sheinfil, A., 1980d, Effect of dietary wheat bran, alfalfa, pectin and carrageenan on plasma cholesterol and fecal bile acid and neutral sterol excretion in rats, *J. Nutr.* **110:**1247–1254.

Reddy, B. S., Mori, H., and Nicolais, M., 1981, Effect of dietary wheat bran and dehydrated citrus fiber on azoxymethane-induced intestinal carcinogenesis in Fischer 344 rats, *J. Natl. Cancer Inst.* **66**:553–557.

Rubino, M. A., Pethica, B. A., and Zuman, P., 1979, The interactions of carcinogens and cocarcinogens with lignin and other components of dietary fiber, in: *Dietary Fibres: Chemistry and Nutrition* (G. Inglett and I. Falkehag, eds.), Academic Press, New York, pp. 251–281.

Slaga, T. J., Sivak, A., and Boutwell, R. K. (eds.), 1978, *Carcinogenesis,* Vol. 2, Raven Press, New York.

Smith-Barbaro, P., Hansen, D., and Reddy, B., 1981, Carcinogen binding to various types of dietary fiber, *J. Natl. Cancer Inst.* **67**:495–497.

Spiller, G. A., 1979, Interaction of dietary fiber with the dietary components: a possible factor in certain cancer etiologies, *Am. J. Clin. Nutr.* **31**(suppl.):S231.

Story, J. A., and Kritchevsky, D., 1979, Bile acid metabolism and fiber, *Am. J. Clin. Nutr.* **31**:199–202.

Talbot, J. M., 1980, The Role of Dietary Fiber in Diverticular Disease and Colon Cancer, Life Sciences Research Office, Federation of American Societies for Experimental Biology, Bethesda, Maryland.

Trowell, H. C., Southgate, D. A. T., Wolever, T. M. S., Leeds, A. R., Gassull, M. A., and Jenkins, D. J. A., 1976, Dietary fiber redefined, *Lancet* **1**:967.

Van Soest, P. J., 1978, Dietary fibers: Their definition and nutritional properties, *Am. J. Clin. Nutr.* **31**:S12–S20.

Walters, R. L., Baird, I. M., and Davis, P. S., 1975, Effect of two types of dietary fibre on faecal steroid and lipid excretion, *Br. Med. J.* **2**:536.

Watanabe, K., Reddy, B. S., Weisburger, J. H., and Kritchevsky, D., 1979, Effect of dietary alfalfa, pectin and wheat bran on azoxymethane or methylnitrosourea-induced colon carcinogenesis in F344 rats, *J. Natl. Cancer Inst.* **63**:141–145.

Wattenberg, L. W., 1978, Inhibition of chemical carcinogenesis, *J. Natl. Cancer Inst.* **50**:11–18.

West, D. W., 1980, An assessment of cancer risk factors in Latter-Day Saints and non-Latter-Day Saints in Utah, in: *Banbury Report 4: Cancer Incidence in Defined Populations* (J. Cairns, J. L. Lyon, and M. Skolnick, eds.), Cold Spring Harbor Laboratory, New York, p. 31.

Wilson, R. B., Hutcheson, D. P., and Wideman, L., 1977, Dimethylhydrazine-induced colon tumors in rats fed diets containing beef fat or corn oil with and without wheat bran, *Am. J. Clin. Nutr.* **30**:176–181.

Wynder, E. L., 1975, The epidemiology of large bowel cancer, *Cancer Res.* **35**:3388–3394.

Wynder, E. L., Kajitani, T., Ishikawa, S., Dodo, H., and Takano, A., 1969, Environmental factors of cancer of the colon and rectum. II. Japanese epidemiological data, *Cancer* **23**:1210.

24

Studies on the Effects of Single Fiber Sources in the Dimethylhydrazine Rodent Model of Human Bowel Neoplasia

HUGH JAMES FREEMAN

1. INTRODUCTION

In North America, colorectal cancer is a commonly diagnosed neoplastic disorder and a major cause of cancer mortality (American Chemical Society, 1977; Schottenfeld and Haas, 1978). Despite improved therapeutic modalities, the prognosis, especially for advanced disease, remains dismal (American Chemical Society, 1977; Schottenfeld and Haas, 1978). Epidemiologic studies suggest, however, that this malignancy is, in part, environmental in origin, and thus potentially preventable (Weisburger *et al.*, 1977). Among several possible agents, dietary factors have been implicated, including excesses of fat, cholesterol, refined carbohydrate, and protein (especially beef) as well as deficiencies of dietary fiber and some trace elements. Because of inherent difficulties in conducting long-term prospective studies on specific dietary components, such as the fiber polymers, in humans, alternative methods have been employed to explore their role, if any, in colon cancer pathogenesis.

HUGH JAMES FREEMAN • University of British Columbia Faculty of Medicine (Gastroenterology) and Cancer Research Center, Vancouver, B.C., Canada V5Z 1L3.

2. CHEMICALLY INDUCED INTESTINAL NEOPLASIA

The chemical induction of intestinal neoplasia in experimental animals was first reported in 1941 by Lorenz and Stewart (1941) using methylcholanthrene, a compound similar in structure to bile acids and cholesterol, and possibly released from naturally occurring fecal steroids by intestinal bacterial metabolism. Subsequently, several structurally different compounds yielding similar results have been used. Substituted biphenyls, for example, are industrial by-products (i.e., dye processing) that appear to be secreted into bowel through the biliary tract. Aflatoxins, present in human foods such as peanut butter, reportedly cause hepatoma and colon cancer in rodents under certain conditions and have been implicated in colon cancer development in two young laboratory investigators working with purified aflatoxin (Deger, 1976). In addition, recent studies indicate that isolated human intestinal mucosal cells may actively metabolize aflatoxin B1, producing mutagenic effects (Freeman and San, 1980). In 1963, Laqueur *et al.* (1963) reported the induction of colon carcinoma in rats following cycad meal ingestion. Later (Laqueur, 1965), it was reported that the active agent in this meal was cycasin, a water-soluble glucoside of a potent carcinogenic hydrazine derivative, methylazoxymethanol. Following this, Druckrey *et al.* (1967) reported the highly selective induction of intestinal tumors in rats with the symmetric hydrazine agent, 1,2-dimethylhydrazine. After administration, this agent is metabolized, resulting in alkylation of DNA, RNA, and cell proteins (Lamont and O'Gorman, 1978). The liver and possibly other tissues, including the colon, may be crucial in the activation of this compound (Lamont and O'Gorman, 1978). After subcutaneous injection, two oxidation reactions result in azomethane and then azoxymethane. The former is volatile and expired via respiration. The latter is another agent commonly used in experimental colon cancer induction and is thought to possibly undergo conjugation with glucuronic acid in the liver followed by biliary excretion and bacterial hydrolysis within the lumen of the bowel. Methylazoxymethanol may result and appears *in vitro* to be unstable at body temperature, decomposing through a series of products to produce a reactive compound capable of methylation of subcellular components (Lamont and O'Gorman, 1978). Other pathways may also be operative and possibly more important, with the agent or one of its metabolites reaching colonic epithelium through the bloodstream (Lamont and O'Gorman, 1978). Other luminal factors seem to play a role since decreased numbers of tumors reportedly occur in germ-free rats treated with dimethylhydrazine, while increased numbers are observed in those administered azoxymethane (Reddy *et al.*, 1975).

While the carcinogenicity of these agents is not specifically known for humans, both synthetic and natural hydrazines are found in the environment as industrial and food contaminants (Toth, 1977; Freeman, 1980). 1,1-Dimethylhydrazine can be detected in tobacco, although no definite relationship between smoking and bowel cancer is established. Hydrazines are found in wild and cultivated mushrooms (i.e., *Gyromitra, Helvella*) and these agents may be responsible for symptoms of gastrointestinal toxicity (i.e., diarrhea) reported to follow mushroom ingestion. Hydrazines are found in rocket propellants and are commercially used in pesticides, herbicides, and blowing agents for plastics. Several hydrazine analogs produce tumors in experimental animals. Finally, some related hydrazines cause DNA damage to human cells cultured *in vitro* under certain conditions (Whiting *et al.*, 1979). Although their role in human colon cancer pathogenesis is far from certain, their potential significance may be broader than simply being used as a convenient agent for experimental tumor induction in rodents (Toth, 1977; Freeman, 1980).

Following administration, tumors occur in both the small and large bowel of the rodent (Freeman, 1980). In the small intestine, most are found near the major bile ducts, providing indirect evidence that some component in bile or pancreatic juice is important. Tumors are seen less often in jejunum and rarely, if ever, in ileum. This distribution closely parallels the site of predilection for human small bowel adenocarcinoma. More commonly, neoplasia occurs in the large bowel. Tumors more often are localized distally in the colon, again paralleling the usual distribution seen in humans. Single or multiple tumors may occur in the same animal, sometimes with coexistent small bowel tumors. Macroscopically, colon tumors appear as polypoid masses or sessile plaques and invasion may occur through the muscle coat. Several forms of invasive carcinoma can occur, although most are well-differentiated adenocarcinomas. Administration of lower doses of carcinogen reportedly leads to an apparent shift to more proximal large bowel. Tumors in proximal colon are often mucinous or colloid type adenocarcinomas similar to the location of this variety of tumor in humans. In this type, extracellular mucus is seen within cystic spaces of the tumor. A rarer signet ring cell type is observed with intracellular mucus displacing the nucleus peripherally. Sarcomas also may be found, but carcinoid type tumors have not been reported. Tumors can produce intussception, obstruction, and bleeding. Protrusion of large tumors through the anal verge may occur (Freeman, 1980). In addition to these pathologic similarities with human colon cancer, these experimental tumors share many of the histochemical (Filipe, 1975), biochemical (Freeman *et al.*, 1978a), and ultrastructural (Barkla and Tutton, 1977) features reported in human large bowel neoplasia.

Thus, the use of these chemical agents can lead to induction of neoplasia in a reliable fashion in the rodent. Moreover, relatively specific and selective action for the bowel is observed. Latency period to tumor induction measures in terms of months and is relatively predictable. Although there are definite species differences in susceptibility for these agents, the morphologic and biologic characteristics of the tumors in this animal model are measureably similar to observations reported in humans. There are, however, differences, particularly in their propensity to induce tumors at other sites, such as the ear canal. Caution, therefore, is warranted in extrapolating results of animal studies to human disease.

3. DIETARY FIBERS

Available information is limited on the precise role of specific dietary components in altering structure and function of normal intestine as well as the pathogenesis of intestinal diseases such as neoplasia. Dietary fiber, a polymeric substance derived from plant cell walls, has been intensely studied in recent years using chemically induced rodent bowel tumor models (e.g., Freeman, 1979). Different types of fiber, including both polysaccharides and nonpolysaccharides, with different chemical compositions and physicochemical properties are evident that may have quite variable actions within the bowel. Although thought to be relatively resistant to human digestive enzymes, some degree of luminal hydrolysis occurs, possibly through the activities of microflora, during transit. Hydrogen, carbon dioxide, water, methane, and short-chain fatty acids may be produced. These may significantly alter intestinal pH, water content, nutrient assimilation, and the energy value of the diet. Stool moisture content, for example, may be significantly altered, depending on the precise kind of fiber polymer being ingested. Hemicellulose adsorbs more water into its structure than cellulose, and both of these polysaccharide polymers adsorb more water than the phenylpropane polymer, lignin. As a result, diets high in polysaccharides generally produce stools with a high moisture content, while diets high in lignin have little effect on stool water content. In a similar fashion, it can be anticipated that there may be differences in their abilities to bind organic compounds and alter intestinal transit, bile acid metabolism, and the intestinal microflora. If these fibers were ingested alone or in combination in various foods, such as wheat bran, quite differing effects on normal bowel function as well as the pathogenesis of various diseases, such as colon neoplasia, might be expected.

4. FIBERS AND INTESTINAL TUMORS

The possible role, if any, that each of these constituent polymers plays in human bowel carcinogenesis is far from established. Epidemiologic studies have implied geographic differences in colon cancer incidence inversely related to the amount of ingested fiber. Although there are some interesting hypotheses that have been popularized (Burkitt, 1971), the limited data available are complicated by several variables, including geographic differences in diagnostic methods, data collection and reporting, and the presence of other diseases. Moreover, there are also measurable differences in the amounts of other food substances, including fat, animal protein, refined sugars, and cholesterol in the diets of people from high- and low-incidence areas of colon cancer. These epidemiologic studies have at least raised questions and provided a framework allowing more critical examination under more carefully controlled conditions in the experimental animal laboratory. To date, studies examining the possible role of differing sources of dietary fiber in experimental colon cancer appear to have provided apparently conflicting results (Wilson *et al.*, 1977; Barbolt and Abraham, 1978; Chen *et al.*, 1978; Fleiszer *et al.*, 1978; Cruse *et al.*, 1978; Freeman *et al.*, 1978b; Watanabe *et al.*, 1978; Bauer *et al.*, 1979). For example, protective effects of wheat bran have been reported by several investigators in various animal species (Wilson *et al.*, 1977; Barbolt and Abraham, 1978; Chen *et al.*, 1978; Fleiszer *et al.*, 1978), whereas others have been unable to reproduce these observations with this food substance as a fiber source (Cruse *et al.*, 1978; Bauer *et al.*, 1979). Wheat bran, however, contains variable amounts of fiber and nonfiber components, depending on the source, mode of preparation (i.e., milling procedures), and moisture content of the bran (Saunders, 1978). The apparent discrepancies reported in earlier studies may reflect, in part, qualitative or quantitative differences in one or more of these dietary constituents (Freeman *et al.*, 1978b). Alternatively, relative differences in food intake and animal nutritional status may be important (Freeman *et al.*, 1978b).

In an additional double-blind study shown in Fig. 1, purified cellulose, fed at a 4.5% level (i.e., approximating the percentage of cellulose measured in a high-fiber food source such as wheat bran) protected nutritionally comparable animals from colonic neoplasia induced with the agent dimethylhydrazine (Freeman *et al.*, 1978b). Interestingly, this protective effect of cellulose appeared to be associated with a shift in tumor distribution from proximal colon to a more distal site (Fig. 2), perhaps reflecting a differential effect of cellulose within the colon or inherent regional differences in the luminal physicochemical environment or colonic mucosa itself (Freeman *et al.*, 1978a,b, 1980b). Differences in

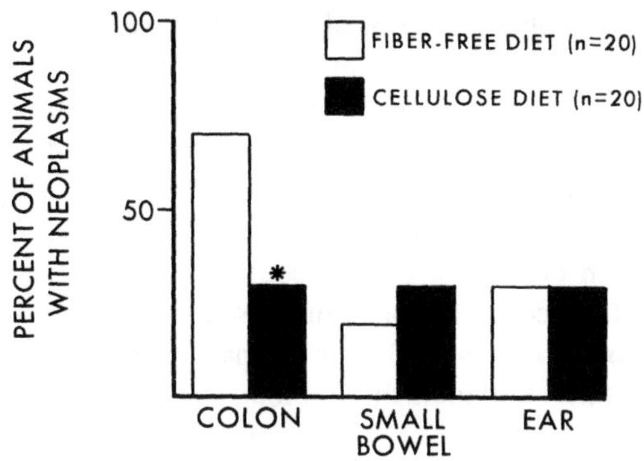

FIGURE 1. Percentage of animals with colon, small bowel, and ear neoplasms. Significant reduction observed only in cellulose-fed animals for colon tumors. *, $p < 0.02$.

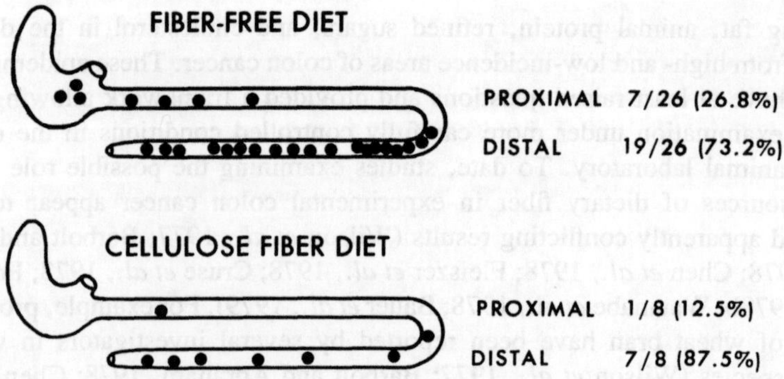

FIGURE 2. Distribution of colon tumors within the large bowel of the fiber-free diet group and the cellulose fiber diet group.

mean colon tumor size and histopathology were not seen. To further examine the role of cellulose in this model, differing dosage levels and schedules of cellulose administration were compared to the effects of feeding identical amounts of pectin as the sole sources of fiber (Freeman *et al.*, 1980a). Rats housed in a carcinogen-containment laboratory were randomly divided into six groups (each containing nine controls and 36 carcinogen-treated animals) and fed one of five diets of similar physical appearance. The diets were designed so that the ratio of calories derived from fat, protein, and carbohydrate were equivalent. Initially, two groups received a fiber-free control diet and then the diet of one of these was changed to 4.5% cellulose after completion of carcinogen injections. The other four groups were given the same diet for the entire study as follows: 4.5% cellulose, 9.0% cellulose, 4.5% pectin, and 9.0% pectin. The injection protocol consisted of weekly subcutaneous administration of 1 mM EDTA with and without a 0.5% solution of 1,2-dimethylhydrazine dihydrochloride. Intakes and outputs as well as weights were measured in metabolic cages. Stool weights were greater in both cellulose fiber groups and the 9.0% pectin group compared to the fiber-free control animals as well as the 4.5% pectin group. Other indices, including body weights, were similar.

Necropsies were performed at two time intervals: one immediately after completion of the carcinogen protocol and a second weeks later. Tumors were examined and classified histologically according to the criteria of Ward (1974). In the initial period shown in Table I, the number and percentage of animals with detectable colon neoplasms were reduced only in cellulose-fed animals. Mean colon tumor number was reduced in both cellulose groups compared to controls ($p < 0.05$) or animals ingesting pectin fiber at the 9.0% level ($p < 0.05$). Although the number and percentage of animals with colon tumors as

TABLE I. Tumor Frequency in the First Sacrifice Period

	Colon tumors		Small bowel tumors	
Diet group ($N = 15$)	Animals	Mean per rat	Animals	Mean per rat
Fiber-free diet	10	1.53 ± 0.34	3	0.27 ± 0.15
Fiber-free diet[a]	10	1.60 ± 0.36	4	0.27 ± 0.12
4.5% Cellulose diet	5	0.67 ± 0.27[b]	2	0.13 ± 0.09
9.0% Cellulose diet	4	0.46 ± 0.27[b]	3	0.20 ± 0.11
4.5% Pectin diet	11	1.20 ± 0.24	6	0.40 ± 0.13
9.0% Pectin diet	11	1.40 ± 0.32	5	0.33 ± 0.13

[a] Changed to 4.5% cellulose after carcinogen protocol.
[b] Mean ± SE; p (t vs. control fiber-free diet) < 0.05.

well as mean colonic tumor number appeared to be lower in the 9.0% cellulose group compared to the 4.5% cellulose group, statistical significance was not achieved, nor were differences significant between controls and pectin groups. Differences in small bowel tumors were not observed in this initial period. In the final period shown in Table II, the number and percentage of animals with colon tumors were reduced in cellulose fiber animals. In both cellulose groups, the mean colon tumor number was significantly less compared to controls or either pectin fiber group ($p < 0.05$). Statistical differences were not seen between the 4.5% and 9.0% cellulose groups. Although the mean colon tumor numbers were slightly greater for both pectin groups compared to controls, these differences were not significant. Finally, no differences were observed between the 4.5% and 9.0% pectin groups. For the small bowel, the mean tumor number was significant only for the 9.0% pectin group compared to controls ($p < 0.05$). This failure of pectin to protect animals from colon tumors had been previously reported (Schottenfeld and Haas, 1978) in a study where a 6.5% pectin diet was

TABLE II. Tumor Frequency in the Last Sacrifice Period

	Colon tumors		Small bowel tumors	
Diet group ($N = 21$)	Animals	Mean per rat	Animals	Mean per rat
Fiber-free diet	16	3.33 ± 0.77	5	0.33 ± 0.14
Fiber-free diet–4.5% cellulose diet	11	1.86 ± 0.51	6	0.57 ± 0.23
4.5% Cellulose diet	13	1.67 ± 0.27[a]	9	0.48 ± 0.13
9.0% Cellulose diet	12	1.52 ± 0.34[a]	8	0.62 ± 0.20
4.5% Pectin diet	16	3.52 ± 0.88	11	0.76 ± 0.22
9.0% Pectin diet	13	3.52 ± 0.83	13	1.10 ± 0.32[a]

[a] Mean ± SE; p (t vs. control fiber-free diet) < 0.05.

administered coincident with carcinogen administration. However, in contrast to that study, pectin feeding for the duration of the study at both the 4.5% and 9.0% levels did not statistically enhance colon tumor induction.

The change from a fiber-free diet to a 4.5% cellulose diet is shown in Table II. Although the number of animals with colon tumors and the mean colon tumor number appeared to be lower compared to controls, a statistically significant effect was not seen compared to animals ingesting a fiber-free diet through the entire study. Similarly, this diet manipulation had no effect on small bowel tumors. These results were of some interest, indirectly suggesting that the protective effect of cellulose depends perhaps upon feeding during carcinogen administration. Although the precise action of cellulose *in vivo* requires further elucidation, this fiber may, in part, directly bind organic compounds and enhance fecal elimination (Freeman, 1980; Kay and Strasberg, 1978). Alternatively, cellulose, by virtue of its water-holding capacity, may lead indirectly to an intraluminal dilution of the critical agent (Freeman, 1980). Or, since intestinal flora appear, in part, to be dietary substrate dependent, fibers such as cellulose may alter the numbers or types of microbes or their capacity for metabolic activation of a carcinogen (Kay and Strasberg, 1978; Freeman, 1980, 1981a,b). Finally, fiber ingestion may indirectly influence the actual *in vivo* metabolism of the chemical agent. Although the number of colon tumors and percentage of colon-tumor bearing animals was less in groups fed only cellulose fiber diets, a greater number had tumors in the late compared to early sacrifice periods, suggesting that the mechanism for this cellulose fiber effect also involves a time-dependent delay in neoplasia development. As previously shown, the development of bowel tumors in this model is determined by dosage and duration of carcinogen administration as well as the time of study after the injection protocol (Thurnherr *et al.*, 1973, Toth *et al.*, 1976).

This time-dependent expression of fiber effect may also explain, in part, differing results of others examining the role of bran diets containing variable mixtures of different fibers, including cellulose, in this model (Fleiszer *et al.*, 1978; Cruse *et al.*, 1978). Significantly increased numbers of colon tumors in pectin compared to cellulose fiber-fed animals were observed at both time periods. Furthermore, the yield of small bowel tumors was increased in the 9.0% pectin group. These results provide strong evidence that different sources of dietary fiber administered in identical amounts can lead to striking differences in bowel tumor development. In addition, it has been observed that carrageenan, an algal polysaccharide fiber polymer (Kay and Strasberg, 1978; Freeman, 1980), enhances colonic neoplasia in rats treated with azoxymethane or methylnitrosourea, although dietary energy content and growth curves in that study were not strictly comparable (Watanabe *et al.*, 1978). This differential effect of individual types of fiber may offer a further explanation for apparent discrepancies among earlier studies using variable mixtures of different fiber and nonfiber components.

5. CONCLUSION

What can be learned from such studies? First, it would appear on the basis of several published reports using this animal model that some fiber components may be protective whereas others may have no effect or may even enhance the development of small or large bowel neoplasia. Although it would be premature to extrapolate directly to the human situation, perhaps the time has come to examine more critically the evidence for the "fiber hypothesis" and suggest that it may need to be refined. Second, most of the information available from these animal models is largely descriptive. We now need to address more precisely the possible molecular mechanisms and biologic actions of different individual fiber polymers within the alimentary tract and their role in diseases, such as bowel carcinogenesis.

ACKNOWLEDGMENTS. Supported by research grants from the National Cancer Institute of Canada, British Columbia Health Care Research Foundation, and Medical Research Council of Canada.

REFERENCES

American Chemical Society, 1977, Cancer statistics 1977, *Chem. Abstr.* **27**:26.

Barbolt, T. A., and Abraham, R., 1978, The effect of bran on dimethylhydrazine-induced colon carcinogenesis in the rat, *Proc. Soc. Exp. Biol. Med.* **157**:656.

Barkla, D. H., and Tutton, P. J., 1977, Surface changes in the descending colon of rats treated with dimethylhydrazine, *Cancer Res.* **37**:262.

Bauer, H. G., Asp, N. G., Oste, R., Dahlqvist, A., and Fredlund, P. E., 1979, Effect of dietary fiber on the induction of colorectal tumors and fecal B-glucuronidase activity in the rat, *Cancer Res.* **39**:3752.

Burkitt, D. P., 1971, Epidemiology of cancer of the colon and rectum, *Cancer* **28**:3.

Chen, W. F., Patchefsky, A. S., and Goldsmith, H. S., 1978, Colonic protection from dimethylhydrazine by a high fiber diet, *Surg. Gynecol. Obstet.* **147**:503.

Cruse, J. P., Lewin, M. R., and Clark, C. G., 1978, Failure of bran to protect against experimental colon cancer in rats, *Lancet* **2**:1278.

Deger, G. E., 1976, Aflatoxin—Human colon carcinogenesis?, *Ann. Intern. Med.* **85**:204.

Druckrey, H., Preussmann, R., Matzies, F., and Ivankovic, S., 1967, Selektive Erzeugung von Darmkrebs bei Ratten durch 1,2-Dimethylhydrazin, *Naturwissenschaften* **54**:285.

Filipe, M. I., 1975, Mucous secretion in rat colonic mucosa during carcinogenesis induced by dimethylhydrazine. A morphological and histochemical study, *Br. J. Cancer* **32**:60.

Fleiszer, D., Murray, D., MacFarlane, J., and Brown, R., 1978, Protective effect of dietary fibre against chemically-induced bowel tumours in rats, *Lancet* **2**:552.

Freeman, H. J., 1979, Dietary fiber and colonic neoplasia, *Can. Med. Assoc. J.* **121**:291.

Freeman, H. J., 1980, Experimental animal studies in colonic carcinogenesis and dietary fiber, in: *Medical Aspects of Dietary Fiber* (G. A. Spiller and R. M. Kay, eds.), Plenum Medical, New York, p. 83.

Freeman, H. J., 1981a, Effects of dietary fibers on fecal bacterial counts in colon carcinogen treated rats, *Clin. Res.* **29**:32.

Freeman, H. J., 1981b, Effects of dietary fibers on fecal B-glucuronidase in colon carcinogen treated rats, *Clin. Res.* **29**:32.

Freeman, H. J., and San, R. H. C., 1980, Use of unscheduled DNA synthesis in freshly isolated human intestinal mucosal cells for carcinogen detection, *Cancer Res.* **40**:3155.

Freeman, H. J., Kim, Y., and Kim, Y. S., 1978a, Glycoprotein metabolism in normal proximal and distal rat colon and changes associated with 1,2-dimethylhydrazine-induced colonic neoplasia, *Cancer Res.* **38**:3385.

Freeman, H. J., Spiller, G. A., and Kim, Y. S., 1978b, A double-blind study on the effect of purified cellulose dietary fiber on 1,2-dimethylhydrazine-induced rat colonic neoplasia, *Cancer Res.* **38**:2912.

Freeman, H. J., Spiller, G. A., and Kim, Y. S., 1980a, A double-blind study on the effects of differing purified cellulose and pectin fiber diets on 1,2-dimethylhydrazine-induced rat colonic neoplasia, *Cancer Res.* **40**:2661.

Freeman, H. J., Lotan, R., and Kim, Y. S., 1980b, Application of lectins for detection of goblet cell glycoconjugate differences in proximal and distal colon of the rat, *Lab. Invest.* **42**:405.

Holloway, W. D., Tasman-Jones, C., and Lee, S. P., 1978, Digestion of certain fractions of dietary fiber in humans, *Am. J. Clin. Nutr.* **31**:927.

Kay, R. M., and Strasberg, S. M., 1978, Origin, chemistry, physiological effects and clinical importance of dietary fibre, *Clin. Invest. Med.* **1**:9.

Lamont, J. T., and O'Gorman, T. A., 1978, Experimental colon cancer, *Gastroenterology* **75**:1157.

Laqueur, G., 1965, The induction of intestinal neoplasms in rats with the glycoside cycasin and its aglycone, *Virchows Arch. (Pathol. Anat.)* **340**:151.

Laqueur, G. L., Mickelson, O., Whiting, M. G., Kurland, L. T., 1963, Carcinogenic properties of nuts from *Cycas circinalis* indigenous to Guam, *J. Natl. Cancer Inst.* **31**:919.

Lorenz, E., and Stewart, H. L., 1941, Intestinal carcinoma and other lesions in mice following oral administration of 1,2,5,6-dibenzanthracene and 20-methylcholanthrene, *J. Natl. Cancer Inst.* **1**:17.

Reddy, B. S., Narisawa, T., Wright, P., Vukusich, D., Weisburger, J. H., and Wynder, E. L., 1975, Colon carcinogenesis with azoxymethane and dimethylhydrazine in germ-free rats, *Cancer Res.* **35**:287.

Saunders, R. M., 1978, Wheat bran: Composition and digestibility, in: *Topics in Dietary Fiber Research* (G. A. Spiller and R. G. Amen, eds.), Plenum Press, New York, p. 43.

Schottenfeld, D., and Haas, J. F., 1978, Epidemiology of colorectal cancer, in: *Gastrointestinal Tract Cancer* (M. Lipkin and R. A. Good, eds.), Plenum Press, New York, p. 207.

Shipley, E. A., 1978, Dietary fiber content of foods, in: *Topics in Dietary Fiber Research* (G. A. Spiller and R. G. Amen, eds.), Plenum Press, New York, p. 203.

Thurnherr, N., Deschner, E. E., Stonehill, E. H., and Lipkin, M., 1973, Induction of adenocarcinomas of the colon in mice by weekly injections of 1,2-dimethylhydrazine, *Cancer Res.* **33**:940.

Toth, B., 1977, The large bowel carcinogenic effects of hydrazines and related compounds occurring in nature and the environment, *Cancer* **40**:2427.

Toth, B., Malick, L., and Shimizu, H., 1976, Production of intestinal and other tumors by 1,2-dimethylhydrazine dihydrochloride in mice, *Am. J. Pathol.* **84**:69.

Ward, J. M., 1974, Morphogenesis of chemically induced neoplasms of the colon and small intestine in rats, *Lab. Invest.* **30**:505.

Watanabe, K., Reddy, B. S., Wong, C. Q., and Weisburger, J. H., 1978, Effect of dietary undegraded carrageenan on colon carcinogenesis in F344 rats treated with azoxymethane or methylnitrosourea, *Cancer Res.* **38**:4427.

Weisburger, J. H., Reddy, B. S., and Wynder, E. L., 1977, Colon cancer—Its epidemiology and experimental production, *Cancer* **40**:2414.

Whiting, R. F., Wei, L., and Stich, H. F., 1979, Enhancement by transition metals of unscheduled DNA synthesis induced by isoniazid and related hydrazines in cultured normal and xeroderma pigmentosum human cells, *Mutat. Res.* **62**:505.

Wilson, R. B., Hutcheson, D. P., and Wideman, L., 1977, Dimethylhydrazine-induced colon tumors in rats fed diets containing beef or corn oil with or without wheat bran, *Am. J. Clin. Nutr.* **30**:176.

25

Bile Acids and Human Colorectal Cancer

M. J. HILL

1. INTRODUCTION

Colerectal cancer is a disease which is much more common in western countries than in the newly developing countries of Africa, Asia, and South and Central America, but migrants from countries with a low incidence to those with a high incidence of the disease rapidly attain an incidence of the disease similar to that of their new homeland. This indicates that the factors determining the incidence of the disease within a population are environmental rather than genetic. Environmental factors can be divided into two classes, namely those associated with the physical environment (such as latitude, altitude, climate, and air pollution, those factors which are shared by a local community) and those associated with the cultural environment (the more personal factors in the environment, such as smoking, diet, and personal hygiene). The effects of these two classes can be studied by, for example, investigating religious groups living within a community and with a shared physical environment; such investigations have shown that the disease is associated with the cultural environment rather than the physical environment, and it is now widely accepted that the diet has a key role in the causation of the disease. There is no consensus, however, concerning which dietary item is involved (Table I) and the conclusion must be that there is no simple relationship between consumption of a single class of dietary items and colorectal carcinogenesis. The epidemiology of colorectal cancer has been reviewed extensively many times and from many points of view (Wynder and

M. J. HILL • Bacterial Metabolism Research Laboratory, PHLS Centre for Applied Microbiology and Research, Salisbury SP4 0JG, England.

TABLE I. Relation between Diet and Large Bowel Cancer[a]

	Type of study		
Dietary item	International (more than 20 countries)	International (small number of countries)	Case control
Meat	✓	✓	✓
Animal protein	✓		
Fat	✓	✓	✓
Bound	✓		
Cholesterol	✓		
Fiber			
Protective		✓	✓
Causative		✓	✓
Beer	✓	✓	
Vitamins A and C (protective)		✓	
Milk (protective)		✓	

[a] For references see Hill (1981).

Shigematsu, 1967; Haenszel and Correa, 1973; Hill, 1975; Correa and Haenszel, 1978).

There have been many attempts to relate the ingestion of preformed carcinogens in food and human carcinogenesis; many such carcinogens have been detected in human feces (Table II), but none has been shown to be related to colorectal carcinogenesis (although many are related, as might be expected, to gastric carcinogenesis). We therefore postulated (Aries *et al.*, 1969; Hill *et al.*, 1971) that large bowel cancer was caused by some bacterial metabolite produced *in situ* in the colon from some benign substrate. A wide range of possible substrates has been investigated (Table III), but the ones yielding the best evidence of a role in colorectal carcinogenesis are the bile acids.

TABLE II. Carcinogens Detected in the Human Diet and Their Relation to Human Carcinogenesis

Carcinogen	Suggested target	Reference
Aflatoxin	Liver	Wogan (1975)
Polycyclic aromatic hydrocarbon in smoked food	Stomach	Sigurjonsson (1967)
N-nitroso compounds in fish	Nasopharynx	Fong and Walsh (1971)
Pyrolysis products in broiled food	Stomach	Sugimura (1979)
Food colors (e.g., butter yellow, Brown FK)	?	
Plant flavenoids	?	Sugimura (1979)

TABLE III. Possible Substrates for the Production of Carcinogens *in Situ* in the Colon by Bacteria[a]

Diet source	Substrate	Product
Meat, protein	Tryptophan	Range of metabolites
	Tyrosine	Phenol and *p*-cresol
	Methionine	Ethionine
	Basic amino acids	Cyclic secondary aminos for *N*-nitrosation
Meat, fat	Bile acids	Co-carcinogenic metabolites
	Cholesterol	Inactivated metabolites
	Lecithin (choline)	Dimethylamine for *N*-nitrosation

[a] From Hill (1977).

2. BILE ACIDS

2.1. Evidence for a Role for Bile Acid Metabolites in Colorectal Carcinogenesis

The bile acids were first suggested as possible substrates for the bacterial production of carcinogens or co-carcinogens because: (1) their fecal concentration is related to the intake of dietary fat (Cummings *et al.*, 1978) and dietary cereal

TABLE IV. Population Studies Relating Fecal Bile Acid Concentration to the Incidence of Large Bowel Cancer

Study	Result	Reference
1. Six populations later extended to nine in four continents	Good correlation between large bowel cancer incidence and FBA concentration	Hill *et al.* (1971)
2. Various racial groups in New York	Good correlation	Wynder and Reddy (1974)
3. Three income groups in Hong Kong	Good correlation	Crowther *et al.* (1976)
4. Comparison of Kuopio, Finland, and New York	Good correlation	Wynder and Reddy (1978)
5. Comparison of Kuopio, Finland, and Copenhagen	No correlation	IARC Working Party (1977)
6. Comparison of urban Finns, rural Finns, urban Danes, and rural Danes	Good correlation	IARC Working Party (1982)
7. Comparison of black and white South Africans	Good correlation	Antonis and Bersohn (1962)[a]

[a] This is an interpretation of the results of a study not related to large bowel carcinogenesis.

fiber (Cummings *et al.*, 1976), both of which have been correlated with the incidence of colorectal cancer; and (2) they are structurally related to a group of known carcinogens, the cyclopenta[*a*]phenanthrenes (Coombs and Croft, 1973).

As a result, the bile acids have been subjected to extremely intense investigation. There have been many studies correlating the risk of large bowel cancer in a population with the mean fecal bile acid (FBA) concentration in that population (Table IV), all except one of which have shown a good correlation. When the three recent studies in Scandinavia are taken together they show a remarkable congruence (Fig. 1) and provide good evidence of a relationship.

There have been a number of case control studies and these have not been as unanimous in their conclusions (Table V), to some extent because they set

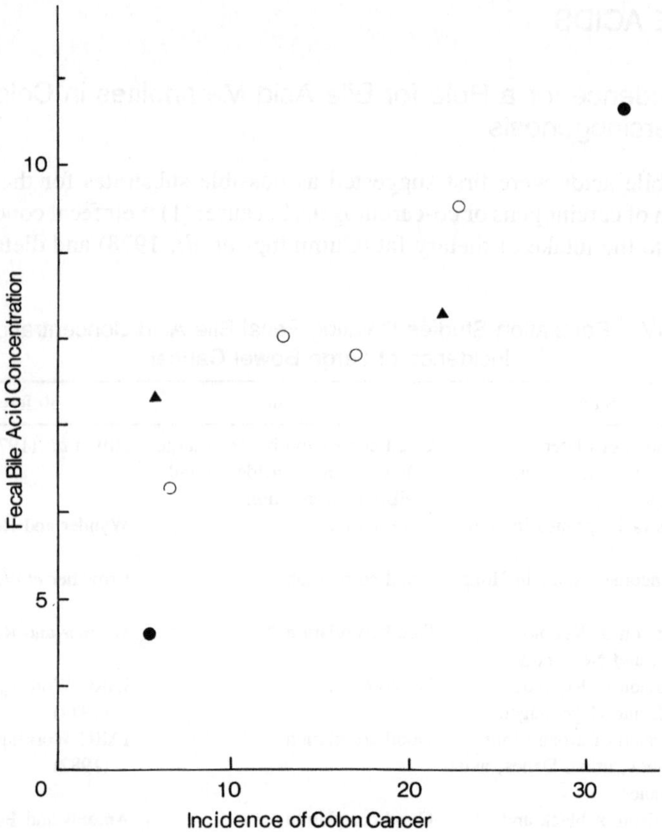

FIGURE 1. The relationship between fecal bile acid (FBA) concentration and the incidence of colorectal cancer in Scandinavian populations. Data from (▲) IARC Working Party (1977), (●) Wynder *et al.* (1978), and (○) IARC Working Party (1982).

TABLE V. Case Control Studies of the Relation between Large Bowel Cancer and Fecal Bile Acid Concentration

Number and type of cases	Number and type of controls	FBA in cases/ FBA in controls	Reference
Colorectal cancer (44)	Nonmalignant bowel disease (99)	>1	Hill *et al*. (1975)
Colon cancer (31)	Normal healthy persons (34)	>1	Wynder and Reddy (1978)
Colorectal cancer (29)	Normal healthy persons (13)	<1	Blackwood *et al*. (1978)
Colorectal cancer	Normal healthy persons	1	Mudd *et al*. (1978)
Colorectal cancer	Nongastrointestinal cancer and normal healthy persons	1	Moskowitz *et al*. (1979)

out to study different things. Wynder and Reddy (1978) showed that the mean FBA concentration in a group of large bowel cancer was higher than that in a group of controls; in contrast Mudd *et al.* (1978) and Moskowitz *et al.* (1979) found no difference in FBA concentration between cases and controls. Hill *et al.* (1975) and Blackwood *et al.* (1978) set out to study a somewhat different hypothesis—that the risk factor in colorectal carcinogenesis was the combination of high FBA concentration and the presence of certain clostridia; as part of their studies, Hill *et al.* found FBA concentration results similar to those of Reddy *et al.*, while Blackwood *et al.* agreed with Mudd *et al.* and Moskowitz *et al.* The findings concerning clostridia will be discussed later.

In addition to these investigations of humans, there have been vast numbers of studies of an animal model using the specific colon carcinogen dimethylhydrazine (or the related compounds azomethane or azoxymethane). These studies provide very strong support for a role for bile acids in the colorectal carcinogenesis produced in this model (Table VI). However, it is important to add a cautionary note. The dimethylhydrazine model is a model of colorectal carcinogenesis caused by a western-style (high-fat, low-fiber) diet; the bile acid theory was formulated to explain a relationship between dietary fat and bowel carcinogenesis. Consequently, support for a role for bile acids in colorectal carcinogenesis from this model is the result of a circular argument. Second, dimethylhydrazine gives large numbers of ear duct and duodenal tumors as well as colorectal tumors; in that respect it is not a good model of human colorectal cancer.

TABLE VI. Evidence from Animal Studies of a Role for Bile Acids in Colorectal Carcinogenesis

Manipulation	Effect on FBA concentration	Effect on colorectal carcinogenesis
Bile diversion to cecum	Increase	Increase
Administration of cholestyramine	Increase	Increase
Rectal instillation of deoxycholic and lithocholic acids	Increase	Increase
Diversion of fecal stream	Decrease	Decrease
Vivonex diet	Decrease	Decrease
Diet change:		
Added fat	Increase	Increase
Added meat	Increase	Increase
Added pectin	Increase	Increase
Added metamucil	Increase	Increase
Added bran	Decrease	Decrease
Added lactulose	Decrease	Decrease

In summary, there is good evidence that bile acids act as co-carcinogens in the dimethylhydrazine model; the question is, how good is this as a model of human colorectal carcinogenesis?

2.2. Identity of the Relevant Bile Acid Metabolites

Although the number of possible fecal bile acids is more than 800 (considering only the saturated ones, but including ethyl, methyl, sulfate, and acetyl esters), interest has focused on the products of two bacterial enzymes with respect to colorectal carcinogenesis. In the dimethylhydrazine model in rats deoxycholic and lithocholic acids are promoters when administered by rectal instillation (Table VII) and by a variety of other tests. In the studies of populations by Hill *et al.* (1971) the incidence of colorectal cancer correlated better with the fecal concentration of deoxycholic acid than with the total FBA concentration, while the reverse was true in the case control study (Hill *et al.*, 1975). Deoxycholic acid and lithocholic acid are both secondary bile acids produced by the action of bacterial 7α-dehydroxylase on the primary bile acids cholic and chenodeoxycholic acids, respectively; Mastromarino *et al.* (1976) showed that the fecal activity of this enzyme was higher in colorectal cancer patients than in control persons and we have confirmed this (Jivraj and Hill, unpublished results).

The second enzyme of interest is 3-oxo-cholyl-4-dehydrogenase, which produces unsaturated bile acids with a 4-en-3-one configuration from 5β-3-oxo substrates. This enzyme has an absolute requirement for a suitable hydrogen acceptor, and in our studies of more than 1200 strains of intestinal bacteria only a group of lecithinase-negative clostridia producing butyric acid as a major end product of glucose fermentation were able to carry out the reaction. The organisms able to carry out this nuclear dehydrogenation reaction do not belong to

TABLE VII. Carcinogenicity and Promoting Activity of Bile Acids

Bile acid	Action	Model	Reference
Deoxycholic acid	Mutagenicity	*Drosophila*	
	Co-mutagenicity	Modified Ames test	Silverman and Andrews (1977)
	Co-carcinogenicity	Skin painting in croton oil	Salaman and Roe (1956)
			Cook *et al.* (1940)
	Co-carcinogenicity	Rectal instillation	Narisawa *et al.* (1974)
Lithocholic acid	Co-carcinogenicity	Rectal instillation	Narisawa *et al.* (1974)
	Co-mutagenicity	Modified Ames test	Silverman and Andrews (1977)
	Mutagenicity	Cell transformation test	Kelsey and Pienta (1979)
Apocholic acid	Carcinogenicity	Skin painting	Lacassagne *et al.* (1961)
Bisnor Δ^5-choleric acid	Carcinogenicity	Skin painting	Lacassagne *et al.* (1966)

a single species of *Clostridium,* but include *C. tertium, C. butyricum, C. paraputrificum,* and *C. indolis* and are referred to collectively as nuclear dehydrogenating clostridia (NDC), to distinguish them from the vast majority that dehydrogenate only hydroxyl groups. In our studies of various population groups we found no differences in the mean numbers of *all* clostridia in feces of populations with wide ranges of colorectal cancer incidence (Hill *et al.*, 1971) and no differences between cases of colorectal cancer and controls. However, when only the NDC are considered, there is a good correlation between carriage rate and the incidence of the disease (Hill and Drasar, 1974) and there is a difference in carriage rate between bowel cancer cases and controls (Hill *et al.*, 1975; Blackwood *et al.,* 1978). The assay of NDC is very simple (Goddard *et al.*, 1975) and is the only way to determine their presence; classical bacteriology is of no help since the NDC do not belong to a single species and not all members of a given species have Δ^4-dehydrogenase activity. Consequently any study in which the clostridia are speciated but not tested for Δ^4-dehydrogenase activity (e.g., Finegold *et al.*, 1975) should be interpreted with caution, while those in which only total clostridial counts were determined (e.g., Moskowitz *et al.,* 1979) confirm the findings of Hill *et al.* (1971, 1975) concerning total clostridia but have no relevance to a study of NDC.

Our current working hypothesis involves both Δ^4-dehydrogenase and 7α-dehydroxylase in the production of bile acids with a 4,6-dien-3-one configuration (Hill 1975, 1977, 1980). Since the Δ^4-dehydrogenase has an absolute requirement for a suitable hydrogen acceptor in feces, identified as phylloquinone (Fernandez and Hill, 1978), we are studying the value of a 4-component discriminant (fecal bile acid concentration, carriage of NDC, 7α-dehydroxylase activity, and vitamin K concentration) in studies of colorectal carcinogenesis.

3. THE ADENOMA–CARCINOMA SEQUENCE

Studies by histopathologists have demonstrated (Morson, 1974) that almost all (if not all) colorectal carcinomas arise in preexisting adenomas, and that there is an adenoma–carcinoma sequence (Fig. 2) involving (1) adenogenesis, (2) adenoma growth, (3) the development of increasingly severe dysplasia, and ultimately (4) carcinogenesis. The evidence in favor of the adenoma–carcinoma sequence has been summarized by Morson (1974) and the evidence for the sequence as outlined in Fig. 2 has been summarized by Hill *et al.* (1978).

If we believe that bile acid metabolites are important in colorectal carcinogenesis, then they must be involved in one or more stages of the adenoma–carcinoma sequence. To determine in which, if any, of the stages they are involved, we have carried out fecal analyses in patients with colorectal adenomas and in those with chronic ulcerative colitis.

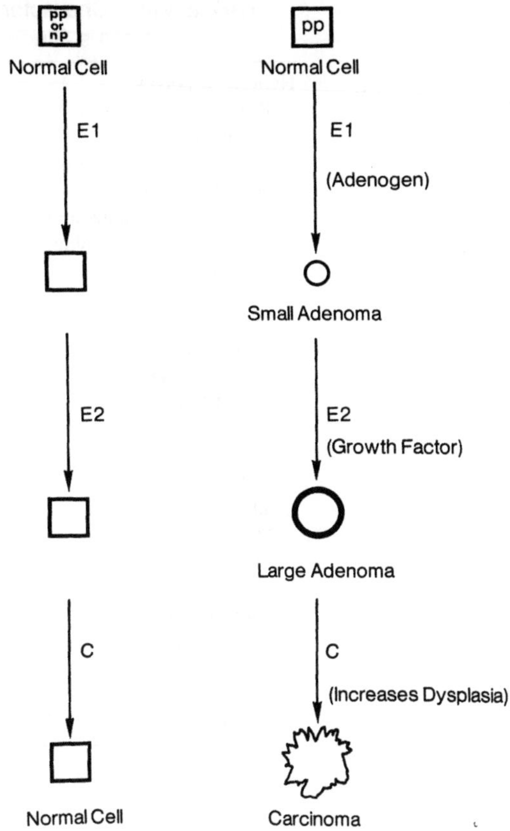

FIGURE 2. The adenoma–carcinoma sequence. The gene *p* is an autosomal recessive gene conferring adenoma-proneness, while *n* is its normal variant.

3.1. Patients with Colorectal Adenomas

We have analyzed fecal samples from 134 patients from whom polyps were removed during sigmoidoscopy or colonoscopy at the outpatients clinic of St. Mark's Hospital. Of these, 114 were adenomas and 20 were nonadenomatous polyps (Table VIII.) The subsite distribution, with only 17% from the right colon (cecum + ascending + transverse colon), reflects the fact that only a small proportion of the polyps were obtained at colonoscopy.

If bile acids are involved in adenogenesis, then we would expect the FBA concentration to be higher in the adenoma patients than in normal healthy control persons; in fact there was no difference (Table IX). However, the mean FBA concentration increased with adenoma size, as did the percentage carrying NDC, indicating that the bile acid metabolite produced by NDC might be involved in adenoma growth. The FBA concentration and percentage carrying NDC also

TABLE VIII. Characteristics of the Polyps Studied

Number of polyps	
Total	134
Adenomas	114
Nonadenomatous polyps	20
Subsite distribution	
Cecum + ascending colon	4
Transverse + descending colon	11
Sigmoid colon	34
Rectum	39
Multiple	26
Size distribution	
<5 mm diameter	26
5–10 mm	29
10–15 mm	19
15–20 mm	16
>20 mm	24
Grade of dysplasia	
Mild	75
Moderate	21
Severe	13
Focal carcinoma	5
Adenoma type	
Tubular	64
Villous	9
Tubulovillous	30

TABLE IX. Fecal Analyses in Adenoma Patients Compared with Normal Healthy Controls

	FBA concentration	Percent carrying NDC
Normal healthy controls	8.3	38%
All adenomas	8.7	38%
Nonadenomatous polyps	6.9	38%
Adenomas		
5 mm diameter	6.8	6%
5–10 mm diameter	8.2	39%
10–15 mm diameter	8.3	59%
15–20 mm diameter	9.3	33%
20 mm diameter	11.3	60%

increased with increasing degree of epithelial dysplasia, but when this was corrected for adenoma size (large adenomas are more likely to show severe dysplasia) there was no relation between fecal analyses and the degree of epithelial dysplasia.

3.2. Patients with Ulcerative Colitis

Patients who have had ulcerative colitis with total colonic involvement for more than 10 years have an increased risk of colorectal carcinogenesis (MacDougall, 1964; Lennard-Jones *et al.*, 1977). A common treatment is prophylactic colectomy, but at St. Mark's Hospital there is a group of more than 100 patients who have had colitis involving the whole colon for more than 10 years and have elected to retain their colon. These patients attend the outpatients clinic regularly for sigmoidoscopy or colonoscopy, when biopsies are taken for histopathological examination for epithelial dysplasia. We have attempted to obtain a fecal specimen from each patient at each outpatient visit for analysis for FBA concentration and NDC carriage. During the (to date) seven years of followup of these patients, seven of the 102 in the study have had a total colectomy because their colitis deteriorated; these patients are, of course, lost to the study. A further nine had severe epithelial dysplasia diagnosed in their biopsy specimens and, since this is a precancerous lesion (Morson and Pang, 1967), these patients were treated by total colectomy; of these seven were found to have Dukes A carcinoma when the excised colon was closely examined. The mean FBA concentration and percentage carrying NDC in those treated surgically for severe dysplasia (i.e., precancer) was much higher than those in the group of colitics treated as a whole (Table X). At this stage in the study the number of patients who have developed precancer or cancer is too small for the fecal analyses to be statistically significantly different, but they indicate that, in this group of patients, the fecal bile acid metabolite produced by the NDC may be

TABLE X. Fecal Analyses in Patients Who Have Had Ulcerative Colitis for More Than 10 Years

	Number of patients	Mean FBA concentration	Percent carrying NDC
All patients	102	7.7	29
Those treated by colectomy			
For severe colitis	7	7.4	33
For precancer	9	10.0	63

involved in the progression from inflammation to dysplasia and then to carcinoma.

4. CONCLUSIONS

The position concerning the role of bile acids in colorectal carcinogenesis is far from settled. There have been enough negative results to indicate that the relationship is unlikely to be simple and that bile acids are unlikely to be the sole cause of colorectal cancer. On the other hand, there have been enough positive studies to indicate that a role for bile acids cannot be dismissed lightly. In my laboratory we are investigating a multistage hypothesis, and have preliminary evidence for bile acids playing a part in determining the rate of adenoma growth. We still have no clues on the etiology of the precurer adenoma—a key factor in colorectal carcinogenesis.

ACKNOWLEDGMENT. The work of my laboratory on large bowel cancer is financially supported by the Cancer Research Campaign, to whom I acknowledge my gratitude.

REFERENCES

Antonis, A., and Bersohn, I., 1962, The influence of diet on faecal lipids in South African white and Bantu prisoners, *Am. J. Clin. Nutr.* **11**:142–55.

Aries, V. C., Crowther, J. S., Drasar, B. S., Hill, M. J., and Williams, R. E. O., 1969, Bacteria and the etiology of cancer of the large bowel, *Gut* **10**:334–5.

Blackwood, A., Murray, W. R., Mackay, C., and Calman, K., 1978, Faecal bile acids and the clostridia in the etiology of colorectal cancer and breast cancer, *Br. J. Cancer* **38**:175.

Cook, I. W., Kennaway, E. L., and Kennaway, N. M., 1940, Production of tumours in mice by deoxycholic acid, *Nature* **145**:627.

Coombs, M. M., and Croft, C. J., 1973, Correlation between carcinogenicity and chemical structure in the cyclopenta[*a*]phenanthrenes, *Cancer Res.* **33**:832.

Correa, P., and Haenszel, W., 1978, Epidemiology of large bowel cancer, in: *Advances in Cancer Research*, Vol. 26 (G. Klein and S. Weinhouse, eds.), Academic Press, New York, pp. 1–141.

Crowther, J. S., Drasar, B. S., Hill, M. J., MacLennan, R., Magnin, D., Peach, S., and Teoh-Chan, C. H., 1976, Faecal steroids and bacteria and large bowel cancer in Hong Kong by socio-economic groups, *Br. J. Cancer* **34**:191–198.

Cummings, J. H., Hill, M. J., Jenkins, D. J., Pearson, J. R., and Wiggins, H. S., 1976, Changes in fecal composition and colonic function due to cereal fiber, *Am. J. Clin. Nutr.* **29**:1468–1473.

Cummings, J., Wiggins, H., Jenkins, D., Houston, H., Drasar, B., Hill, M., and Jivraj, T., 1978, The influence of different levels of dietary fat intake on faecal composition, microflora and gastrointestinal transit time, *J. Clin. Invest.* **61**:953.

Fernandez, F., and Hill, M. J., 1978, A faecal hydrogen acceptor for clostridial 3-oxo-steroid Δ^4-dehydrogenase, *Trans. Biochem. Soc.* **6**:376–377.

Finegold, S. M., Flora, D., Attebury, H., and Sutter, V., 1975, Fecal bacteriology of colonic polyp patients and control patients, *Cancer Res.* **35**:3407–3417.

Fong, Y. Y., and Walsh, E. O., 1971, Carcinogenic nitrosamines in Cantonese salt-dried fish, *Lancet* **2**:1032.

Goddard, P., Fernandez, F., West, B., Hill, M. J., and Barnes, P., 1975, The nuclear dehydrogenation of steroids by intestinal bacteria, *J. Med. Microbiol.* **8**:429–435.

Haenszel, W., and Correa, P., 1973, Cancer of the large intestine. Epidemiological findings, *Dis. Colon Rectum* **16**:371–7.

Hill, M. J., 1975, The etiology of colon cancer, *Crit. Rev. Toxicol* **4**:31–82.

Hill, M. J., 1977, Bacterial metabolism, in: *Topics in Gastroenterology 5* (S. C. Truelove and E. Lee, eds.), Blackwell, Oxford, p. 45.

Hill, M. J., 1980, The aetiology of colorectal cancer, in: *Recent Advances in Gastrointestinal Pathology* (R. Wright, ed.), Saunders, London, pp. 297–310.

Hill, M. J., 1981, Metabolic epidemiology of large bowel cancer, in: *Gastrointestinal Cancer* (J. J. DeCosse and P. Sherlock, eds.), Martinus Nijhoff, The Hague, pp. 187–226.

Hill, M. J., and Drasar, B. S., 1974, in: *Anaerobic Bacteria: Role in Disease* (A. Balows, R. M. DeHaan, V. Dowell, and L. Guze, eds.), Thomas, Springfield, Illinois, pp. 119–133.

Hill, M. J., Drasar, B. S., Aries, V. C., Crowther, J. S., Hawksworth, G. M., and Williams, R. E. O., 1971, Bacteria and aetiology of cancer of large bowel, *Lancet* **1**:95–100.

Hill, M. J., Drasar, B. S., Williams, R. E. O., Meade, T. W., Cox, A. G., Simpson, J. E. P., and Morson, B. C., 1975, Faecal bile acids and clostridia in patients with cancer of the large bowel, *Lancet* **1**:535–538.

Hill, M. J., Morson, B. C., and Bussey, H. J. R., 1978, Etiology of adenoma–carcinoma sequence in large bowel, *Lancet* **1**:245–247.

IARC Working Party, 1977, Dietary fibre, transit time, faecal bacterial, steroids and colon cancer in two Scandinavian populations, *Lancet* **2**:207–211.

IARC Working Party, 1982, Diet and faecal analyses in 4 Scandinavian populations, *Nutr. Cancer* (in press).

Kelsey, M. I., and Pienta, R. J., 1979, Transformation of hamster embryo cells by cholesterol-α-epoxide and lithocholic acid, *Cancer Lett.* **6**:143–149.

Lacassagne, A., Buu-Hoi, N. P., and Zajdela, F., 1961, Carcinogenic activity of apocholic acid, *Nature* **190**:1007–8.

Lacassagne, A., Buu-Hoi, N. P., and Zajdela, F., 1966, Carcinogenic activity *in situ* of further steroid compounds, *Nature* **209**:1026–7.

Lennard-Jones, J., Morson, B. C., Ritchie, J. K., Shove, D., and Williams, C. B., 1977, Cancer in colitis: Assessment of the individual risk by clinical and histological criteria, *Gastroenterology* **73**:1280–1289.

MacDougall, I. P. M., 1964, Clinical identification of those cases of ulcerative colitis most likely to develop cancer of the bowel, *Dis. Colon Rectum* **7**:447–450.

Mastromarino, A., Reddy, B. S., and Wynder, E. L., 1976, Metabolic epidemiology of colon cancer: Enzymic activity of the fecal flora, *Am. J. Clin. Nutr.* **29**:1455–1460.

Morson, B. C., 1974, The polyp–cancer sequence in the large bowel, *Proc. R. Soc. Med.* **67**:451–457.

Morson, B. C., and Pang, L. S. C., 1967, Rectal biopsy as an aid to cancer control in ulcerative colitis, *Gut* **8**:423–434.

Moskowitz, M., White, C., Barnett, R. N., Stevens, S., Russell, E., Vargo, D., and Floch, M. H., 1979, Diet, faecal bile acids and neutral steroids in carcinoma of the colon, *Dig. Dis. Sci.* **24**:746–751.

Mudd, D. G., McKelvey, S. T., Sloan, J. M., and Elmore, D. T., 1978, Faecal bile acid concentrations in patients at increased risk of large bowel cancer, *Acta. Gastroenterol. Belg.* **41**:241–4.

Narisawa, T., Magadia, N., Weisburger, J., and Wynder, E. L., 1974, Promoting effect of bile acids on colon carcinogenesis after intrarectal instillation of *N*-methyl-*N*-nitro-*N*-nitrosoguanidine in rats, *J. Natl. Cancer Inst.* **53**:1093–1097.

Salaman, M. H., and Roe, F. J. C., 1956, Further tests for tumour-initiating activity: *N,N*-di-(2-chloroethyl)-*p*-aminophenylbutyric acid (CB 1348) as initiator of skin tumour formation in the mouse, *Br. J. Cancer* **10**:363–379.

Sigurjonsson, J., 1967, Occupational variations in mortality from gastric cancer in relation to dietary differences, *Br. J. Cancer* **21**:651.

Silverman, S. J., and Andrews, A. W., 1977, Bile acids: Co-mutagenic activity in the *Salmonella* mammalian microsome mutagenicity test, *J. Natl. Cancer Inst.* **59**:1557.

Sugimura, T., 1979, Naturally occurring genotoxic carcinogens, in: *Naturally Occurring Carcinogens, Mutagens, and Modulators of Carcinogenesis* (E. C. Miller *et al.*, eds.), Japanese Science Society Press, Tokyo, pp. 241–261.

Wogan, G. N., 1975, Dietary factors and special epidemiological situations of liver cancer in Thailand and East Africa, *Cancer Res.* **35**:3499–3502.

Wynder, E. L., and Reddy, B. S., 1974, Metabolic epidemiology of colorectal cancer, *Cancer* **34**:801–806.

Wynder, E. L., and Reddy, B. S., 1978, Etiology of cancer of the colon, in: *Colon Cancer* (E. Grundmann, ed.), Gustav Fischer Verlag, Stuttgart, pp. 1–14.

Wynder, E. L., and Shigematsu, T., 1967, Environmental factors of cancer of the colon and rectum, *Cancer* **20**:1520–1561.

26

Postulated Mechanisms whereby Fiber May Protect against Large Bowel Cancer

JOHN H. CUMMINGS and W. J. BRANCH

1. INTRODUCTION

Ten years ago Burkitt (1971) suggested, on the basis of his observations of the lifestyle of western and of rural African peoples, that increased large bowel cancer risk was due in part to lack of dietary fiber. He went on to postulate a mechanism whereby fiber might protect against the development of this cancer. The suggestion was that fiber, by increasing the amount of feces passed, led to a reduction in the concentration of potentially carcinogenic substances in the bowel, to a reduced transit time through the colon and therefore reduced contact time between the carcinogen and the mucosa, and to alterations in bacterial metabolism by fiber in favor of less carcinogen production.

Despite major advances in the understanding of dietary fiber metabolism over the past ten years and similar strides which have been made in large bowel cancer research, this hypothesis of Burkitt's has yet to be superseded.

2. INCREASED FECAL BULK—THE DILUTION HYPOTHESIS

The suggestion that by increasing the bulk of feces, and presumably the volume of colonic contents, fiber dilutes potentially noxious substances in the

JOHN H. CUMMINGS and W. J. BRANCH • Dunn Clinical Nutrition Centre, Old Addenbrookes Hospital, Cambridge CB2 1QE, England.

bowel is a reasonable one and appeals to common sense. However, when one examines the experimental evidence for this theory, it would appear to be oversimplified. First, the carcinogens or promoters of human large bowel cancer are unknown. Therefore, the protective role of dietary fiber can be assessed only indirectly, by looking at its effects on metabolism within the bowel. Second, it is evident from physiological studies that fiber is a mixture of substances with varying effects on the large intestine. It is therefore not possible to generalize about fiber–carcinogen interactions. However, the consumption of some types of dietary fiber undoubtedly leads to a fall in the concentration of some substances in feces.

Figure 1 shows the concentration of an inert substance (plastic pellets) in the feces of three groups of six subjects while eating a typical western-type diet, containing approximately 20 g of dietary fiber, and then the same diet with the

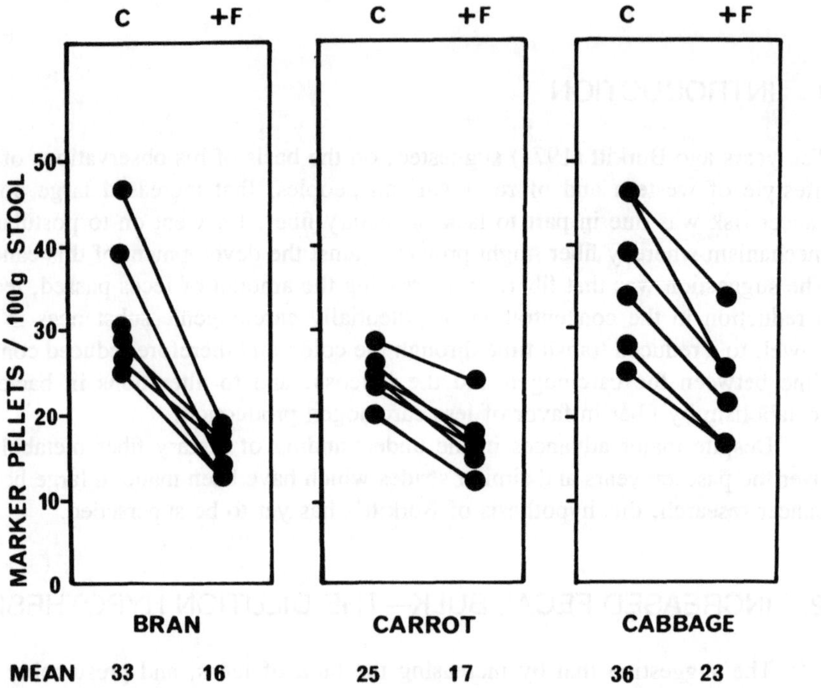

FIGURE 1. Concentration of radioopaque pellets in the feces of three groups of healthy subjects while eating a controlled western-type diet or the same diet with the addition of 18 g of fiber from bran or cabbage or 20 g from carrot. Each subject took ten pellets with each meal throughout each of two 3-week dietary periods.

addition of 18 g of fiber from either bran, cabbage, or carrot. Ten pellets were given to each subject with every meal (30/day) throughout each of the two 3-week dietary periods (Cummings *et al.*, 1978). Bran, cabbage, and carrot affect colonic function in different ways (Stephen and Cummings, 1980). Bran resists digestion in the colon and increases fecal bulk largely by virtue of its physical presence in the stools. By contrast, cabbage and carrot are almost totally digested by the colonic microflora and increase fecal bulk by stimulating microbial cell growth. Both these mechanisms, however, lead to a significant reduction in marker concentration, which is proportional to the size of the increase in stool weight. Bran halves the concentration of inert marker while doubling fecal output, while cabbage and carrot reduce marker concentration by around 50% and increase daily stool output by 60%. How much these changes reflect the concentration of substances in the rest of the colon is unknown, but it is likely that in the left side of the colon there is significant dilution.

Inert substances are, of course, unlikely to be important in colonic carcinogenesis. Bile acid concentration in feces, however, has been implicated in the colon cancer story. Hill and colleagues have shown that fecal bile acid concentrations are increased both in subjects with carcinoma of the colon (Hill *et al.*, 1975) and in those populations that experience an increased risk of large bowel cancer (Hill *et al.*, 1971). Figure 2 shows bile acid excretion in the feces of six subjects while taking a standard western-type diet and then the same diet this time with the addition of approximately 32g of dietary fiber from wheat bran. Part A of the figure indicates that the total fecal bile acid excretion increased in these subjects, from 199 to 279 mg/day, but parts B and C show that if bile acid excretion is expressed as a concentration, either of fecal solids excretion or as a concentration in the whole stool, then there is a significant fall with the addition of bran. The convention of expressing bile acid excretion per gram of fecal solids has been adopted in colon cancer studies (Hill *et al.*, 1975) and appears to relate to large bowel cancer risk. Concentrations of bile acids greater than 6 mg/g fecal solids signify an increased risk. However, this is a notional concentration, since bile acids are particularly insoluble substances. Their actual concentration in the aqueous phase of colonic contents in these circumstances is unknown. It would, however, be of great value to know the concentration and type of bile acid present in this phase of the stool, and the effect of fiber on it.

A carcinogen, or tumor promoter, effective in the large bowel is likely to be present in solution in the colon. The aqueous phase of colonic contents may be sampled using the dialysis bag technique described by Wrong *et al.* (1965). If the concentrations of water-soluble substances in feces are measured, it is clear that no general rule for the effect of dietary fiber can be made. Two readily identifiable groups of substances are present in solution: electrolytes, such as sodium and potassium, and products of bacterial metabolism, such as ammonia and short-chain fatty acids.

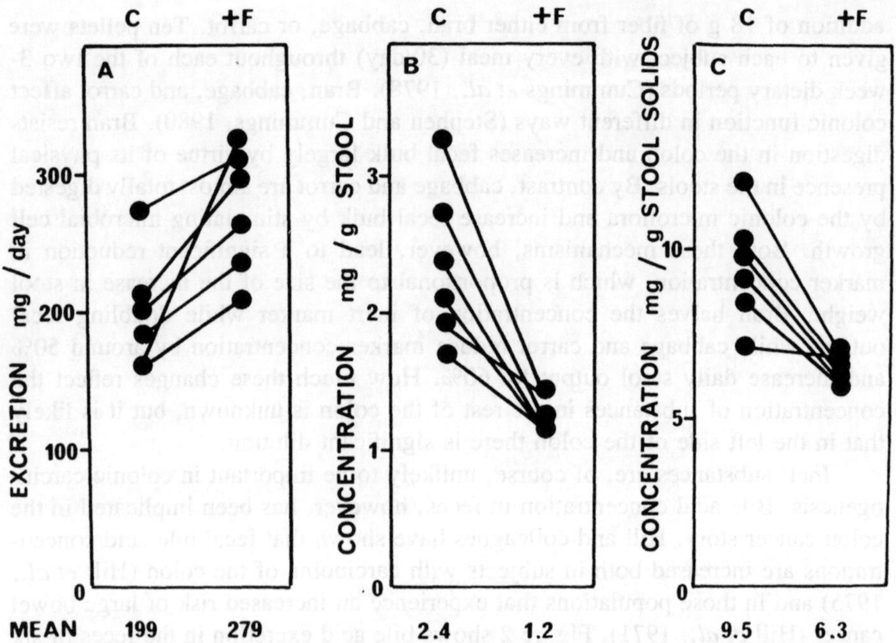

FIGURE 2. (A) Fecal bile acid excretion (mg/day) in six healthy subjects while eating a controlled western-type diet and then the same diet with 32 g added wheat fiber. Each dietary period was for 3 weeks and feces were analyzed for the third week of each period. (B) Fecal bile acid concentrations expressed as mg/g whole stool from the same study as in (A). (C) Fecal bile acid concentrations expressed as mg/g stool solids from the same study as in (A). (From Cummings *et al.*, 1976.)

Table I shows sodium and potassium concentrations in the feces of four subjects while they were taking a controlled diet or the same diet with the addition of 32 g of dietary fiber from wheat. This dietary change was achieved by adding bran and All Bran and by the substitution of wholemeal for white bread and bran biscuits for plain biscuits. It is apparent from these data that sodium concentrations are unchanged by this dietary manipulation despite a threefold increase in stool weight in these subjects and a similar change in the concentration of inert marker. Potassium concentrations increased significantly in the fecal dialysate. (Dietary potassium intake increased from 95 to 123 mmol/day in this particular study.) In general, sodium and potassium concentrations are little affected by changes in dietary fiber intake (Branch and Cummings, unpublished data). Although the concentration of these substances in the stool is known to be related to total stool weight, the changes in fecal output produced by fiber are too small to be associated with large changes in electrolyte concentrations. It is only where stool weights rise to those seen in diarrhea that significant alterations in concentration are likely to be seen. In the context of the western-type diet, sodium and potassium concentration are little altered by fiber.

The main anions in the stool are short-chain fatty acid (SCFA) anions, acetate, propionate, and butyrate. These are present in fecal water in concentrations of around 70–100 mM and are the direct end products of breakdown of dietary fiber by the colonic microflora. One might expect that the addition of fiber to a diet would lead to a rise in their concentration in fecal water. However, as is seen from Table I, this is not the case. SCFA concentrations remain unchanged when wheat fiber is added to the diet and when other types of fiber are added (Branch and Cummings, unpublished data). These findings are in keeping with studies in animals. SCFA concentrations fall during fasting, however, and there may be changes in the molar ratios of the three different fatty acids with diet.

Concentrations of SCFA depend not only on their production, but also on their absorption, which is rapid from the human colon (McNeil *et al.*, 1978), and utilization. The balance between these processes keeps their relative concentration constant. However, production rates and absorption of SCFA from the colon may be important in tumorigenesis.

Ammonia, another product of bacterial metabolism in the human colon, does change its concentration when fiber is added to the diet. Figure 3 shows that when a group of subjects add 18 g of dietary fiber from bran to their diet, total fecal nitrogen excretion increases but ammonia concentration in fecal dialysate falls. Similar changes can be seen with largely metabolized dietary fiber sources, such as cabbage. The concentration of ammonia in fecal dialysate is dependent on production from urea and from protein by bacteria, absorption across the mucosa, and incorporation into bacterial protein. It is probably this

TABLE I. Concentrations of Various Substances in Feces of Four Subjects Taking a Control Diet or Control plus Added Wheat Fiber[a]

	Control diet	+ Wheat fiber	
Acetate, mM	31.7 ± 2.8	39.5 ± 2.5	
Propionate, mM	13.2 ± 1.3	12.3 ± 0.9	
Butyrate, mM	12.8 ± 1.0	14.3 ± 1.2	
Sodium, mM	8.6 ± 0.6	8.7 ± 0.1	
Potassium, mM	112.6 ± 4.7	128.4 ± 3.9	($p < 0.02$)
Stool weight, g/day	81.0 ± 11.0	210.0 ± 9.0	($p < 0.001$)
Marker concentration, number of pellets per 100 g feces	40.0 ± 5.0	14.0 ± 0.6	($p < 0.02$)

[a] From Cummings *et al.* (1979). Control diet: fiber content 22 g/day. The same diet with added wheat fiber: fiber content 53 g/day. Values are ± SEM.

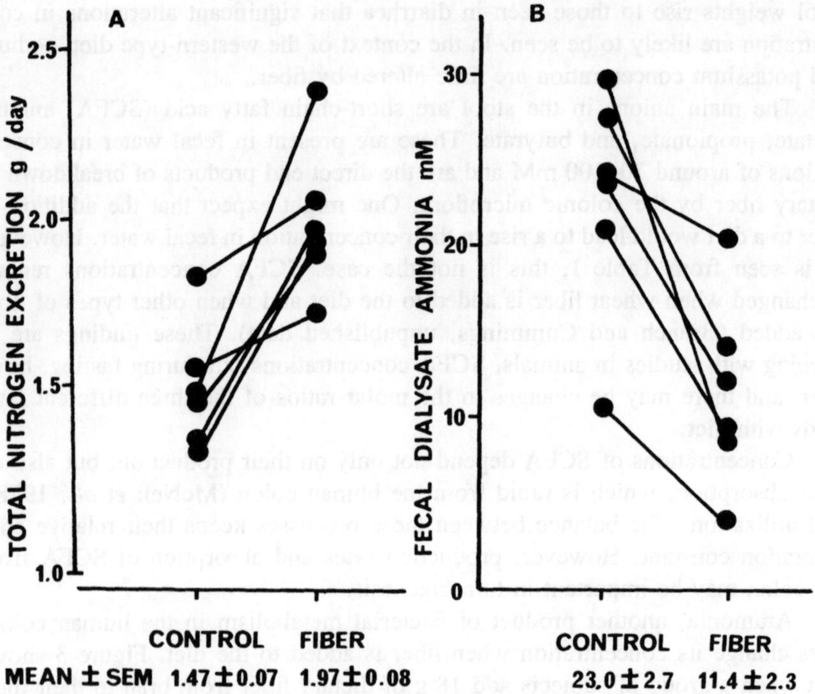

FIGURE 3. Fecal nitrogen excretion and average fecal dialysate ammonia concentration in six healthy subjects eating a controlled western-type diet for three weeks and then the same diet with the addition of 18 g of dietary fiber from wheat bran. (From Cummings *et al.*, 1981.)

latter pathway that leads to an overall fall in ammonia concentrations when fiber is added to the diet, since dietary fiber is known to stimulate microbial cell growth in the colon (Stephen and Cummings, 1979, 1980).

It is, therefore, very difficult to generalize about the diluting effect of dietary fiber on substances in the large bowel, particularly since most studies are of feces, which may not reflect changes in the cecum and right colon. Other factors, in addition to the simple physical bulking properties of fiber, may also influence the concentration of substances in the stool. These include absorption from the colon, bacterial metabolism, and perhaps also transit time.

3. THE EFFECT OF TRANSIT TIME

Increasing dietary fiber intake has been shown in many studies to speed up the rate of passage of material through the intestine. In general the magnitude of this change is closely related to the effect which a particular fiber exerts on fecal output. Fibers such as bran, which produce the largest changes in stool output, lead to the greatest reduction in transit time. However, it is important to ask whether transit itself, independently, determines colonic function, and if so, can this be related to protection against large bowel cancer as Burkitt has suggested?

Studies in man indicate that the digestibility of fiber is influenced by transit time. When the breakdown of cellulose from mixed sources in the diet is correlated with transit time in individuals, then a relationship emerges. In those subjects with a transit time of more than about 50 hr cellulose digestion during passage through the gut is maximum at around 80%, and no matter how long transit becomes, there is no further metabolism of the cellulose. When, however, transit time falls below 50 hr, then the digestion of cellulose also falls, until in subjects with transit of around 24 hr it is only about one-third to one-half digested (see Fig. 3, Chapter 2).

This finding agrees well with the work of Van Soest and his group (Van Soest, 1975) on the digestion of cellulose in ruminants. They have shown that cellulose digestion requires a considerable period of time, more than for digestion of other cell wall polysaccharides. Van Soest suggests that there are two phases in cellulose digestion, a rapid one lasting about 15 hr and subsequently a slow one which may go on for a further 24–48 hr. He also indicates in his studies (Van Soest, 1973) that a small fraction of cell wall cellulose may be indigestible. As far as the breakdown of cellulose is concerned, therefore, the time material spends in the gut is of importance.

This has been confirmed in man by Stephen (1980), who looked at the digestion of total dietary fiber in subjects on controlled diets whose transit time was either speeded up with senna or slowed down with codeine phosphate. Speeding up transit time through the gut in these subjects significantly reduced dietary fiber digestibility.

Transit time is also important in determining other colonic events. In subjects on a given diet of constant composition, fecal ammonia concentrations can be shown to be related to transit time (Cummings, 1978). Moreover, it has been shown that part of the variation in urine volatile phenol excretion, a bacterial breakdown product of tyrosine, can be related to transit time in individuals (Cummings *et al.*, 1979). Those subjects with the slowest transit time show the greatest urine phenol excretion, suggesting that greater microbial protein breakdown is occurring where transit is prolonged. Transit time is also an important determinant of rumen bacterial cell yields from carbohydrate (Hespell, 1979), and in man has been shown to influence fecal microbial cell output on subjects on a constant diet (Stephen, 1980). In addition, Davignon *et al.* (1968) have shown that colonic turnover time (which is equivalent to transit time) can be related to the degree of steroid degradation in the colon in hyperlipidemic subjects.

These data therefore all suggest that the time which material spends in the gut is important in determining metabolism, but that for some substances, such as cellulose, there may be a transit time beyond which further changes in metabolism are not evident. If the relationship between transit time and metabolism holds for materials other than cellulose, then the importance of transit will only be seen in studies where changes are occurring over this critical period of 24–48 hr.

Two studies of transit in relation to large bowel cancer have been done since Burkitt's original observation in the African. Both that of Glober *et al.* (1974) in Japanese and Japanese migrants to Hawaii, and the IARC Intestinal Microecology group (1977), in Scandinavians, have failed to demonstrate a relationship between incidence of bowel cancer and transit.

4. BACTERIAL METABOLISM

The stimulation in microbial growth which is seen when dietary fiber is fed or when transit through the gut is speeded up has a number of important implications. One relates to changes in nitrogen metabolism, since it is currently thought that a number of potential mutagens in the colon contain nitrogen (Bruce *et al.*, 1977). Another is the production of SCFA from fermentation and their possible antitumor properties.

4.1. Nitrogen Metabolism

When dietary fiber intake is increased, this results in an increase in fecal nitrogen excretion. Early investigators attributed this to the presence of undigested protein in feces. While undigested protein may contribute a small proportion of total nitrogen excretion, the major fraction is associated with the bacterial mass (Stephen and Cummings 1979). Animal studies have shown most fecal nitrogen is associated with bacteria, and on feeding indigestible carbohydrates, bacterial nitrogen increases (Mason, 1969; Mason and Palmer, 1973).

Although free amino acids occur in the colon, the majority of nitrogen available for bacterial use is present as ammonia. Ammonia is generated by microbial action on urea, on protein, or other nonurea sources, although the quantitative contribution from each source is unknown. After its production, ammonia may be absorbed and transported to the liver, excreted in feces, or incorporated into bacterial cells in the large bowel. Incorporation into bacteria is dependent on the availability of energy-yielding carbohydrate for microbial cell growth. When dietary fiber intakes are increased, the availability of carbohydrate for microbial fermentation is increased and the effect of this is to reroute nitrogen from ammonia into microbial protein synthesis. As a result of this, although fecal nitrogen excretion increases, ammonia concentrations fall (Fig. 3) (Cummings *et al.*, 1981).

Ammonia has been shown to effect the metabolism and the morphology of cells and may promote tumor incidence (Visek, 1972, 1978). Low concentrations (5–10 mM) alter intermediary metabolism and DNA synthesis, reduce the lifespan of intestinal cells, and induce faster cell turnover. In general, ammonia is more toxic for healthy than for transformed cells and thus a high concentration of ammonia in the bowel lumen may select for neoplastic growth. Ammonia, by increasing cell turnover, increases the probability of genetic damage occurring in the presence of oncogenic agents. Dividing cell populations are more susceptible to chemical carcinogenesis (Warwick, 1971) and the concept of nonspecific injury and cancer promotion is long-standing in medicine. Experimental evidence relating large bowel injury and dimethylhydrazine-induced tumors (Pozharisski, 1975) shows that cell kinetics is of fundamental importance during chemical carcinogenesis.

Further evidence relating ammonia, cell turnover, and cancer is seen in ureterosigmoidostomy patients. These patients have very high luminal ammonia concentrations (up to 100 mM) (McConnell *et al.*, 1979) and greatly increased risk of developing tumors distal to the site of the implantation (Tank *et al.*, 1973).

Bruce *et al.* (1977) have isolated a mutagenic *N*-nitroso compound from human feces, and carcinogenic *N*-nitroso derivatives may be synthesized in the bowel from nitrite (Tannenbaum *et al.*, 1978). Nitrite is present in the diet or may be formed from dietary or endogenous nitrate. Although the evidence for endogenous nitrate synthesis continues to be debated (Green *et al.*, 1981), bacterial synthesis of nitrate from ammonia has been demonstrated both in culture and natural ecosystems. The reduction in ammonia concentration that occurs on a high-fiber diet may indirectly reduce the formation of *N*-nitroso compounds in the bowel lumen. Inhibition of nitrite production may be due also to the low luminal pH that occurs on fiber feeding, since bacterial nitrification has an alkaline pH optimum.

4.2. Hydrogen Ion Concentration

An association of high dietary fiber intakes with a low fecal pH has been noted by Walker *et al.* (1979). The pH in the colon may fall as a consequence of fermentation of complex carbohydrate, as it does when subjects take the disaccharide lactulose (Bown *et al.*, 1974), although studies with poorly digested fiber such as bran have failed to show pH changes in feces (Stephen and Cummings, 1981). The pH may play an important role in absorption from the colon and in the regulation of cell growth.

The use of alkaline media has been reported by Pikovski (1954) to provide selective growth conditions for epithelial tumor cells in tissue culture. Cancer cells can grow over a wide pH range and are not susceptible to the growth-inhibitory effects of pH variation (Ceccarini and Eagle, 1971). These authors also demonstrated that contact inhibition between cells is pH dependent. Changing the pH in culture of embryo cells from 6.6 to 7.6 results in a marked increase in DNA synthesis and in cell growth (Rubin, 1971). Other evidence indicates that cell permeability increases in alkaline media.

Luminal pH may therefore influence the growth and metabolism of colonic epithelial cells. The significance of these observations is increased by the finding that patients with large bowel cancer have higher fecal pH values than controls (MacDonald *et al.*, 1978). Subjects on low-fiber diets therefore may provide an environment in the large bowel selective for the growth of abnormal cells.

Low luminal pH may also be protective against the effects of bacterial metabolites. Ammonia is trapped in the bowel at acid pH and absorption into epithelial cells reduced (Brown *et al.*, 1975). Bile acid degradation is also pH dependent. In an acid environment 7α-dehydroxylation is inhibited, possibly through a direct effect on the enzyme or the poor solubility of these compounds in acid conditions. Tumor-promoting activity of bile acids has been demonstrated using the MNNG rat model (Narisawa *et al.*, 1974). Diversion of bile from the bowel lumen results in a pronounced fall in epithelial cell turnover, suggesting that these compounds can also influence cell division and thus may increase carcinogenic risk (Deschner and Raicht, 1979).

4.3. Short-Chain Fatty Acids

Polysaccharide degradation in the colon generates, in theory at least, 200 mmol of SCFA daily, comprising 120 mmol acetate, 50 mmol propionate, and 30 mmol butyrate. The concentrations of these fatty acids found in human feces are about 45, 12, and 13 mM, respectively (Cummings, 1981). SCFA, in addition to affecting luminal pH, also alter colonic metabolism.

Butyrate has been shown to modify the metabolism, structure, and function of a wide range of cell types. It is an important substrate for energy metabolism in the human isolated colonocyte (Roediger, 1980), and in the distal colon is metabolized apparently in preference to glucose. Butyrate has both stimulatory and inhibitory effects on a wide range of enzymes (Prasad, 1980). It induces the accumulation of acetylated histones in cultured cells and may stabilize chromatin structure during cell division. Low concentrations of butyrate have been shown *in vitro* to reduce DNA synthesis and suppress cell proliferation both in normal and malignant cells (Hagopian *et al.*, 1977). The concentration of butyrate required to modulate growth and influence cell morphology is well within the range found in the human colonic lumen. Suppression of *in vitro* neoplastic characteristics in Syrian hamster cells (Leavitt *et al.*, 1978) and reversal of virus-induced dedifferentiation (Leder and Leder, 1975) by butyrate have been shown. Butyrate at a concentration of 3 mM can revert the morphology of rat hepatoma cells to a more normal appearance (Borenfreund *et al.*, 1980) and this change is associated with the cells becoming anchorage dependent for growth, a characteristic of nontransformed cells. Transformed cells generally produce less fibronectin than normal cells and lack the surface matrix. In virus-treated rat kidney cells, this loss of fibronectin production may be reversed with butyrate (Hayman *et al.*, 1980).

The relevance of these studies to colon cancer in the human is speculative, but as the primary source of butyrate is fermentation of carbohydrates in the colon, a protective role for fiber against bowel cancer through butyrate production may be postulated.

5. CONCLUSION

Although epidemiological studies indicate that dietary fiber is protective against large bowel cancer, the exact mechanism for this protection cannot be stated with certainty as yet. Adding fiber to the diet increases the bulk of colonic content and dilutes some, but not all, constituents. Fiber has major effects on colonic metabolism through the microbial flora. Fermentation produces SCFA, of which butyrate has been shown in experimental studies to be antineoplastic. The stimulation of microbial growth by fiber reroutes nitrogen in the bacterial protein and may protect against the formation of carcinogenic nitrogen-containing substances such as nitrosomines and ammonia. At the present time the significance of a shortened transit time has not been confirmed as protective against bowel cancer, although experimental studies show that transit may control metabolic events, such as fiber breakdown. Overall, the impact of fiber on colonic function is clear, but its direct relevance to carcinogenesis remains to be clarified.

REFERENCES

Borenfreund, E., Schmid, E., Bendich, A., and Franke, W. W., 1980, Constitutive aggregates of intermediate sized filaments of the vimentin and cytokeratin type in cultured hepatoma cells and their dispersal by butyrate, *Exp. Cell Res.* **127**:215–235.

Bown, R. L., Gibson, J. A., Sladen, G. E., Hicks, B., and Dawson, A. M., 1974, Effects of lactulose and other laxatives on ileal and colonic pH as measured by a radiotelemetry device, *Gut* **15**:999–1004.

Bown, R. L., Gibson, J. A., Fenton, J. C. B., Snedden, W., Clark, M. L., and Sladen, G. E., 1975, Ammonia and urea transport by the excluded human colon, *Clin. Sci. Mol. Med.* **48**:279–287.

Bruce, W. R., Varghese, A. J., Furrer, R., and Land, P. C., 1977, A mutugen in the feces of normal humans, in: *Origins of Human Cancer* (H. H. Hiatt, J. D. Watson, and J. A. Winsten, eds.), Cold Spring Harbor Conferences on Cell Proliferation, Vol. 4, Book C, Cold Spring Harbor Laboratory, Cold Spring Harbor, New York, pp. 1641–1646.

Burkitt, D. P., 1971, Epidemiology of cancer of the colon and rectum, *Cancer* **28**:3–13.

Ceccarini, C., and Eagle, H., 1971, pH as a determinant of cellular growth and contact inhibition, *Proc. Natl. Acad. Sci.* **68**:229–233.

Cummings, J. H., 1978, Diet and transit through the gut, *J. Plant Foods* **3**:83–95.

Cummings, J. H., 1981, Short chain fatty acids in the human colon, *Gut* **22**:763–779.

Cummings, J. H., Hill, M. J., Jenkins, D. J. A., Pearson, J. R., and Wiggins, H. S., 1976, Changes in fecal composition and colonic function due to cereal fiber, *Am. J. Clin. Nutr.* **29**:1468–73.

Cummings, J. H., Southgate, D. A. T., Branch, W., Houston, H., Jenkins, D. J. A., and James, W. P. T., 1978, Colonic response to dietary fiber from carrot, cabbage, apple, bran and guar gum, *Lancet* **1**:5–8.

Cummings, J. H., Hill, M. J., Bone, E. S., Branch, W. J., and Jenkins, D. J. A., 1979, The effect of meat protein and dietary fiber on colonic function and metabolism. Part II. Bacterial metabolites in feces and urine, *Am. J. Clin. Nutr.* **32**:2094–2101.

Cummings, J. H., Stephen, A. M., and Branch, W. J., 1981, Implications of dietary fiber breakdown in the human colon, in: *Banbury Report No. 7, Gastrointestinal Cancer: Endogenous Factors* (R. Bruce, S. Tannenbaum, and P. Correa, eds.) Cold Spring Harbor Laboratory, Cold Spring Harbor, New York, pp. 71–81.

Davignon, J., Simmonds, W. J., and Ahrens, E. H., 1968, Influences of chromic oxide as an internal standard for balance studies in formula-fed patients and for assessment of colonic function, *J. Clin. Invest.* **47**:127–138.

Deschner, E. E., and Raicht, R. F., 1979, The influence of bile on the kinetic behaviour of colonic epithelial cells of the rat, *Gastroenterology* **76**:1120.

Glober, G., Klein, K. C., Moore, W. O., and Abba, B. C., 1974, Bowel transit-times in two populations experiencing similar colon cancer risks, *Lancet* **2**:80–81.

Green, L. C., Tannenbaum, S. R., and Goldman, P., 1981, Nitrate synthesis in the germ free and conventional rat, *Science* **212**:56–58.

Hagopian, H. K., Riggs, M. G., Swartz, L. A., and Ingram, V. M., 1977, Effect of *n*-butyrate on DNA synthesis in chick fibroblasts and Hela cells, *Cell* **12**:855–860.

Hayman, E. G., Engvall, E., and Ruoslahti, E., 1980, Loss of fibronectin in cell surface matrix of transformed rat kidney cells and its restoration with butyrate, *Proc. Am. Assoc. Cancer Res.* **21**:A1117.

Hespell, R. B., 1979, Efficiency of growth by ruminal bacteria, *Fed. Proc.* **38**:2707–2712.

Hill, M. J., Crowther, J. S., Drasar, B. S., Hawkesworth, G., Aries, V., and Williams, R. E. O., 1971, Bacteria and aetiology of cancer of large bowel, *Lancet* **1**:95–100.

Hill, M. J., Drasar, B. S., Williams, R. E. O., Meade, T. W., Cox, A. G., Simpson, J. E. P., and Morson, B. C., 1975, Fecal bile acids and clostridia in patients with cancer of the large bowel, *Lancet* **1**:535–538.

IARC Intestinal Microecology Group, 1977, Dietary fiber, transit time, fecal bacteria, steroids and colon cancer in two Scandinavian populations, *Lancet* **2**:207–211.

Leavitt, J., Barrett, J. C., Crawford, B. D., and Ts'o, P. O. P., 1978, Butyric acid suppression of the *in vitro* neoplastic state of Syrian hamster cells, *Nature* **271**:262–265.

Leder, A., and Leder, P., 1975, Butyric acid, a potent inducer of erythroid differentiation in cultural erythroleukemic cells, *Cell* **5**:319–322.

MacDonald, I. A., Webb, G. R., and Mahony, D. E., 1978, Fecal hydroxysteroid dehydrogenase activities in vegetarian Seventh-Day Adventists, control subjects and bowel cancer patients, *Am. J. Clin. Nutr.* **31**:S233–238.

McConnell, J. B., Murison, J., and Steward, W. K., 1979, The role of the colon in the pathogenesis of hyperchloraemic acidosis in ureterosigmoid anastomosis, *Clin. Sci.* **57**:305–312.

McNeil, N. I., Cummings, J. H., and James, W. P. T., 1978, Short chain fatty acid absorption by the human large intestine, *Gut* **19:**819–822.

Mason, V. C., 1969, Some observations on the distribution and origin of nitrogen in sheep feces, *J. Agric. Sci. Camb.* **73:**99–111.

Mason, V. C., and Palmer, R., 1973, The influence of bacterial activity in the alimentary canal of rats on fecal nitrogen excretion, *Acta Agric. Scand.* **23:**141–150.

Narasawa, T., Magadia, N. E., Weisburger, J. H. and Wynder, E. L., 1974, Promoting effects of bile acids on colon carcinogenesis after intrarectal installation of N-methyl-N^1-nitro-N-nitrosoguanidine in rats, *J. Natl. Cancer Inst.* **53:**1093–1097.

Pikovski, M. A., 1954, Differentiated mammary gland tumor tissue grown in an alkaline medium *in vitro, Exp. Cell Res.* **7:**52–57.

Pozharisski, K. M., 1975, The significance of nonspecific injury for colon carcinogenesis in rats, *Cancer Res.* **35:**3824–3830.

Prasad, K. N., 1980, Butyric acid: A small fatty acid with diverse biological functions, *Life Sci.* **27:**1351–1358.

Roediger, W. E. W., 1980, Role of anaerobic bacteria in the metabolic welfare of the colonic mucosa in man, *Gut* **21:**793–798.

Rubin, H., 1971, pH and population density in the regulation of animal cell multiplication, *J. Cell Biol.* **51:**686–702.

Stephen, A. M., 1980, Dietary Fiber and Human Colonic Function, Ph.D. Thesis, University of Cambridge, Cambridge, England.

Stephen, A. M., and Cummings, J. H., 1979, The influence of dietary fiber on fecal nitrogen excretion in man, *Proc. Nutr. Soc.* **38:**141A.

Stephen, A. M., and Cummings, J. H., 1980, Mechanism of action of dietary fiber in the human colon, *Nature* **284:**283–284.

Stephen, A. M., and Cummings, J. H., 1981, The effect of wheat fiber on fecal pH in man, *Gastroenterology* **80:**1294.

Tank, E. S., Karsch, D. N., and Lapides, J., 1973, Adenocarcinoma of the colon associated with uretosigmoidostomy, *Dis. Colon Rectum* **16:**300–304.

Tannenbaum, S., Tett, D., Young, V., Land, P., and Bruce, W., 1978, Nitrite and nitrate are formed by endogenous synthesis in the human intestine, *Science* **200:**1487–1489.

Van Soest, P. J., 1973, The uniformity and nutritive availability of cellulose, *Fed. Proc.* **32:**1804–1808.

Van Soest, P. S., 1975, Physico-chemical aspects of fiber digestion, in: *Digestion and Metabolism in the Ruminant* (I. W. McDonald and A. C. I. Warner, eds)., University of New England Publishing Unit, Armidale, Australia, pp. 351–365.

Visek, W. J., 1972, Effects of urea hydrolysis on cell life-span and metabolism, *Fed. Proc.* **31:**1178–1193.

Visek, W. J., 1978, Diet and cell growth modulation by ammonia, *Am. J. Clin. Nutr.* **31:**S216–S220.

Walker, A. R. P., Walker, B. F., and Segal, I., 1979, Fecal pH value and its modification by dietary means in South African black and white school children, *S. A. Med. J.* **55:**495–498.

Warwick, G. P., 1971. Effect of the cell cycle on carcinogenesis, *Fed. Proc.* **30:**1760–1765.

Wrong, O., Metcalfe-Gibson, A., Morrison, R. B. I., Ng, S. T., and Howard, A. V., 1965, *In vivo* dialysis of feces as a method of stool analysis. I. Techniques and results in normal subjects, *Clin. Sci.* **28:**357–75.

Index

MIX
Papier aus verantwortungsvollen Quellen
Paper from responsible sources
FSC® C105338

If you have any concerns about our products,
you can contact us on
ProductSafety@springernature.com

In case Publisher is established outside the EU,
the EU authorized representative is:
Springer Nature Customer Service Center GmbH
Europaplatz 3, 69115 Heidelberg, Germany

Printed by Libri Plureos GmbH
in Hamburg, Germany